AF303070

Numerical Approaches to Spatial Correlations in
Strongly Interacting Fermion Systems


Dissertation

zur Erlangung des Doktorgrades

des Fachbereichs Physik

der Universität Hamburg


vorgelegt von
Hartmut Hafermann
aus
Leer (Ostfriesland)


Hamburg
2009

**Bibliografische Information der Deutschen Nationalbibliothek**

Die Deutsche Nationalbibliothek verzeichnet diese Publikation in der Deutschen Nationalbibliographie; detaillierte bibliografische Daten sind im Internet über http://dnb.d-nb.de abrufbar.

1. Aufl. - Göttingen: Cuvillier, 2010
  Zugl.: Hamburg, Univ., Diss., 2009

  978-3-86955-347-4


| | |
|---|---|
| Gutachter der Dissertation | Prof. Dr. Alexander I. Lichtenstein |
| | Prof. Dr. Michael Potthoff |
| | Prof. Dr. Jim K. Freericks |
| Gutachter der Disputation | Prof. Dr. Alexander I. Lichtenstein |
| | Prof. Dr. Philipp Werner |
| Datum der Disputation | 14. Juli 2009 |
| Vorsitzender des Prüfungsausschusses | PD Dr. Alexander Chudnovskiy |
| Vorsitzender des Promotionsausschusses | Prof. Dr. Robert Klanner |
| Dekan der Fakultät für Mathematik, Informatik und Naturwissenschaften | Prof. Dr. Heinrich Graener |

Umschlaggestaltung: Henning Labuhn

Meiner Familie
und Inga.

There is a theory which states that if ever anyone discovers exactly
what the Universe is for and why it is here, it will instantly disappear
and be replaced by something even more bizarre and inexplicable.
There is another theory which states that this has already happened.

Douglas Adams (1952-2001)

# Kurzzusammenfassung

Gegenstand der vorliegenden Arbeit ist die Entwicklung numerischer Methoden zur Beschreibung räumlicher Korrelationen in stark wechselwirkenden fermionischen Systemen. Die vor kurzem entwickelten zeitkontinuierlichen Quanten Monte Carlo Methoden zur Lösung des Quantenstörstellenproblems bilden die Grundlage dieser Verfahren. Es wird gezeigt wie Korrelationsfunktionen in imaginärer Zeit sowie Zweiteilchenvertexfunktionen berechnet werden können.

Eine erste Anwendung bildet das zweilagige Hubbard Modell auf dem Bethe-Gitter. Es wird im Rahmen der dynamischen Molekularfeldtheorie numerisch exakt gelöst. Die dynamische Suszeptibilität sowie Spinkorrelationen werden berechnet. Das System weist unter anderem einen Phasenübergang zum Singlett-Isolator auf, welcher mit der Ausbildung einer Ladungs- sowie Spin-Lücke einhergeht. Für eine weitere Anwendung wird das Anderson Störstellenproblem als Modell einer Kondo Störstelle auf metallischem Substrat herangezogen. Die räumlichen Spin- und Ladungskorrelationen der nichtwechselwirkenden Leitungselektronen lassen sich mittels der Vertexfunktion der Störstelle ohne weitere Näherung bestimmen. Es wird gezeigt, dass die räumliche Variation der lokalen Badkorrelationen eine klare Signatur der Kondo Abschirmwolke enthält. Diese ist somit prinzipiell über Tunnelmikroskopie Experimente nachweisbar.

Der Hauptteil der Arbeit behandelt die Methode dualer Fermionen zur Lösung quantenmechanischer Vielteilchenprobleme. Hierauf aufbauend wird ein effizientes Verfahren zur approximativen Lösung des Quantenstörstellenproblems vorgestellt, welches eine direkte und stabile analytische Fortsetzung dynamischer Größen auf die reelle Achse ermöglicht.

Die Anwendung auf Gitterprobleme erlaubt, ausgehend von der dynamischen Molekularfeldtheorie (DMFT), räumliche Korrelationen zu berücksichtigen. Im zweidimensionalen Hubbard Modell führt dies gegenüber DMFT zu einer qualitativen Änderung des Metall-Isolator Übergangs durch kurzreichweitige antiferromagnetische Korrelationen. Das Verfahren wird auf Mehrorbitalsysteme verallgemeinert und die Diagrammregeln für die Störungstheorie werden abgeleitet. Es wird ein flexibles Konzept zur objektorientierten Implementierung von störungstheoretischen Verfahren entwickelt. Eine neuartige Cluster-Methode wird vorgeschlagen und getestet. Ferner wird gezeigt, wie verallgemeinerte Suszeptibilitäten berechnet werden können.

Ein weiteres wesentliches Resultat ist die erhaltende Leiternäherung für duale Fermionen in Anlehnung an die Fluktuationsaustauschnäherung. Diese berücksichtigt Effekte langreichweitiger Korrelationen und reproduziert die Pseudolücke im Einteilchenspektrum. Es wird gezeigt, dass die Methode im Gegensatz zur konventionellen Fluktuationsaustauschnäherung insbesondere im Fall starker Elektronenkorrelationen anwendbar ist. Basierend auf der Leiternäherung wird ein Verfahren vorgeschlagen, das in Verbindung mit Bandstrukturmethoden erstmals die realistische Beschreibung von korrelierten Materialien unter Einbeziehung langreichweitiger Korrelationen erlaubt.

# Abstract

This thesis presents efficient numerical approaches to spatial correlations in strongly interacting fermion systems. The recently proposed continuous-time quantum Monte Carlo algorithms provide the basis for most of these approaches and are presented in detail. It is shown how imaginary-time correlation functions and two-particle vertex functions are computed with these methods.

A first application is the two-plane Hubbard model on the Bethe lattice in infinite dimensions. Within dynamical mean-field theory it is solved numerically exactly. The dynamical susceptibility and spin-correlations of the model are computed. It exhibits a phase transition to a singlet-insulator phase, which is accompanied by the simultaneous formation of a spin- and charge-gap.

For a second application, a Kondo impurity on a metallic substrate is modeled by an Anderson impurity model. The spatial spin and charge correlations of the noninteracting conduction electron bath can be obtained from the vertex function of the impurity model without invoking any approximations. It is shown that the spatial variation of the local bath correlation function exhibits a clear signature of the Kondo screening cloud, which is therefore accessible to scanning tunneling microscopy experiments.

The main part of the thesis is concerned with the dual fermion approach to the solution of the quantum many-body problem, which is introduced in detail. The method is employed to devise an efficient solver for the Anderson impurity model. It is free of statistical errors and allows a direct and stable analytical continuation of dynamical quantities to the real axis.

The application of the dual fermion approach to quantum lattice models allows to account for spatial correlations beyond dynamical mean-field theory. It is shown that the inclusion of short-range antiferromagnetic correlations in the two-dimensional Hubbard model leads to a qualitative modification of the metal-insulator transition. The dual fermion formalism is generalized to the multiorbital case and the diagram rules of the perturbation theory are derived. A flexible concept for the object-oriented implementation of combined numerical and diagrammatic techniques is developed. Based hereupon, a novel cluster-method is implemented and tested. It is further shown how generalized susceptibilities are obtained within this approach.

Another important result is the conserving infinite-order ladder approximation, which accounts for fluctuation-exchange corrections to the dual self-energy. It allows to capture the effects of long-range correlations. In contrast to the fluctuation-exchange approximation, it is shown that this method is applicable in particular for the case of strong electronic correlations. It is further demonstrated that it reproduces the pseudo-gap in the single-particle spectrum. A scheme based on the ladder approximation is proposed, which, in conjunction with band structure methods, for the first time allows a realistic and computationally feasible description of correlated materials including the effects of long-range correlations beyond dynamical mean-field theory.

# Contents

# Chapter 1

# Introduction

Condensed matter physics is probably the field in physics with the greatest impact on everyday life. Well-known examples are the invention of the transistor or the application of the GMR-effect in modern hard-disk drives. Notwithstanding its practical use, condensed matter physics is also a fascinating subject.

This is particularly true for so-called strongly correlated electron systems. The interplay of many degrees of freedom in many-body quantum systems with strong interactions leads to emergent behavior and to very diverse phenomena such as the fractional quantum Hall effect [1], itinerant electron magnetism, heavy-fermion [2, 3] and non-Fermi-liquid behavior, the Kondo effect [4] and high-temperature superconductivity [5, 6]. Concepts from other fields of physics have their counterpart in condensed matter physics. Examples comprise massless Dirac fermions in graphene, the realization of asymptotic freedom in the Kondo effect and spontaneous symmetry breaking in magnetic or superconducting materials.

The term "correlations" is often used as a synonym for interacting, without acknowledging its actual meaning; it emphasizes the fact that electrons, e.g. in a crystal, do not move independently. Whether or not an electron will hop to a neighboring site or orbital (considering only processes that are compatible with the Pauli principle) will strongly depend on whether the site is already occupied, due to the Coulomb repulsion. This real space picture is in strong contrast to the reciprocal space picture of independent, nearly free electrons moving in the background potential of the nuclei as described by Bloch waves. Strong electronic correlations are therefore expected in materials containing atoms with open d- or f-shells, which often exhibit remarkable properties. Examples are the transition metals V, Fe, Co, Ni, Cu and their oxides such as the Mott insulator NiO and the possible singlet insulator $VO_2$ [7]. The $CuO_2$ planes determine the properties of the high-temperature cuprate superconductors. Recently superconductivity has been observed in iron-based compounds, the iron-pnictide superconductors [8]. Materials containing rare earth or actinide ions with partially filled f-shells can exhibit heavy fermion behavior with effective masses thousand times larger than the electron mass.

The Coulomb repulsion due to the confinement of the electrons in these orbitals is of the order of the kinetic energy. This often results in a delicate balance between localization and delocalization. As a consequence, these materials are very sensitive to externally controllable parameters, such as pressure and doping, and are therefore promising candidates for applications. In addition, the huge degeneracy in the spin sector at low energies leads to a large number of possible ground states. Quantum fluctuations, in particular singlet correlations, play an important role and might favor exotic states [3]. The interplay between low energy spin fluctuations and the fermionic excitations is believed to ultimately lead to the phenomenon of high temperature superconductivity [6]. As of today, applications arising from an increased understanding of strongly correlated materials are only beginning to be explored [9]. The development of reliable theoretical tools to calculate the material specific properties of strongly correlated materials therefore remains one of the primary concerns of modern theoretical condensed matter physics.

The description of correlated materials is theoretically challenging. The framework of density functional theory (DFT) provides a quantitative description with remarkably accurate predictions (apart from some shortcomings, such as the band-gap problem) for simple metals, semiconductors and band-insulating materials. However, the failure of approximations to the DFT, such as the local density approximation (LDA), to capture the physics of strong correlations, renders a realistic description of correlated systems difficult. A classic example is the Mott insulator NiO which is erroneously predicted to be metallic in absence of magnetic long-range order. There have been attempts to improve upon the LDA, such as the LDA+U approach [10] and the GW approximation [11]. While introducing a wavevector dependence into the self-energy, the applicability of the GW approximation is still restricted to systems with relatively weak correlations. The static nature of the LDA+U describes the formation of Hubbard bands for systems with orbital order, but ignores dynamical correlations and fails to reliably predict the temperature dependent phase diagram. [12].

A quantitative understanding has only become possible in recent years. This is not primarily due to a strong increase of computational resources, but the development of new theoretical concepts and methods. The main difficulty lies in the fact that two very different energy scales play an important role in the redistribution of the spectral weight. The description of the low energy physics requires a very accurate understanding of the high-energy excitations and their feedback on low energy excitations. Dynamical mean-field theory (DMFT) has been a major step forward in the understanding of correlated materials [13]. This is particularly true for the physics of the Mott transition in fermionic systems [14]. DMFT for the first time allowed a consistent description of both the low-energy coherent features – the long-lived quasiparticle excitations – and the incoherent high-energy excitations due to the Coulomb repulsion, acting on short timescales. The former give rise to the quasiparticle peak at the Fermi level, while the latter manifest themselves in the broad Hubbard bands at high energies, giving rise to the well-known three peak structure of the spectrum [15].

DMFT maps the lattice problem to a local quantum impurity problem embedded in an electronic bath subject to a self-consistency condition. It becomes exact in the limit of infinite coordination number, where quantum fluctuations become purely local. Its applicability is justified when the physics is dominated by strong local interactions and the phases of the fermions are averaged over a large number of bonds connecting a given site to the lattice (up to 12 for fcc and hcp lattices), so that memory effects in the bath can be neglected [16, 17].

The success of DMFT to capture the local temporal quantum fluctuations was exploited to merge it with a material specific theory in the LDA+DMFT approach [18, 19]. This approach has been applied to a variety of systems (for a review, see [12]) and for example correctly predicted the temperature dependence of the local moment in the transition metals iron and nickel [20]. The Curie temperatures were however overestimated since DMFT completely neglects spatial correlations and cannot capture the reduction of the critical temperature due to the presence of long-wavelength magnons. For short-range spatial correlations, which e.g. are present in systems with strong dimerization, this problem was successfully overcome through cluster extensions to DMFT. An early application of cluster DMFT to a real material was the case of $Ti_2O_3$ [21] where single-site DMFT does not reproduce the insulating state stabilized by nonlocal Coulomb interactions and strong chemical bonding between Ti-Ti pairs. Intersite correlations between titanium atoms were also found to be important in cluster calculations of the low-dimensional quantum spin system TiOCl [22]. The metal-insulator transition in $VO_2$ was attributed to the formation of dynamical singlets between neighboring Vanadium atoms [7].

At present, however, no computationally feasible approaches are available to reliably predict the properties of correlated materials where the correlations are manifestly long-ranged. In iron, evidence was found for a quasiparticle mass enhancement due to electron-magnon coupling in angular resolved photoemission spectroscopy (ARPES) experiments [23]. The same renormalization effects are expected to play a crucial role in the $CuO_2$ planes of the cuprates [24, 25]. Moreover, long-range correlations are expected in low-dimensional systems such as nanostructures, or quasi-one-dimensional organic conductors, such as TTF-TCNQ [26].

Driven by the shortcomings of DMFT, many efforts have been undertaken to go beyond the mean-field description. An expansion in $1/z$, where $z$ is the coordination number, does not converge as the action depends in a non-analytic way on the coordination number [27]. DMFT has been supplemented with a momentum dependent self-energy by including spin and charge density wave-like fluctuations through a Gaussian random field [28]. Building on earlier work on strong-coupling expansions for the Hubbard model [29, 30], a general framework to perform a systematic cumulant expansion around DMFT even considering non-local Coulomb interaction was developed in Ref. [31]. Attempts to include the effect of short-range correlations were primarily based on various cluster generalizations of DMFT [32, 33, 34, 35]. These approaches break translational invariance either in real or momentum space explicitly and might artificially favor states which order at some finite wave-vector. The range of the correlations included is determined by the cluster size. While these approaches constitute a systematic way to improve upon DMFT, they become computationally highly demanding for large cluster sizes. Effects caused by long-wavelength fluctuations and correlations related with a narrow region of reciprocal space, such as the vicinity of Van Hove singularities [36], can hardly be taken into account by cluster approaches. Alternatives are finite-size quantum Monte Carlo simulations. These are however hampered by the fermionic sign problem at low temperatures and away from half-filling [37]. As with exact diagonalization methods, they do not operate in the thermodynamic limit.

Recent developments have led to approaches which combine numerical and analytical (diagrammatic) methods to overcome the abovementioned problems and to include long-range correlations beyond the dynamical mean-field description [38, 39, 40]. These are based on straightforward diagrammatic extensions of DMFT. In an alternative approach, a systematic expansion around DMFT is transformed into a standard diagrammatic technique in terms of auxiliary, so-called dual fermions [41]. As detailed in the present work, the dual fermion approach is a promising method for including the effects of spatial correlations into quantum simulations of real materials.

## 1.1  Structure and Scope

This thesis is devoted to the development and application of theoretical approaches to strongly interacting fermion systems for which spatial correlations are relevant. The thesis is structured as follows:

The basic concepts and models in condensed matter theory, which are relevant to this thesis, are briefly introduced in chapter 2. The approaches and applications presented in subsequent chapters require the computation of two-particle correlation functions and $n$-particle Green functions, or the associated vertex functions of an Anderson impurity model. For the most part, the recently developed so-called continuous-time solvers are employed for this purpose. These provide a numerically exact solution of the impurity problem, i.e. without systematic errors due to imaginary time discretization. They are able to handle general interactions, including the full Coulomb vertex. Two complementary approaches are discussed in detail in chapter 3. These are based on an expansion of the partition function in either the interaction (weak-coupling approach) or the impurity-bath hybridization (strong-coupling expansion) and the collection of diagrams into fermionic determinants. It is shown how Green functions and higher-order correlation functions are computed on Matsubara frequencies or in imaginary time.

In chapter 4, these solvers are applied to the two-plane Hubbard model on the Bethe lattice in infinite dimensions. The model is solved exactly within cluster dynamical mean-field theory. The short-range nearest-neighbor correlations lead to the formation of a singlet-insulating state, as revealed by the calculation of the spin correlation functions. The dynamical susceptibility is computed and shows the formation of a concomitant spin-gap.

Chapter 5 is devoted to the study of spatial correlations around a Kondo impurity, as can be probed by scanning-tunneling microscopy experiments. The system is modeled by an Anderson impurity exerted to a bath of noninteracting electrons. For this model, the spin and charge correlation functions as a function of distance from the impurity can be obtained exactly from the knowledge of the two-particle impurity vertex function. It is shown that the spatial variation of the correlations contains a signature of the Kondo screening cloud.

The dual fermion approach is introduced in chapter 6. This novel, perturbative method combines numerical with diagrammatic techniques. The introduction of auxiliary fermions through a continuous Hubbard-Stratonovich transformation allows to formulate a Feynman-type perturbation expansion and to approximate the solution to a many-body problem through a perturbation of a nontrivial, albeit numerically solvable reference problem. When applied to quantum lattice models, it allows to incorporate the effect of spatial correlations beyond dynamical mean-field theory. The relation of the approach to DMFT is established and numerical results are shown for illustration. The application to the two-dimensional Hubbard model reveals that the metal-insulator transition is greatly affected by spatial correlations. The leading order diagrammatic correction captures the effect of antiferromagnetic short-range correlations and leads to a strong reduction of the critical interaction and a qualitative modification of the transition lines compared to DMFT.

As part of this thesis, the dual fermion approach is analyzed in detail and generalized in various respects. The general matrix formulation is derived and implemented. This allows to perform multiorbital and cluster calculations. A detailed derivation is given in the appendix. The extension to symmetry broken phases is discussed and it is shown how the approach can be applied to real systems in conjunction with realistic band structure calculations. The dual fermion approach has been implemented in a way to ensure a high degree of flexibility and reusability of the code. The concepts that have guided the object-oriented C++ implementation are shortly outlined in chapter 7.

The cluster dual fermion approach is introduced as a perturbative extension to cluster DMFT in chapter 8. As a benchmark, it is applied to the one-dimensional Hubbard model and compared to results from density matrix renormalization group (DMRG) and variational cluster approach (VCA) calculations.

In chapter 9 an efficient approximate technique for solving the Anderson impurity problem based on the dual perturbation theory is presented. It is based on a diagrammatic expansion around a reference system which is solvable by exact diagonalization and is hence free of statistical errors. This application is not related to spatial correlations, but it provides a deeper understanding of the nature of the dual perturbation theory. It is established that the corresponding perturbation expansion approaches standard perturbation theory in the limit of small interaction and strong hybridization, whereas it becomes equivalent to the hybridization expansion of the Green function near the atomic limit. As a generic test case, it is applied to the Anderson impurity model embedded in an electronic bath with semielliptical density of states. Good agreement is found with numerically exact quantum Monte Carlo results, even at lower temperatures. The solver is applied in the Kondo regime to elucidate the role of the reference system. It is further shown that spectral properties can be directly obtained on the real axis, without the need of numerical analytical continuation. First results are compared to those obtained from analytical continuation using Padé approximants.

In chapter 10, it is shown how generalized susceptibilities can be calculated in the dual fermion approach. This allows to detect possible instabilities of a system towards ordered phases, such as spin density wave states. It is shown that in the DMFT limit, the dual fermion approach to calculate susceptibilities remarkably is exactly equivalent to the DMFT approach. The pairing susceptibility in the two-dimensional Hubbard model is considered as an application. The eigenfunction corresponding to the leading eigenvalue of the Bethe-Salpeter equation resembles the d-wave symmetry of the order parameter. In agreement with previous studies, the dominant contribution to pairing is found to originate from the triplet (magnetic) spin channel of the particle-hole vertex.

In chapter 11, the dual fermion approach is supplemented with a ladder approximation to the dual self-energy. The ladder dual fermion approach (LDFA) includes diagrams from the particle-hole fluctuation channels similar to the fluctuation-exchange-approximation (FLEX). It is demonstrated that spin fluctuations responsible for the formation of the pseudogap are included through the LDFA.

The convergence properties of the approach are also investigated. It is shown that a perturbation theory in terms of dual fermions does not constitute an uncontrolled approximation, but rather enhances the convergence properties compared to straightforward diagrammatic extensions of DMFT. It is demonstrated that in marked contrast to FLEX, the LDFA is applicable even for strong coupling. The performance of the dual fermion approach, in particular the LDFA, is assessed by comparison with numerical results from lattice Monte Carlo simulations for the two-dimensional Hubbard model.

Finally, a computationally efficient approach for the study of quantum lattice models is proposed. It combines the exact diagonalization method for the solution of the reference problem with the LDFA.

General conclusions and an outlook for the application of the methods developed in this thesis are provided in chapter 12.

# Chapter 2

# Models and Concepts

In condensed matter physics, one is faced with the intricate task of finding an appropriate description of systems consisting of a huge number of particles. Owing to the quantum mechanical nature of the problem and neglecting relativistic effects, we are faced with the task of solving the many-particle Schrödinger equation

$$\tilde{H}\,\Psi(\mathbf{x}_1\ldots\mathbf{x}_N) = E\,\Psi(\mathbf{x}_1\ldots\mathbf{x}_N)\,, \tag{2.1}$$

where the number of the particles $N$ is of the order of the Avogadro constant $N_A = 10^{23}$. In a real solid, the electrons interact with each other and move in the field of the nuclei, which in turn interact among themselves and oscillate about their equilibrium positions. As always, we have to employ physical intuition, which here tells us that the fact that the nuclei are much heavier than the electrons, $m_e/M_n \approx 10^{-4} - 10^{-6}$, will cause the electrons to move rather independently of the motion of the nuclei. In other words, the motion of electrons and nuclei is decoupled, which is true up to terms of order $m_e/M_n$. If we are only interested in the excitations of the electronic system, we may regard the nuclei as frozen in their equilibrium positions and solve the electronic problem in a static background potential of the nuclei. This is the essence of the Born-Oppenheimer approximation, which leaves us with the Hamiltonian of the electronic system subject to an external potential $V$:

$$\tilde{H} = \tilde{H}_0 + \tilde{V}_{\text{ee}} = \sum_{i=1}^{N}\left(-\frac{\hbar^2}{2m}\nabla_{\mathbf{x}}^2 + V_{\text{ext}}(\mathbf{x}_i)\right) + e^2 \sum_{i<j}\frac{1}{\left|\mathbf{x}_i - \mathbf{x}_j\right|}\,. \tag{2.2}$$

What we neglect are the excitations of the system of the nuclei – the phonons – and their interaction with electrons, but we still have the freedom to introduce phonons and the electron-phonon coupling perturbatively afterwards[1]. By distinguishing core and

---

[1]The fact that the electron-phonon interaction is a small perturbation does not imply that it has a small effect. It may break symmetries and lift the associated selection rules, as for example recently observed in the case of phonon mediated tunneling into graphene [42].

valence electrons, the problem is further simplified. The former are tightly bound to the nuclei and do not participate in bonding. The Hamiltonian (2.2) then describes the system of valence electrons moving in the external potential of the ion cores. Despite these simplifications, an exact solution of the many-particle Schrödinger equation is impossible and one has to resort to approximations. Two main routes have evolved to deal with the Hamiltonian (2.2): the density functional theory and a description in terms of quantum lattice models.

## 2.1 Density Functional Theory

The density functional theory (DFT) is a very successful *ab-initio* theory to describe the band features of weakly correlated materials. Its applications range from the description of molecules to solids, within quantum chemistry and solid state physics and it is widely used in science and industry. A thorough introduction can e.g. be found in Ref. [43].

The theory is based on the two Hohenberg-Kohn theorems. The first theorem states that for an interacting electron system in an external potential $V_{\text{ext}}$, this potential is uniquely determined (up to a constant) by the ground state density $n_0(\mathbf{x})$. This statement is nontrivial, for it implies that the ground state properties of the interacting $N$-particle system, including the many-body wavefunction, are uniquely determined through the ground state density only. Instead of dealing with the full wavefunction which depends on $3N$ particle coordinates, it is sufficient to use the density. A recipe of how to calculate the ground-state density is provided by the second Hohenberg-Kohn theorem.

In the formulation by Levy and Lieb, a functional of the density is constructed on the space of densities that can be represented by $N$-body wavefunctions. It is defined as the minimum of the expectation values of the exact kinetic and electron-electron interaction, taken over the class of wavefunctions that yield the density $n(\mathbf{x})$: $F[n] := \min_{\Psi_n}\langle\Psi|T + V_{\text{ee}}|\Psi\rangle$. By the Ritz variational principle, the ground-state density $n_0(\mathbf{x})$ minimizes $F[n]$ and the functional of the total energy is given by

$$E[n] = F[n] + \int d\mathbf{x}\, v_{\text{ext}}(\mathbf{x})n(\mathbf{x}) \,. \tag{2.3}$$

In order to find the ground state density of the system, one therefore needs to solve the constrained variational problem

$$\frac{\delta E[n]}{\delta n(\mathbf{x})} = \mu \,, \tag{2.4}$$

where $\mu$ plays the role of a Lagrangian multiplier (chemical potential) which fixes the particle number $N = \int d\mathbf{x}\, n(\mathbf{x})$. In practice, this is achieved by considering a noninteracting auxiliary system in an effective potential described by the Schrödinger equation

$$\left(-\frac{\hbar^2}{2m}\nabla^2 + v_{\text{eff}}(\mathbf{x})\right)\psi_i(\mathbf{x}) = \epsilon_i\,\psi_i(\mathbf{x}) \,, \tag{2.5}$$

where the effective potential is chosen such that the Kohn-Sham orbitals $\psi_i$ reproduce the density of the interacting many-body system:

$$n(\mathbf{x}) = \sum_{i=1}^{N} |\psi_i(\mathbf{x})|^2 \; . \tag{2.6}$$

This yields the correct particle number given that the Kohn-Sham orbitals fulfill the orthonormality constraint $\langle \psi_i | \psi_j \rangle = \delta_{ij}$, $i, j = 1, \ldots, N$. In order to identify the effective potential, the energy functional is written in the form

$$E[n] = T_0[n] + E_{\mathrm{H}}[n] + E_{\mathrm{ext}}[n] + E_{\mathrm{xc}}[n] \; , \tag{2.7}$$

where the kinetic energy functional $T_0$ of the noninteracting system, the Hartree functional $E_{\mathrm{H}}$ and the corresponding potential $v_{\mathrm{H}}$ are known explicitly. The exchange correlation functional contains the many-particle effects. Variation gives

$$\frac{\delta E[n]}{\delta n} \frac{\delta n}{\delta \psi^*} = \frac{\delta T_0}{\delta \psi^*} + (v_{\mathrm{H}} + v_{\mathrm{ext}} + v_{\mathrm{xc}}) \frac{\delta n}{\delta \psi^*} = 0 \; . \tag{2.8}$$

Solving (2.5) is hence equivalent to (2.4) provided that $v_{\mathrm{eff}} = v_{\mathrm{H}} + v_{\mathrm{ext}} + v_{\mathrm{xc}}$. The Kohn-Sham eigenvalues play the role of Lagrange-multipliers which ensure the orthonormality constraints. Since the effective potential depends on the density and also determines the density through the solution of the Kohn-Sham equations, these have to be solved self-consistently.

The energy functional (2.7) is known except for the exchange-correlation functional. Knowing $E_{\mathrm{xc}}[n]$ would imply that the ground state properties of all interacting electron systems were known. Various approximations to the exchange-correlation functional have been devised, the most common being the local density- (LDA) or generalized gradient approximations (GGA). The LDA (GGA) are known to over(under)-estimate bond strengths. The local spin-density approximation includes the electron spin and a relativistic formulation of DFT has been developed. The available electronic structure codes mainly differ in the construction of the basis set and potential used to solve the Kohn-Sham equations. Pseudopotential methods allow a very fast solution of large systems. In the augmented wave methods partial solutions within atomic spheres and interstitial regions have to be matched such that the solution is continuously differentiable at the boundary for all energies. The related computational effort could be significantly reduced by linearizing the matching condition in the LMTO formalism introduced by Anderson [44]. Recent augmented wave codes are very accurate, but the construction of the basis set (APW, FLAPW) in these codes (WIEN2k, FLEUR) is rather involved. A compromise between speed and accuracy is the projector augmented wave method implemented in VASP.

## 2.2   Quantum Lattice Models

Methods of quantum field theory are widely used in condensed matter physics. They provide a convenient way to treat systems consisting of many particles. This includes cases where the particle number is allowed to fluctuate, as in superconductivity. The difficulty to construct and deal with (anti-)symmetrized (fermionic) bosonic many-particle wave functions for identical particles is circumvented by introducing operators that create or annihilate a particle in a given state. The proper (fermion) boson statistics are imposed through the (anti-)commutation relations of the field operators:

$$[\psi_\alpha(\mathbf{x}), \psi_\beta(\mathbf{x}')]_{(\pm)} = 0 , \qquad [\psi_\alpha(\mathbf{x}), \psi_\beta^\dagger(\mathbf{x}')]_{(\pm)} = \delta_{\alpha\beta}\delta(\mathbf{x} - \mathbf{x}') . \tag{2.9}$$

Instead of considering the many-particle Schrödinger equation $H|\psi\rangle = E|\psi\rangle$ in a particular representation, such as the position representation (2.1), with a fixed number of particles, the Hamiltonian itself is expressed in terms of field operators. This corresponds to replacing the wavefunction by an operator acting on a quantum field with a fluctuating particle number and is referred to as second quantization. As described in standard textbooks, the Hamiltonian is expressed in terms of field operators as

$$H = \sum_{\alpha\beta} \int d\mathbf{x}\, \psi_\alpha^\dagger(\mathbf{x}) \left( -\frac{\hbar^2}{2m}\nabla_\mathbf{x}^2 + V_{\text{ext}}(\mathbf{x}) \right) \psi_\alpha(\mathbf{x})$$

$$+ \frac{1}{2}\sum_{\alpha\beta\gamma\delta} \int d\mathbf{x} \int d\mathbf{x}'\, \psi_\alpha^\dagger(\mathbf{x})\psi_\beta^\dagger(\mathbf{x}')V(\mathbf{x} - \mathbf{x}')\psi_\delta(\mathbf{x}')\psi_\gamma(\mathbf{x}) . \tag{2.10}$$

A representation which is more convenient for practical calculations can be obtained by considering Wannier functions. These are localized around a given lattice site $\mathbf{X}_i$ and can be chosen as

$$\phi_{i\alpha}(\mathbf{x} - \mathbf{X}_i) = \frac{1}{\sqrt{N}} \sum_\mathbf{k} e^{-i\mathbf{k}\mathbf{X}_i}\psi_{\mathbf{k}\alpha}(\mathbf{x}) , \tag{2.11}$$

where the $\psi_{\mathbf{k}\alpha}(\mathbf{x})$ are Bloch functions of the noninteracting Hamiltonian. The index $\alpha$ labels spins and orbitals. Defining the (creation) annihilation opearator of a Wannier state as $(c_{i\alpha}^\dagger)\, c_{i\alpha}$, the field operators can be expressed in the form $\psi_\alpha(\mathbf{x}) = \sum_i \phi_{i\alpha}(\mathbf{x})c_{i\alpha}$ and the Hamiltonian takes the form of a lattice model

$$H = \sum_{ij}\sum_{\alpha\beta} t_{ij}^{\alpha\beta} c_{i\alpha}^\dagger c_{j\beta} + \frac{1}{2}\sum_{ijkl}\sum_{\alpha\beta\gamma\delta} U_{ijkl}^{\alpha\beta\gamma\delta} c_{i\alpha}^\dagger c_{j\beta}^\dagger c_{l\delta}c_{k\gamma} , \tag{2.12}$$

where the matrix elements are given by

$$t_{ij}^{\alpha\beta} = \int d\mathbf{x}\, \phi_\alpha^*(\mathbf{x} - \mathbf{X}_i)\left( -\frac{\hbar^2}{2m}\nabla^2 + V_{\text{ext}}(\mathbf{x}) \right)\phi_\beta(\mathbf{x} - \mathbf{X}_j) ,$$

$$U_{ijkl}^{\alpha\beta\gamma\delta} = e^2 \int d\mathbf{x} \int d\mathbf{x}' \frac{\phi_\alpha^*(\mathbf{x} - \mathbf{X}_i)\phi_\beta^*(\mathbf{x}' - \mathbf{X}_j)\phi_\gamma(\mathbf{x} - \mathbf{X}_k)\phi_\delta(\mathbf{x}' - \mathbf{X}_l)}{|\mathbf{x} - \mathbf{x}'|} . \tag{2.13}$$

The second quantization representation is essentially a reformulation of the original problem and the task of reliably calculating the properties of the general Hamiltonian (2.12) is theoretically challenging. Instead of attempting a full diagonalization of the problem, one concentrates on the causal single-particle (and sometimes two-particle) Green function,

$$G_{\alpha\beta}(\mathbf{x} - \mathbf{x}', \tau - \tau') := -\langle T_\tau c_\alpha(\mathbf{x}, \tau) c_\beta^\dagger(\mathbf{x}', \tau') \rangle \,. \tag{2.14}$$

which takes the place as the central quantity similar to the density in DFT. The remainder of this thesis is mainly concerned with finding approximations to the Green function, or the self-energy. In reciprocal space, the latter is defined through $G(\mathbf{k}, i\omega)^{-1} = i\omega + \mu - h_\mathbf{k} - \Sigma(\mathbf{k}, i\omega)$. Here $h_\mathbf{k}$ is the Fourier transform of the hopping, the bare dispersion. The self-energy contains information about the electronic correlations. Spatial correlations – which are of particular interest here – manifest themselves in an explicit wavevector dependence of the self-energy. Approximate results should be examined to whether a feature is inherent to the model or an artefact of the particular approximation. This is difficult for the complex Hamiltonian (2.12). For the study of the basic physical mechanisms, simpler model Hamiltonians are traditionally used. These are derived from the general Hamiltonian by reducing the number of matrix elements to the dominant contributions, often assuming a short-range Coulomb interaction.

A prominent example obtained in this way is the single-band Hubbard model (or multi-orbital generalizations thereof). In a tight-binding approximation, the hopping matrix element in (2.13) is taken with respect to the atomic orbitals, which have little overlap with their neighbors. Therefore, the matrix elements are usually restricted to nearest-neighbor hopping $t$ and next-nearest-neighbor hopping $t'$. Due to screening, the local intraatomic matrix elements $U_{iiii}^{\alpha\beta\gamma\delta}$ are expected to strongly dominate. Hubbard proposed to restrict the Coulomb matrix to these elements [45, 46]. For a single band-model, the resulting Hamiltonian reads

$$H = \sum_{\langle ij\rangle\sigma} t_{ij} c_{i\sigma}^\dagger c_{j\sigma} + U \sum_i n_{i\uparrow} n_{i\downarrow} \,, \tag{2.15}$$

where $t_{ij} = t$ if $i$ is a nearest neighbor of $j$, $t_{ij} = t'$ for next-nearest-neighbors and zero otherwise. Despite the approximate nature of the model it is still under current active research. An exact (Bethe-Ansatz) solution is available only for the one-dimensional model [47, 48], which is a Luttinger liquid [49]. The two-dimensional square lattice model is a minimal model to study the phenomenon of high-temperature superconductivity. It is believed to capture the low energy physics of the cuprates [50]. Another aspect is the Mott transition [14, 9] which arises due to the interplay between the kinetic energy and Coulomb repulsion. The orbital selective Mott transition was studied in the two-orbital model [51]. The model was investigated on the Kagomé lattice to elucidate the effect of frustration [52]. The abundance of numerical results make the Hubbard model an ideal benchmark system for testing new theoretical approaches.

## 2.3 Coherent State Path Integral

For the most part in this thesis, the coherent state path integral formalism will be used and is shortly outlined in the following. The notation is adapted from [53]. The fermion coherent state is defined as

$$|\eta\rangle = e^{-\sum_\alpha \eta_\alpha c_\alpha^\dagger}|0\rangle \,, \tag{2.16}$$

where $|0\rangle$ is the vacuum of the Fock space and $\alpha$ labels the single-particle states. The coherent states are eigenstates of the annihilation operator,

$$c_\alpha|\eta\rangle = \eta_\alpha|\eta\rangle \,, \tag{2.17}$$

where it follows from the anticommutation relations of the operators that the eigenvalues must inhere this property: the $\eta_\alpha$'s are anticommuting Grassmann numbers. In contrast to boson coherent states, which most closely resemble classical behavior, fermion coherent states have no analogous physical interpretation. This is related to the fact that they do not belong to the physical fermion Fock space. The coherent states nevertheless form an overcomplete set in the physical fermion Fock space with closure relation

$$\int d\eta_\alpha^* d\eta_\alpha e^{-\sum_\alpha \eta_\alpha^* \eta_\alpha}|\eta\rangle\langle\eta| = \mathbb{1} \,, \tag{2.18}$$

which allows one to expand any physical fermion state in terms of coherent states. The partition function in the grandcanonical ensemble can be expressed in terms of coherent states as

$$\mathcal{Z} = \mathrm{Tr}\, e^{-\beta(H-\mu N)} = \sum_n \langle n|e^{-\beta(H-\mu N)}|n\rangle = \int d\eta_\alpha^* d\eta_\alpha e^{-\sum_\alpha \eta_\alpha^* \eta_\alpha}\langle -\eta|e^{-\beta(H-\mu N)}|\eta\rangle \,, \tag{2.19}$$

where the minus sign in $\langle -\eta|$ stems from the anticommutativity of the Grassman numbers. The partition function may be viewed as the trace over the imaginary time evolution operator $\mathcal{U} = e^{-\beta(H-\mu N)}$, which is obtained from its real time counterpart by a Wick rotation, i.e. the time axis is rotated in the complex time plane by letting $it \to \tau$. The path integral represents the sum over all imaginary time trajectories. Its construction is based on a formal discretization of the imaginary time interval from 0 to $\beta$ into $L$ slices of width $\epsilon = \beta/L$. Inserting the closure relation $L-1$ times, the partition function in the continuum limit $L \to \infty$ can be written

$$\mathcal{Z} = \lim_{L\to\infty} \int e^{-\sum_\alpha \eta_\alpha^* \eta_\alpha}\langle -\eta|\left(e^{-\epsilon(H-\mu N)}\right)^L|\eta\rangle \prod_{l=0}^L d\eta_{\alpha l}^* d\eta_{\alpha l}$$

$$= \lim_{L\to\infty} \int e^{-\sum_\alpha \eta_{\alpha L}^* \eta_{\alpha 0}} e^{-\sum_{l=0}^L \sum_\alpha \eta_{\alpha l}^* \eta_{\alpha l}}\langle \eta_L|e^{-\epsilon(H-\mu N)}|\eta_{L-1}\rangle \ldots \langle \eta_1|e^{-\epsilon(H-\mu N)}|\eta_0\rangle \prod_{l=0}^L d\eta_{\alpha l}^* d\eta_{\alpha l} \,. \tag{2.20}$$

In order to evaluate the matrix elements, the operators are taken in normal ordered form, with all annihilation operators appearing to the right of the creation operators. If $H$ is normal ordered, the exponential is normal ordered up to an error of order $\epsilon^2$ (as obvious from an expansion of the exponential), which vanishes in the continuum limit. Thus a matrix element evaluates to $\langle \eta_l | e^{-\epsilon H[c^\dagger, c]} | \eta_{l-1} \rangle = e^{\sum_\alpha \eta^*_{\alpha l} \eta_{\alpha l-1} - \epsilon H[\eta^*_{\alpha l}, \eta_{\alpha l-1}]} + O(\epsilon^2)$. Note that for a Hubbard-type interaction $U n_\uparrow n_\downarrow = +U c^\dagger_\uparrow c^\dagger_\downarrow c_\downarrow c_\uparrow$ and the creators act on a time slice an instant later than the annihilators.

Using the fact that the trace imposes antiperiodic boundary conditions, $\eta_i = -\eta_f$ and $\eta^*_L = -\eta_f$, $\eta_0 = \eta_i$, where $i$ and $f$ label initial and final states, the discrete expression for the action is given by

$$S(\eta^*, \eta) = \epsilon \sum_{l=1}^{L} \left( \sum_\alpha \eta^*_{\alpha l} \left\{ \frac{(\eta_{\alpha l} - \eta_{\alpha l-1})}{\epsilon} - \mu \eta_{\alpha l-1} \right\} + H[\eta^*_{\alpha l}, \eta_{\alpha l-1}] \right). \qquad (2.21)$$

In the continuum limit, the finite difference is symbolically replaced by a derivative, which yields the partition function in its final form

$$\mathcal{Z} = \int_{\eta_f = -\eta_i} e^{-\int_0^\beta d\tau \{ \sum_\alpha \eta^*_\alpha(\tau)[\partial_\tau - \mu] \eta_\alpha(\tau) + H[\eta^*_\alpha(\tau), \eta_\alpha(\tau)] \}} \mathcal{D}[\eta^*_\alpha(\tau), \eta_\alpha(\tau)]. \qquad (2.22)$$

The measure denotes integration over all (antiperiodic) trajectories. In the remainder of this thesis, the coherent state path integral will be denoted in the short hand notation

$$\mathcal{Z} = \int e^{-S[c^*, c]} \mathcal{D}[c^*, c], \qquad (2.23)$$

where $c$ and $c^*$ are used to label the Grassmann fields. While the action formulation will be used for the most part, the Hamiltonian formulation (the first equality in Eq. 2.19) is equivalent and can be used instead.

When defining averages, e.g. for Green's function, operators are interlaced between coherent states according to the time slice on which they are defined. Hence the path integral by construction represents time ordered products. Therefore time ordering will not explicitly be indicated when denoting expectation values in terms of path integrals. Notwithstanding the fact that the Green function defined in terms of operators (2.14) is not normal-ordered, the Green function will be denoted in terms of the path integral as

$$G_{12} := -\langle c_1 c_2^* \rangle \equiv \int c_2^* c_1 e^{-S[c^*, c]} \mathcal{D}[c^*, c], \qquad (2.24)$$

where the index 1 comprises frequency, momentum, spin and orbital indices.

## 2.4   Anderson Impurity Model

Nowadays, the Anderson impurity model is widely used in condensed matter physics. The model has originally been introduced by P. W. Anderson to study impurities with open d-shells, such as iron, cobalt and nickel, embedded in a metallic host and the associated local moment formation [54, 55]. The model assumes that electrons interact only locally on the impurity (the d-electrons). The impurity can exchange electrons with noninteracting electron bands (the conduction electrons). As shown below, the conduction electrons can be integrated out exactly, rendering the problem a purely local one. One then speaks of the impurity exerted to a "bath" of free electrons.

Apart from the original purpose, the model has further been employed to describe a variety of systems, ranging from atoms on metallic substrates [56] to quantum dots, in and out-of equilibrium [57] and to study the Kondo effect [4]. It has experienced revived interest with the advent of dynamical mean-field theory [15, 12], where the interacting lattice problem is approximately mapped to a local quantum impurity problem supplemented with a self-consistency condition (see Sec. 2.6). The dual fermion approach introduced in chapter 6 is also based on an impurity problem.

The Anderson model is a quantum many-body problem. The Hilbert space grows exponentially with an increasing number of on-site degrees of freedom. This renders a solution of the multiorbital model, which for example arises in the context of LDA+DMFT, a highly nontrivial task. The versatile applications have brought about a variety of solvers for the problem, and after nearly half a century past the original formulation, new solvers are still being developed. Rigorous methods such as the exact-diagonalization method [58] and Monte Carlo techniques provide numerically essentially exact solutions. A frequently used variant is the Hirsch-Fye method described in section 3.2. Recently, a new class of so-called continuous-time quantum Monte Carlo solvers has emerged. These are described in detail in chapter 3. In chapter 9, an approximate solver based on a perturbation expansion around the exact diagonalization solution is proposed.

In order to cover the following applications, the Hamiltonian of the Anderson impurity model is written in the following general form:

$$H = \sum_{k\alpha} \epsilon_{k\alpha} f^{\dagger}_{k\alpha} f_{k\alpha} + \sum_{k\alpha\beta} V_{k\alpha\beta} c^{\dagger}_{\alpha} f_{k\beta} + V^{*}_{k\beta\alpha} f^{\dagger}_{k\alpha} c_{\beta} + \sum_{\alpha} E_{\alpha} c^{\dagger}_{\alpha} c_{\alpha} + H_{\mathrm{loc}}[c^{\dagger}, c] . \qquad (2.25)$$

Greek letters are used for the combined index $\alpha \equiv \{m\sigma\}$, which labels spin and orbital degrees of freedom, respectively. Except when otherwise stated, it will be silently assumed that all quantities are diagonal in spin-space. The first term in (2.25) describes a gas of noninteracting fermions with dispersion $\epsilon_{k\alpha}$ where $k$ labels the single-particle states in band $\alpha$. The second term describes the processes of bath fermions hopping on or off the impurity with matrix amplitudes $V_{k\alpha\beta}$, that allow transitions from any band to any of the impurity levels $E_{\alpha}$. For applications within DMFT, let $E_{\alpha} \equiv -\mu$ (Eq. 2.55), where $\mu$ is the chemical potential. $H_{\text{loc}}$ is the local Hamiltonian for the impurity. For most applications, it is sufficient to write it in the form

$$H_{\text{loc}}[c^\dagger, c] = \sum_{\alpha\beta} t_{\text{loc}}^{\alpha\beta} c_\alpha^\dagger c_\beta + \frac{1}{2} \sum_{\alpha\beta\gamma\delta} U_{\alpha\beta\gamma\delta} c_\alpha^\dagger c_\beta^\dagger c_\delta c_\gamma , \tag{2.26}$$

which allows to account for interorbital transitions and the full Coulomb matrix. For example, the rotationally invariant Hamiltonian (2.98) in Sec. 2.9 has this form.

Repeating the steps in the preceding section, the partition function for the model can be written

$$\mathcal{Z} = \int e^{-S[c^*,c;f^*,f]} \mathcal{D}[c^*, c; f^*, f] , \tag{2.27}$$

where the imaginary time action is given by

$$S = -\int_0^\beta d\tau \int_0^\beta d\tau' \left\{ \sum_{k\alpha\beta} f_{k\alpha}^*(\tau) G_{f\,k\alpha\beta}^{-1}(\tau - \tau') f_{k\beta}(\tau') - \sum_{k\alpha\beta} \left[ c_\alpha^*(\tau) V_{k\,\alpha\beta} f_{k\beta}(\tau') \right. \right.$$
$$\left. \left. + f_{k\alpha}^*(\tau) V_{k\beta\alpha}^* c_\beta(\tau') \right] + \sum_{\alpha\beta} c_\alpha^*(\tau) G_{0\,\alpha\beta}^{-1}(\tau - \tau') c_\beta(\tau') \right\} + \int_0^\beta d\tau S_{\text{loc}}[c^*(\tau), c(\tau)] .$$

$$\tag{2.28}$$

The Green functions are defined by

$$[\partial_\tau + \epsilon_{k\alpha}] G_{f\,k\alpha\beta}(\tau - \tau') = -\delta_{\alpha\beta}\delta(\tau - \tau') \tag{2.29}$$

and

$$[\partial_\tau + E_\alpha] G_{0\,\alpha\beta}(\tau - \tau') = -\delta_{\alpha\beta}\delta(\tau - \tau') .$$

As usual, for fermions, these are subject to the antiperiodic boundary condition $G_{\alpha\beta}(0^-) = -G_{\alpha\beta}(\beta)$. Since the bath is noninteracting, the bath degrees of freedom described by $f$ and $f^*$ can be integrated out exactly from Eq. 2.28. This is facilitated by a frequently used Gaussian identity for Grassmann variables. For the present case it is written in the form (cf. appendix A.1)

$$\int \exp(-f_i^* a_{ij} f_j + c_i^* b_{ij} f_j + f_i^* \bar{b}_{ij} c_j) \mathcal{D}[f^*, f] = (\det a) \exp(c_i^* [b a^{-1} \bar{b}]_{ij} c_j) . \tag{2.30}$$

The resulting action reads

$$S_{\text{eff}} = -\sum_{\alpha\beta} \int_0^\beta d\tau \int_0^\beta d\tau' c_\alpha^*(\tau) \mathcal{G}_{\alpha\beta}^{-1}(\tau - \tau') c_\beta(\tau') + \int_0^\beta d\tau S_{\text{loc}}[c^*(\tau), c(\tau)] . \tag{2.31}$$

The determinant that arises by application of (2.30) is insignificant since it cancels in the calculation of expectation values (it corresponds to a constant shift of the Hamiltonian).

Once the bath degrees of freedom are integrated out, the "effective medium" Green function $\mathcal{G}(\tau - \tau')$ contains retardation effects: electrons can hop onto the impurity at time $\tau'$ and return to the bath at some later time $\tau$. In Fourier space, this function is given by

$$\mathcal{G}_{0\,\alpha\beta}^{-1}(i\omega) = (i\omega - E_\alpha)\delta_{\alpha\beta} - \Delta_{\alpha\beta}(i\omega) , \tag{2.32}$$

where the hybridization function of the bath,

$$\Delta_{\alpha\beta}(i\omega) = \sum_{k\gamma} \frac{V_{k\alpha\gamma} V_{k\beta\gamma}^*}{i\omega - \epsilon_{k\gamma}} \tag{2.33}$$

has been introduced. For the impurity solvers discussed in chapter 3, the impurity problem is used in this form. The hybridization function and the local Hamiltonian fully specify a given problem.

## 2.5 Exact Diagonalization

In the exact diagonalization method for the impurity problem, the bath described by the hybridization function is represented by a finite number of bath levels (orbitals). The full Hamiltonian for the system consisting of impurity and bath is evaluated in a suitable basis and diagonalized exactly. Some details, which are relevant for the superperturbation impurity solver presented in chapter 9, are given here.

Assume a system of $N_i$ orbitals on the impurity and $N_b$ bath states. In the occupation number basis, the many-particle Hilbert space is spanned by the vectors

$$|n_{1_i}^\uparrow, \ldots, n_{N_i}^\uparrow, n_{1_b}^\uparrow, \ldots, n_{N_b}^\uparrow\rangle |n_{1_i}^\downarrow, \ldots, n_{N_i}^\downarrow, n_{1_b}^\downarrow, \ldots, n_{N_b}^\downarrow\rangle , \tag{2.34}$$

where the occupation number is $n_i^\sigma = 0, 1$ for fermions. In principle, one can evaluate the Hamiltonian in this basis and diagonalize the resulting matrix. The diagonalization and evaluation of Green functions becomes unfeasible already for a small number of states, because the dimension $d$ of the Hilbert space increases with the number of orbitals exponentially as $2^{2(N_i+N_b)}$. For a single impurity orbital and six bath orbitals, the number of states is $2^{14} = 16384$. The memory requirement for the full matrix using double precision (16 byte) complex numbers is 4 GB.

It is obviously vital to take the symmetries of the system into account. In the present example, the total spin $\langle S_z \rangle$ and the total particle number $\langle N \rangle = \langle N_\uparrow \rangle + \langle N_\downarrow \rangle$ are conserved and the corresponding operators commute with the Hamiltonian:

$$[N_\uparrow, H] = [N_\downarrow, H] = [S_z, H] = 0 . \tag{2.35}$$

It follows that the Hamiltonian does not mix states with different particle number and different values of the total spin. These quantum numbers define a symmetry sector of the Hamiltonian. The Hamiltonian matrix takes a blockdiagonal structure so that it is possible to diagonalize it in each of the subspaces separately. This reduces the computational effort and memory requirements considerably (for example from $O(d^2)$ to $O(m(d/m)^2)$ for $m$ blocks of equal size). Depending on the problem, further symmetries may be present, such as translational symmetry and associated momentum conservation on a periodic cluster.

In exact diagonalization, the imaginary time Green function is evaluated as follows:

$$G_{\alpha\beta}(\tau) = -\langle c_\alpha(\tau)c_\beta^\dagger(0)\rangle = -\frac{1}{\mathcal{Z}}\mathrm{Tr}[e^{-\beta H}c_\alpha(\tau)c_\beta^\dagger(0)] \ . \tag{2.36}$$

Using the eigenbasis of the Hamiltonian, inserting the closure relation and writing the time dependence of the Heisenberg operators explicitly, this becomes

$$G_{\alpha\beta}(\tau) = -\frac{1}{\mathcal{Z}}\sum_{lm} e^{-\beta E_l}\langle l|e^{\tau H}c_\alpha e^{-\tau H}|m\rangle\langle m|c_\beta^\dagger|l\rangle = -\frac{1}{\mathcal{Z}}\sum_{lm} e^{-\beta E_l}e^{\tau(E_l-E_m)}\langle l|c_\alpha|m\rangle\langle m|c_\beta^\dagger|l\rangle \ ,$$

$$\tag{2.37}$$

where the partition sum is $\mathcal{Z} = \sum_l \exp(-\beta E_l)$. Fourier transform to Matsubara frequencies yields

$$G_{\alpha\beta}(i\omega) = -\frac{1}{\mathcal{Z}}\sum_{lm}\langle l|c_\alpha|m\rangle\langle m|c_\beta^\dagger|l\rangle e^{-\beta E_l}\int_0^\beta d\tau e^{\tau(i\omega+E_l-E_m)}$$

$$= \frac{1}{\mathcal{Z}}\sum_{lm}\frac{\langle l|c_\alpha|m\rangle\langle m|c_\beta^\dagger|l\rangle}{i\omega_n + E_l - E_m}\left(e^{-\beta E_l} + e^{-\beta E_m}\right) \ , \tag{2.38}$$

where $\exp(-\beta E_l)\{\exp[\beta(E_l - E_m)]+1\} = \exp(-\beta E_l)+\exp(-\beta E_m)$ and $\exp(i\beta\omega_n) = -1$ by definition of the Matsubara frequencies. The evaluation of the Green function requires to compute the trace over a product of operator matrices $C_{lm} := \langle l|c|m\rangle$. The transformation which block-diagonalizes the Hamiltonian also brings these matrices into a block structure. For example, the creation operator $c_\uparrow^\dagger$, will only have nonzero matrix elements between an initial state with a given spin-projection $S_z$ and particle number $N$ and a final state with spin $S_z + \frac{1}{2}$ and $N + 1$ particles. This allows to simplify the computation of the trace. Before performing the trace over all states $l$ in a given symmetry sector, one determines whether the trace is nonzero. This is done by checking if there is a "path", connecting a given symmetry sector with itself: The $c$-operator connects a state in symmetry sector $i$ with some other state in sector $j$ and the contribution to the trace is nonzero only if $c^\dagger$ connects sector $j$ with $i$. The benefit of using this procedure becomes significant for the calculation of the two-particle Green function and is also used for the computation of the trace which arises in the strong-coupling continuous-time solver discussed in Sec. 3.4. At low temperatures one may further exploit the fact that high-energy states do not contribute to the Green function. A cutoff energy $E_l$ may be defined by $\exp(-\beta E_l) < \epsilon$ where $\epsilon$ is some small number which, e.g., is taken to be equal to the effective machine precision. If both eigenvalues $E_l$, $E_m$ in (2.38) are above the thus defined cutoff, the contribution to the Green function need not be evaluated. Note that evaluation of the single-particle Green function therefore only requires knowledge of the low-energy part of the spectrum.

## 2.6 Dynamical Mean-Field Theory

Dynamical mean-field theory (DMFT) is by now a well-established description of strongly correlated electron systems [15, 12]. While the focus will be on Hubbard-like models in this thesis, DMFT has also been applied to a variety of lattice models, such as the Kondo-lattice model, the periodic Anderson model and the Falicov-Kimball model, including systems out of equilibrium.

DMFT may be viewed as a quantum analog for classical mean-field theories, which provides an intuitive understanding of the approach. It was realized that the construction of an approximation in analogy to the classical case results in a nontrivial mean-field theory, which fully takes into account the local, time dependent quantum fluctuations. As a consequence, the mean field cannot be represented by a single number, but rather by a time dependent field and hence the name dynamical.

DMFT is often introduced using the cavity construction. The underlying idea is to focus on a given site of the lattice and evaluate the effective field that couples to a local observable caused by all other sites on average. For the Ising model, this construction leads to the familiar Weiss mean-field result for the magnetization. Neglecting spatial variations of the magnetization allows to express the effective field caused by the surroundings in terms of the local quantity being sought, which can in turn be computed from the knowledge of the effective field. Since neither the expectation value of the local observable nor the effective field are known a priori, the problem needs to be solved self-consistently (in analogy to the familiar graphical solution of the transcendent equation in case of the Ising model).

### 2.6.1 Cavity Construction

The cavity construction of DMFT is outlined in the following. It is derived here in a general multiorbital form for Hubbard-like models, such as (2.15). For the derivation it is useful to decompose the action at site $i$ into the atomic part and a bond term containing the hopping:

$$S_i^{\mathrm{at}} = \sum_{\alpha\beta} \int_0^\beta d\tau \left\{ c_{i\alpha}^*(\tau)[\partial_\tau - \mu]c_{i\beta}(\tau) + S_{\mathrm{loc}}[c_i^*(\tau), c_i(\tau)] \right\}$$

$$S_{ij}^{\mathrm{t}} = \int_0^\beta d\tau \sum_{\alpha\beta} \left[ t_{ij}^{\alpha\beta} c_{i\alpha}^*(\tau)c_{j\beta}(\tau) + t_{ji}^{\beta\alpha} c_{j\beta}^*(\tau)c_{i\alpha}(\tau) \right] . \tag{2.39}$$

Focussing on site 0, the lattice action allows for the decomposition $S = S_0^{\mathrm{at}} + S^{(0)} + \Delta S$, where

$$S^{(0)} = \sum_{i\neq 0} S_i^{\mathrm{at}} + \sum_{\substack{\langle ij \rangle \\ i,j\neq 0}} S_{ij}^{\mathrm{t}} \quad \text{and} \quad \Delta S = \sum_j S_{0j}^{\mathrm{t}} . \tag{2.40}$$

$S^{(0)}$ denotes the action for the lattice in the presence of the cavity. It contains all bonds (sum over pairs $\langle ij \rangle$), except those coupling to site 0. $\Delta S$ contains the bonds connecting site 0 to the lattice. A given site is isolated from the rest of the lattice together with its bonds, thus creating a cavity. By integrating out the remainder of the lattice, the effect of all other sites is collected into the dynamical mean-field. The effective dynamics for the site is governed by an effective action obtained by integrating out the degrees of freedom on all sites except for site 0,

$$\frac{1}{\mathcal{Z}_{\text{eff}}} e^{-S_{\text{eff}}[c_0^*, c_0]} = \frac{1}{\mathcal{Z}} \int e^{-S[c_i^*, c_i]} \mathcal{D}[c_{i\neq0}^*, c_{i\neq0}] . \tag{2.41}$$

To this end, the partition function is rewritten using the above decomposition of $S$ and expanding in $\Delta S$:

$$\begin{aligned}
\mathcal{Z} &= \int e^{-S_0^{\text{at}}} \mathcal{D}[c_0^*, c_0] \int e^{-S^{(0)}} \sum_{n=0}^{\infty} \frac{1}{n!} (\Delta S)^n \mathcal{D}[c_{i\neq0}^*, c_{i\neq0}] \\
&= \int e^{-S_0^{\text{at}}} \mathcal{D}[c_0^*, c_0] \mathcal{Z}^{(0)} \sum_{\substack{n=0 \\ n \text{ even}}}^{\infty} \frac{1}{n!} \langle (\Delta S)^n \rangle^{(0)} ,
\end{aligned} \tag{2.42}$$

where the average is over the cavity action $\langle \ldots \rangle^{(0)} = (1/\mathcal{Z}^{(0)}) \int e^{-S^{(0)}} \ldots \mathcal{D}[c_{i\neq0}^*, c_{i\neq0}]$, which is nonzero only for processes conserving the number of particles on the cavity lattice and hence odd terms vanish in this expansion. The lowest nontrivial order reads

$$\frac{1}{2!} \langle (\Delta S)^2 \rangle^{(0)} = \sum_{\alpha\beta} \sum_{\gamma\delta} \int_0^\beta d\tau \int_0^\beta d\tau' c_{0\alpha}^*(\tau) \sum_{ij} t_{0i}^{\alpha\beta} t_{j0}^{\gamma\delta} \langle c_{i\beta}(\tau) c_{j\gamma}^*(\tau') \rangle^{(0)} c_{0\delta}(\tau') . \tag{2.43}$$

Note that the factor $1/n!$ cancels by expanding the product $(\Delta S)^n$ and collecting equivalent terms (the combinatorics is essentially the same as for the derivation of the dual potential in appendix A.3). Raising the sum to the power of an exponential is done by taking its logarithm. By the linked-cluster theorem, this corresponds to replacing the sum of diagrams in the average over the cavity action by the sum of all connected diagrams. As a result, the effective action can be written in the form

$$\begin{aligned}
S_{\text{eff}} = S_0^{\text{at}} + \sum_{n=0}^{\infty} \sum_{i_1 \ldots j_n} \sum_{\alpha_1\beta_1 \ldots \gamma_n\delta_n} & \int_0^\beta d\tau_1 \int_0^\beta d\tau_1' \ldots \int_0^\beta d\tau_n \int_0^\beta d\tau_n' \, t_{0i_1}^{\alpha_1\beta_1} t_{j_10}^{\gamma_1\delta_1} \ldots t_{0i_n}^{\alpha_n\beta_n} t_{j_n0}^{\gamma_n\delta_n} \times \\
& \times c_{0\alpha_1}^*(\tau_1) c_{0\delta_1}(\tau_1') \ldots c_{0\alpha_n}^*(\tau_n) c_{0\delta_n}(\tau_n') \, G_{i_1, j_1 \ldots i_n j_n}^{(0)\beta_1 \ldots \gamma_n}(\tau_1, \ldots, \tau_n; \tau_1', \ldots, \tau_n') ,
\end{aligned} \tag{2.44}$$

which involves the $2n$-point connected correlation functions $G_{i_1, j_1 \ldots i_n j_n}^{(0)\beta_1 \ldots \gamma_n}(\tau_1, \ldots, \tau_n') :=$ $(-1)^n \langle c_{i_1\beta_1}(\tau_1) c_{j_1\gamma_1}^*(\tau_1') \ldots c_{i_n\beta_n}(\tau_n) c_{j_n\gamma_n}^*(\tau_n') \rangle_{\text{con}}^{(0)}$ of the cavity action. Up to now this is a reformulation of the problem which is still unsuitable for practical calculations. An important observation is that the effective action drastically simplifies in the limit of infinite dimensions. It was realized by Metzner and Vollhardt [16], that the hopping amplitude has to be scaled with $1/\sqrt{z}$ (with $z$ being the coordination number; $z = 2d$ for the hypercubic lattice) to yield a nontrivial model in this limit. At the lowest nontrivial order, the fact that $G_{ij}^{(0)}$ is proportional to $t^{|i-j|}$, where $|i - j|$ is the Manhattan distance, implies that $G_{ij}^{(0)}$ falls off at least as $1/z$ (the Manhattan distance of two nearest neighbors of site 0 is at least 2). The prefactor of $G_{ij}^{(0)}$ in (2.44) is also of order $1/z$. When summed over all sites $i,j$, this term therefore turns out to be of order unity. By the same reasoning, the higher order terms are shown to fall off at least as $1/z$. As a result, only the lowest-order contribution survives in the limit of infinite dimensions. The effective action can hence be written in the form

$$S_{\text{eff}} = -\sum_{\alpha\delta} \int_0^\beta d\tau \int_0^\beta d\tau' c_{0\alpha}^*(\tau)\mathcal{G}_{\alpha\delta}^{-1}(\tau - \tau')c_{0\delta}(\tau') + \int_0^\beta d\tau \tilde{S}_{\text{loc}}[c^*(\tau), c(\tau)] . \quad (2.45)$$

For later purposes, the one-electron part $t_{\text{loc}}$ has been shifted from the local Hamiltonian $H_{\text{loc}} = \sum_{\alpha\beta} t_{\text{loc}}^{\alpha\beta} c_\alpha^* c_\beta$ to $\mathcal{G}$, which in analogy to the classical case is referred to as the Weiss field. It reads

$$\mathcal{G}_{\alpha\delta}^{-1}(i\omega) = (i\omega + \mu)\delta_{\alpha\delta} - t_{\text{loc}}^{\alpha\delta} - \sum_{ij} t_{0i}^{\alpha\beta} G_{ij}^{(0)\beta\gamma}(i\omega)t_{j0}^{\gamma\delta} . \quad (2.46)$$

This expression takes a particularly simple form on the Bethe lattice with nearest-neighbor hopping (Fig. 4.1). One needs to evaluate the sum corresponding to a hopping from site 0 (orbital $\alpha$) to any site $i$ (and orbital $\beta$), propagation on the cavity lattice described by $G^{(0)}$ from $i$ to any other site $j$ ($\beta$ to $\gamma$) and hopping back to site 0 into orbital $\delta$. When site 0 is removed from this lattice, the nearest neighbors of this site are disconnected, so that $G_{ij}^{(0)} = G_{ii}^{(0)}\delta_{ij}$. Since in the limit of infinite coordination number removing one site from this lattice does not change the Green function, one may further identify $G^{(0)}$ with $G$. Rescaling $t$ as $t \to t/\sqrt{z}$ and performing the sum over the $z$ neighbors ($t$ only connects nearest neighbors) yields the matrix relation

$$\mathcal{G}^{-1}(i\omega) = (i\omega + \mu)\mathbb{1} - t_{\text{loc}} - t\,G(i\omega)\,t . \quad (2.47)$$

Note that the quantities $t$ and Green functions are matrices. This closes the set of equations. The local observable $G_{\alpha\beta}(\tau - \tau') = -\langle c_\alpha(\tau)c_\beta^*(\tau')\rangle_{S_{\text{eff}}}$ for a given Weiss field is obtained from $S_{\text{eff}}$. Eq. 2.47 in turn relates the Weiss field to a given $G$, which required to relate $G$ to the cavity Green function in (2.46). For a general lattice, this relation reads

$$G_{ij}^{(0)} = G_{ij} - G_{i0}G_{00}^{-1}G_{0j} . \quad (2.48)$$

It was first derived by Hubbard in a different context [59]. For the evaluation of the self-consistency condition (2.46) for the multiband case one may use the prescription by Lichtenstein and Katsnelson [19]. Inserting (2.48) into (2.46) yields

$$\mathcal{G}^{-1}(i\omega) = (i\omega + \mu)\mathbb{1} - t_{\text{loc}} - \sum_{ij}\left[t_{0i}G_{ij}(i\omega)t_{j0} - t_{0i}G_{i0}(i\omega)G_{00}^{-1}(i\omega)G_{0j}(i\omega)t_{j0}\right] . \quad (2.49)$$

Translational invariance will always be assumed. Introducing the Fourier transform and its inverse for Green's function,

$$G_{ij}(i\omega) = \frac{1}{N}\sum_{K} G(\mathbf{k}, i\omega)e^{i\mathbf{k}(\mathbf{x}_i - \mathbf{x}_j)} , \qquad G(\mathbf{k}, i\omega) = \frac{1}{N}\sum_{ij} G_{ij}(i\omega)e^{-i\mathbf{k}(\mathbf{x}_i - \mathbf{x}_j)} \quad (2.50)$$

and similar for the hopping, $h_{\mathbf{k}} = \sum_j t_{j0}e^{-i\mathbf{k}\mathbf{x}_i}$, the sum in (2.49) can be expressed in the form

$$\mathcal{G}^{-1}(i\omega) = (i\omega + \mu)\mathbb{1} - t_{\text{loc}} - \frac{1}{N}\sum_{k} h_{\mathbf{k}}G(\mathbf{k}, i\omega)h_{\mathbf{k}} - \frac{1}{N^2}\sum_{kk'} h_{\mathbf{k}}G(\mathbf{k}, i\omega)G_{00}^{-1}(i\omega)G(\mathbf{k}', i\omega)h_{\mathbf{k}'} .$$
$$(2.51)$$

Introducing the local quantity $\Lambda(i\omega) = i\omega + \mu - t_{\text{loc}} - \Sigma(i\omega)$, one may express $h_{\mathbf{k}}$ as $\Lambda(i\omega) - G^{-1}(\mathbf{k}, i\omega)$ and one finds that, e.g.

$$\frac{1}{N}\sum_{k} h_{\mathbf{k}}G(\mathbf{k}, i\omega) = \frac{1}{N}\sum_{k}\left[\Lambda(i\omega) - G^{-1}(\mathbf{k}, i\omega)\right]G(\mathbf{k}, i\omega) = \Lambda(i\omega)\frac{1}{N}\sum_{k}[G(\mathbf{k}, i\omega) - \mathbb{1}]$$
$$(2.52)$$

and analogously for $(1/N)\sum_{k} G(\mathbf{k}, i\omega)h_{\mathbf{k}}$. The term $(1/N)\sum_{k} G(\mathbf{k}, i\omega) = G_{00}(i\omega)$ is the local Green function. Similarly,

$$\frac{1}{N}\sum_{k} h_{\mathbf{k}}G(\mathbf{k}, i\omega)[\Lambda(i\omega) - G^{-1}(\mathbf{k}, i\omega)] = \frac{1}{N}\sum_{k} h_{\mathbf{k}}G(\mathbf{k}, i\omega)\Lambda(i\omega) - \frac{1}{N}\sum_{k} h_{\mathbf{k}}$$
$$= \Lambda(i\omega)G_{00}(i\omega)\Lambda(i\omega) - \Lambda(i\omega) , \quad (2.53)$$

where (2.52) and $\sum_{\mathbf{k}} h_{\mathbf{k}} = 0$ were used. Inserting this back into (2.51) leads to the final result

$$\mathcal{G}^{-1}(i\omega) = (i\omega + \mu)\mathbb{1} - t_{\text{loc}} - \Lambda(i\omega) + G_{00}^{-1}(i\omega) = \Sigma(i\omega) + G_{00}^{-1}(i\omega) . \quad (2.54)$$

A connection to the Anderson impurity model can be established by considering an effective Hamiltonian

$$H_{\text{eff}} = \sum_{k\alpha}\tilde{\epsilon}_{k\alpha}f_{k\alpha}^{\dagger}f_{k\alpha} + \sum_{k\alpha\beta}\tilde{V}_{k\alpha\beta}c_{\alpha}^{\dagger}f_{k\beta} + \tilde{V}_{k\beta\alpha}^{*}f_{k\alpha}^{\dagger}c_{\beta} - \mu\sum_{\alpha}c_{\alpha}^{\dagger}c_{\alpha} + H_{\text{loc}}[c^{\dagger}, c]. \quad (2.55)$$

According to Sec. 2.4, this Hamiltonian gives rise to an action of the form 2.45, for which the bare Green function of the impurity model reads

$$\mathcal{G}^{-1}(i\omega) = (i\omega + \mu)\mathbb{1} - t_{\text{loc}} - \Delta(i\omega) . \quad (2.56)$$

The hybridization function expressed in terms of the effective parameters $\tilde{\epsilon}_k$ and $\tilde{V}_k$ according to Sec. 2.4 is given by

$$\Delta_{\alpha\beta}(i\omega) = \sum_{k\gamma} \frac{\tilde{V}_{k\alpha\gamma}\tilde{V}^*_{k\gamma\beta}}{i\omega - \tilde{\epsilon}_{k\gamma}} \, . \tag{2.57}$$

The representation for $\Delta(i\omega)$ is not unique; it depends on the particular choice of the effective Hamiltonian for which other choices are possible. Eq. 2.54 can now be fulfilled by determining the effective parameters or the hybridization function $\Delta(i\omega)$ such that

$$G^{-1}_{00}(i\omega) = (i\omega + \mu)\mathbb{1} - t_{\mathrm{loc}} - \Delta(i\omega) - \Sigma(i\omega) \, . \tag{2.58}$$

Comparing this to Eq. 2.56, one sees that the right hand side of (2.58) is equal to the inverse of the interacting Green function of the impurity model $g^{-1}(i\omega)$ (which will always be denoted by lowercase $g$). Note that according to (2.54) the self-energy is determined from the local impurity problem. This equation also allows for an effective medium interpretation [15] in the spirit of the coherent potential approximation. One may envision the impurity embedded in a medium described by the Green function $G_{\mathrm{med}}(\mathbf{k}, i\omega)^{-1} = (i\omega + \mu)\mathbb{1} - h_{\mathbf{k}} - \Sigma(i\omega)$, where the interaction $H_{\mathrm{loc}}$ is added and the self-energy is subtracted on the impurity site. The medium can be integrated out exactly. Identifying the interacting Green function of the impurity model with the local part of the effective medium Green function is the same as identifying $\Sigma(i\omega)$ of the medium with the self-energy of the impurity model and recovers (2.54).

Before proceeding to the practical implementation, it is instructive to adapt a different point of view and consider the structure of the perturbation theory in infinite dimensions. By power-counting in $1/d$, similar to the reasoning above, it follows that if two interaction vertices in a diagram are connected by bare Green functions via at least three paths, only the contribution for which both vertices are located on the same site survives in the infinite-dimensional limit [15]. This is always the case for skeleton diagrams, which are constructed from the vertices and full interacting Green functions as lines. As a result, all diagrams in an expansion of the self-energy in terms of skeleton diagrams collapse to a single site, so that the self-energy becomes local. The same is true for the Luttinger-Ward functional $\Phi$ [60, 61], which is the sum over all vacuum to vacuum skeleton diagrams. The two lowest-order contributions are depicted in Fig. 2.1. All diagrams of this functional collapse to a single site, so that $\Phi$ is a functional of the local Green function only. The self-energy is obtained from the Luttinger-Ward functional by functional derivative with respect to $G$,

$$\Sigma_{ij}(i\omega) = \frac{\delta\Phi}{\delta G_{ji}(i\omega)} \, , \tag{2.59}$$

which diagrammatically corresponds to taking out a line connecting the vertices $i$ and

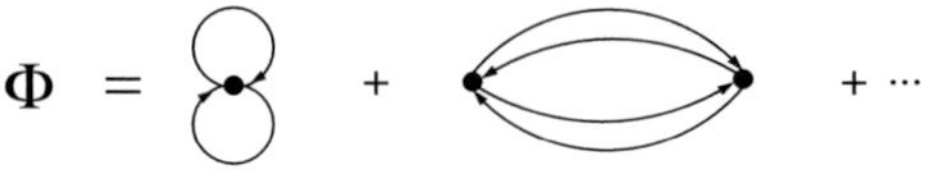

*Figure 2.1:* Lowest-order contributions to the Luttinger-Ward functional $\Phi$.

*j.* In accordance with the foregoing, it follows that the self-energy in the infinite-dimensional limit is local, $\Sigma = \Sigma(i\omega)$. For the perturbation theory in momentum space, the above property translates to ignoring momentum conservation at the vertices. Consider the Laue function, which for a two-leg vertex is defined as

$$\Delta^{\mathrm{L}} := \sum_{\mathbf{x}} e^{i\mathbf{x}(\mathbf{k}_1 - \mathbf{k}_2 + \mathbf{k}_3 - \mathbf{k}_4)} = N\delta_{\mathbf{k}_1 + \mathbf{k}_3, \mathbf{k}_2 + \mathbf{k}_4} \; . \tag{2.60}$$

As noted in Ref. [62], the Laue function in DMFT (for a general $n$-particle vertex) is approximated by a constant, $\Delta^{\mathrm{L}}_{\mathrm{DMFT}} \equiv 1$, which corresponds to completely ignoring momentum conservation at each vertex. As a consequence the momentum sums may be carried out for each propagator independently, which corresponds to replacing it by its local part.

## 2.6.2　Self-Consistency Loop

In general, neither the Weiss field nor the Green function are known a priori and need to be determined self-consistently. In practice, this is done by performing a self-consistency loop, as depicted in Fig. 2.2. The procedure is as follows: Starting from an initial guess for the self-energy, the lattice Green function is obtained as

$$G(\mathbf{k}, i\omega) = \left[(i\omega + \mu)\mathbb{1} - t_{\mathrm{loc}} - h_{\mathbf{k}} - \Sigma(i\omega)\right]^{-1} \; . \tag{2.61}$$

The local lattice Green function is obtained by summing (integrating) over all $\mathbf{k}$-points in the first Brillouin zone. Application of a local Dyson-like equation $\mathcal{G}^{-1}(i\omega) = \Sigma(i\omega) + G^{-1}(i\omega)$ yields the Weiss field, which is the input to the impurity solver. The solver is then employed to calculate the local Green function $g(i\omega)$. From this quantity, the new self-energy can be extracted using Dyson's equation. The calculation of the lattice Green function using the new guess for the self-energy closes the self-consistency loop. A measure of convergence is e.g. provided by evaluating $\| \Sigma(i\omega)_{\mathrm{new}} - \Sigma(i\omega)_{\mathrm{old}} \| :=$ $(1/N_{\mathrm{max}}) \sum_{n=0}^{N_{\mathrm{max}}-1} \sum_{\alpha\beta} \left| \Sigma_{\alpha\beta}(i\omega)_{\mathrm{new}} - \Sigma_{\alpha\beta}(i\omega)_{\mathrm{old}} \right|$, or a corresponding measure involving the local Green function. The iterations are stopped when this measure becomes smaller than some predetermined accuracy $\epsilon$. In order to enhance the convergence of the self-consistency loop and to avoid oscillations about the fixed point of the equations, the new

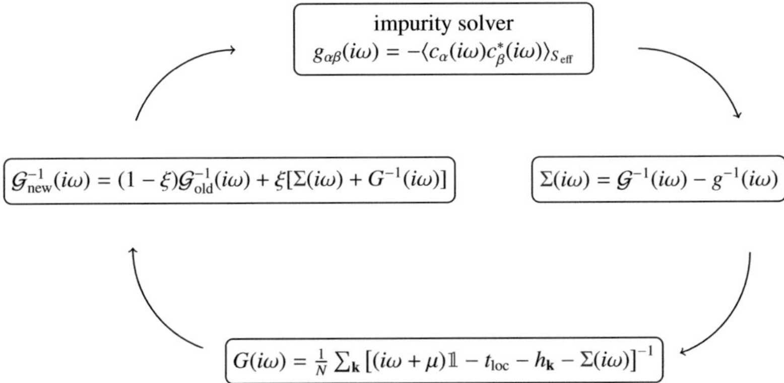



*Figure 2.2:* DMFT self-consistency loop using the self-energy $\Sigma(i\omega)$ as the adjustable quantity. The loop can be started by constructing the local Green function $G(i\omega)$, thereby taking $\Sigma(i\omega) \equiv 0$.

and old guesses for the Weiss field are mixed. In most of the cases encountered so far, a simple linear mixing has proven sufficient:

$$\mathcal{G}_{\text{new}}^{-1}(i\omega) = (1-\xi)\mathcal{G}_{\text{old}}^{-1}(i\omega) + \xi[\Sigma(i\omega) + G^{-1}(i\omega)]$$
$$= \mathcal{G}_{\text{old}}^{-1}(i\omega) + \xi[\Sigma(i\omega) + G^{-1}(i\omega) - \mathcal{G}_{\text{old}}^{-1}(i\omega)] \,. \tag{2.62}$$

In the above equation, $\Sigma(i\omega)$ is the new self-energy. Upon self-consistency the new and old self-energies coincide. Setting $\Sigma_{\text{new}} = \Sigma_{\text{old}}$ in (2.62) hence does not change the fixed point and allows one to write

$$\mathcal{G}_{\text{new}}^{-1}(i\omega) = \mathcal{G}_{\text{old}}^{-1}(i\omega) + \xi[G^{-1}(i\omega) - g_{\text{old}}^{-1}(i\omega)] \,, \tag{2.63}$$

with $g_{\text{old}}^{-1}(i\omega) = \mathcal{G}_{\text{old}}(i\omega) - \Sigma_{\text{old}}(i\omega)$. This explicitly shows that self-consistency is reached when the local part of the lattice Green function $G(i\omega)$ coincides with the Green function $g(i\omega)$ of the impurity model. The mixing parameter $\xi > 0$ controls the ratio of the new and old Weiss fields entering the next iteration. For $\xi$ too small, convergence may be slowed down unnecessarily, while for $\xi$ chosen too large, oscillations about the self-consistent solution may occur.

The choice of the starting guess for the self-energy may be based on a priori knowledge. If nothing is known, the guess $\Sigma(i\omega) \equiv 0$ may serve as the starting point. The initial $\mathcal{G}$ is simply calculated from the local part of the noninteracting Green function of the lattice by virtue of Eq. 2.54.

In general the self-consistency loop is well behaved and typically converges within a few iterations so that $\Sigma(i\omega) \equiv 0$ is a generic choice. In the vicinity of a second-order

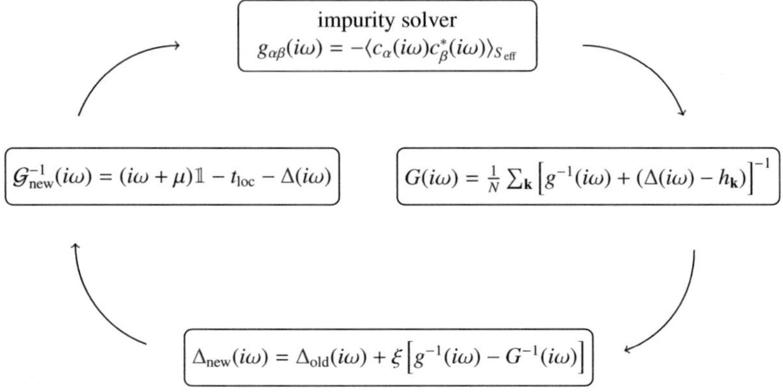



*Figure 2.3:* DMFT self-consistency loop using the hybridization function $\Delta(i\omega)$ as the adjustable quantity. Taking $\Sigma(i\omega) \equiv 0$ and $\Delta_{\text{old}}(i\omega) \equiv 0$ corresponds to $g^{-1}(i\omega) = (i\omega + \mu)\mathbb{1} - t_{\text{loc}}$ and generates the same initial guess for the local Green function as in Fig. 2.2.

transition however, the critical slowing down on the broken-symmetry side decelerates the convergence. It is enhanced by performig successive DMFT calculations with the control parameter of the transition approaching the critical value and by taking the self-consistent result for the self-energy from the previous DMFT calculation as an initial guess for the next. An example where this has been used is given in chapter 4. In the coexistence region of a first-order metal-insulator transition the solution to the DMFT equations is not unique and the solution depends on the initial guess. The coexistence region can be isolated by independently approaching the transition from the metallic and isolating sides, again using the self-consistent solution from the previous calculation as the initial guess. An application can be found in chapter 6.8.

In the self-consistency loop described above, the self-energy is the quantity that is adjusted through the iterations. Here a second formulation is given, which is the natural one for the dual fermion approach introduced in chapter 6. Instead of the self-energy, the hybridization function is adjusted through the iterations. The self-energy need not be explicitly evaluated, since the local part of the lattice Green function, Eq. 2.61, can be expressed in terms of the hybridization function and the bare Hamiltonian as[2]

$$G(i\omega) = \frac{1}{N} \sum_{\mathbf{k}} \left[ g^{-1}(i\omega) + (\Delta(i\omega) - h_{\mathbf{k}}) \right]^{-1} , \tag{2.64}$$

where $g^{-1}(i\omega) = (i\omega + \mu)\mathbb{1} - t_{\text{loc}} - \Delta(i\omega) - \Sigma(i\omega)$. The new guess for the hybridization is obtained as follows:

$$\Delta_{\text{new}}(i\omega) = (1 - \xi)\Delta_{\text{old}}(i\omega) + \xi[(i\omega + \mu)\mathbb{1} - t_{\text{loc}} - \Sigma_{\text{old}}(i\omega) - G^{-1}(i\omega)]$$

$$= \Delta_{\text{old}}(i\omega) + \xi[g_{\text{new}}^{-1}(i\omega) - G^{-1}(i\omega)] \,. \tag{2.65}$$

Here $\xi > 0$ as before and $g_{\text{new}}(i\omega)$ has been inserted in place of $g_{\text{old}}(i\omega)$ in $\Delta_{\text{old}}(i\omega) = (i\omega + \mu)\mathbb{1} - t_{\text{loc}} - \Sigma_{\text{old}}(i\omega) - g_{\text{old}}^{-1}(i\omega)$ for convenience. The two self-consistency loops are obviously equivalent and have the same fixed point, $g = G$.

Note that the self-consistency condition can conveniently be used to generate the initial guess for the hybridization function corresponding to $\Sigma \equiv 0$. It is given by

$$\Delta_{\text{init}}(i\omega) = (i\omega + \mu)\mathbb{1} - t_{\text{loc}} - G_{0\,\text{loc}}^{-1}(i\omega) \,, \tag{2.66}$$

where $G_{0\,\text{loc}}$ is the local part of the lattice Green function (2.61) evaluated for $\Sigma \equiv 0$. Using $\xi = 1$, $\Delta_{\text{old}}(i\omega) \equiv 0$, and constructing $g(i\omega)$ and $G(i\omega)$ for $\Sigma(i\omega) \equiv 0$ in (2.65) is equivalent to (2.66).

## 2.7 Cluster Extensions of DMFT

In quantum cluster approaches (for a review, see Ref. [35]), the lattice problem with infinite degrees of freedom is reduced to a cluster problem with less degrees of freedom. The cluster need not be a physical subsystem of the original lattice [34, 63, 64]. Here the cellular DMFT (CDMFT), the dynamical cluster approximation (DCA) and the variational cluster approximation (VCA) will be shortly outlined. The introduction of the CDMFT provides the basis for the discussion of the cluster dual fermion approach introduced in chapter 8.

The notation is adapted from Ref. [35]. The $d$-dimensional lattice containing $N$ sites is grouped into clusters of linear dimension $L_{\text{c}}$ containing $N_{\text{c}} = L_{\text{c}}^d$ sites. As depicted in Fig. 2.4 a) for $d = 1$, the lattice vectors are decomposed as $\mathbf{x} = \mathbf{X} + \tilde{\mathbf{x}}$, where the vector $\tilde{\mathbf{x}}$ denotes the position of a cluster within the superlattice and $\mathbf{X}$ labels sites within the cluster. The reciprocal space is split into cells accordingly, as show in b). A wave vector in the original lattice is given by $\mathbf{k} = \tilde{\mathbf{k}} + \mathbf{K}$, where $\tilde{\mathbf{k}}$ is a superlattice wavevector and $\mathbf{K}$ is the cluster momentum. The number of clusters or superlattice momenta is $N/N_{\text{c}}$. Tiling the lattice into clusters breaks translational invariance of the original lattice. While superlattice momentum is still conserved, the lattice momentum is conserved up to $\mathbf{K}$ only. The cluster momentum $\mathbf{K}$ has components $K_i = n_i(2\pi/L_{\text{c}})$ and becomes a reciprocal lattice vector, i.e. $\exp(i\mathbf{K}\tilde{\mathbf{x}}) \equiv 1$. The Fourier transform and

---

[2]It may appear as if (2.64) is an implicit relation for determining $g(i\omega)$, since self-consistency requires $g = G$. However, both $g$ and $G$ contain the self energy, which in turn is a functional of the hybridization and needs to be determined through the solution of the impurity problem.

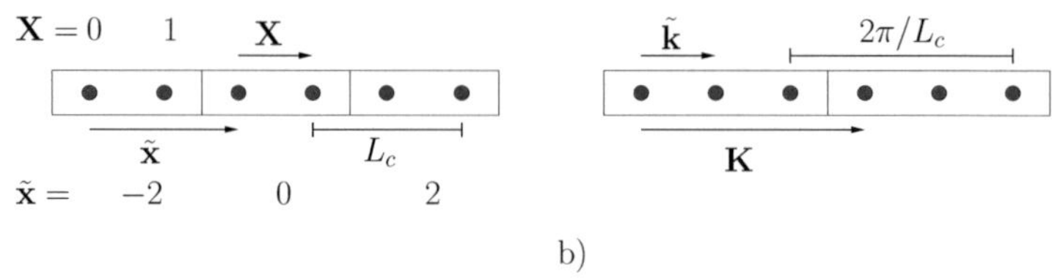


*Figure 2.4:* a) Division of a one-dimensional chain into real-space clusters of size $L_c = 2$. A vector of the original lattice is decomposed as $\mathbf{x} = \tilde{\mathbf{x}} + \mathbf{X}$, where $\tilde{\mathbf{x}}$ points to the origin of the clusters in the superlattice and $\mathbf{X}$ labels sites within a cluster. All vectors are in units of the (original) lattice spacing. b) Corresponding division of the reciprocal space into cells of width $2\pi/L_c$. Points within a cell are connected by superlattice momentum $\tilde{\mathbf{k}}$ and cells are connected by the cluster momentum $\mathbf{K}$.

its inverse of a quantity $G$ are thus given by

$$G(\mathbf{K}, \tilde{\mathbf{k}}) = \sum_{\mathbf{X}\tilde{\mathbf{x}}} G(\mathbf{X}, \tilde{\mathbf{x}})e^{-i[(\mathbf{K}+\tilde{\mathbf{k}})\mathbf{X}+\tilde{\mathbf{k}}\tilde{\mathbf{x}}]} , \qquad G(\mathbf{X}, \tilde{\mathbf{x}}) = \frac{1}{N}\sum_{\mathbf{K}\tilde{\mathbf{k}}} G(\mathbf{K}, \tilde{\mathbf{k}})e^{i[(\mathbf{K}+\tilde{\mathbf{k}})\mathbf{X}+\tilde{\mathbf{k}}\tilde{\mathbf{x}}]} . \quad (2.67)$$

The sum over $\tilde{\mathbf{x}}$ yields the superlattice transform and summing over $\mathbf{X}$ corresponds to the intracluster transform.

Following Ref. [35], CDMFT can be obtained from a locator expansion, which is an expansion in real space around the finite cluster. To this end, the hopping and the self-energy are split into intercluster and intracluster parts,

$$t(\tilde{\mathbf{x}} - \tilde{\mathbf{x}}') = \delta t(\tilde{\mathbf{x}} - \tilde{\mathbf{x}}') + t_c\delta_{\tilde{\mathbf{x}},\tilde{\mathbf{x}}'} ,$$
$$\tilde{\Sigma}(\tilde{\mathbf{x}} - \tilde{\mathbf{x}}', i\omega) = \delta\Sigma(\tilde{\mathbf{x}} - \tilde{\mathbf{x}}', i\omega) + \Sigma_c(i\omega)\delta_{\tilde{\mathbf{x}},\tilde{\mathbf{x}}'} . \qquad (2.68)$$

All quantities are matrices in the cluster sites. The locator expansion around the cluster described by the cluster Green function $g_c(i\omega) = [(i\omega + \mu)\mathbb{1} - t_c - \Sigma_c(i\omega)]^{-1}$ in $\delta t$ and $\delta\Sigma$ reads

$$G(\tilde{\mathbf{x}} - \tilde{\mathbf{x}}', i\omega) = g_c(i\omega)\delta_{\tilde{\mathbf{x}},\tilde{\mathbf{x}}'} + g_c(i\omega)\sum_i [\delta t(\tilde{\mathbf{x}} - \tilde{\mathbf{x}}_i) + \delta\Sigma(\tilde{\mathbf{x}} - \tilde{\mathbf{x}}_i, i\omega)]G(\tilde{\mathbf{x}}_i - \tilde{\mathbf{x}}', i\omega) . \quad (2.69)$$

In the cluster approaches, correlations beyond the extension of the cluster are neglected and the remainder of the system is assumed to be uncorrelated. This corresponds to neglecting the intercluster self-energy $\delta\Sigma$ and allows one to map the lattice problem to a cluster embedded in an uncorrelated host. Due to the translational invariance of the superlattice, Eq. 2.69 is diagonalized with respect to superlattice momenta by Fourier transform and with the approximation $\delta\Sigma = 0$ reads

$$G(\tilde{\mathbf{k}}, i\omega) = g_c(i\omega) + g_c(i\omega)\delta t(\tilde{\mathbf{k}})G(\tilde{\mathbf{k}}, i\omega) = [g_c^{-1}(i\omega) - \delta t(\tilde{\mathbf{k}})]^{-1} . \qquad (2.70)$$

A corresponding relation restricted to the cluster is obtained by averaging this equation over all superlattice momenta,

$$\bar{G}(i\omega) = \frac{N_c}{N} \sum_{\tilde{\mathbf{k}}} G(\tilde{\mathbf{k}}, i\omega) = \frac{N_c}{N} \sum_{\tilde{\mathbf{k}}} [g_c^{-1}(i\omega) - \delta t(\tilde{\mathbf{k}})]^{-1} . \tag{2.71}$$

This step is referred to as coarse-graining. It corresponds to neglecting the phase factors $e^{i\tilde{\mathbf{k}}\tilde{\mathbf{x}}}$ on the vertices of self-energy diagrams which are associated with the position of the cluster in the original superlattice. The cluster Green function contains the self-energy $\Sigma_c(i\omega)$, which can therefore be determined as a functional of the coarse-grained Green function $\bar{G}(i\omega)$ from the solution of an impurity model. Since $g_c(i\omega)$ is independent of $\tilde{\mathbf{k}}$, it is possible to determine a local function $\Delta(i\omega)$ such that

$$\bar{G}(i\omega)^{-1} = g_c^{-1}(i\omega) - \Delta(i\omega) , \tag{2.72}$$

which defines the hybridization function. In order to establish the relation to an impurity model, one defines the excluded cluster Green function $\mathcal{G}^{-1}(i\omega) = \bar{G}^{-1}(i\omega) + \Sigma_c(i\omega)$ as the bare Green function to $\bar{G}$. By comparison with (2.72) it follows that

$$\mathcal{G}^{-1}(i\omega) = (i\omega + \mu)\mathbb{1} - t_c - \Delta(i\omega) . \tag{2.73}$$

The interacting Green function of the impurity model $g(i\omega)$ is hence related to the cluster Green function $g_c(i\omega)$ by

$$g^{-1}(i\omega) = \mathcal{G}^{-1}(i\omega) - \Sigma_c(i\omega) = g_c^{-1}(i\omega) - \Delta(i\omega) , \tag{2.74}$$

so that Eq. 2.72 is seen to be equivalent to requiring that the coarse-grained Green function $\bar{G}$ be equal to the Green function $g$ of the effective cluster impurity model. In particular, Eq. 2.71 takes the form

$$\bar{G}(i\omega) = \frac{N_c}{N} \sum_{\tilde{\mathbf{k}}} \left\{ g^{-1}(i\omega) + [\Delta(i\omega) - \delta t(\tilde{\mathbf{k}})] \right\}^{-1} . \tag{2.75}$$

These equations clearly resemble those in Sec. 2.6.2. While here they have been obtained from the locator expansion, they are covered by the derivation in Sec. 2.6.1. Using the local degrees of freedom to label sites within a cluster, instead of (or in addition to) orbitals and identifying the intra- and inter-cluster parts $t_c$ and $\delta t$ of the hopping with $t_{loc}$ and $h_{\mathbf{k}}$, the self-consistency loops in Sec. 2.6.2 are seen to be general enough to accommodate an implementation of the CDMFT equations[3]. Examples of CDMFT calculations can be found in chapters 4 and 8.

---

[3] In the notation of Sec. 2.6.1, $N$ should then be replaced the number of clusters $N/N_c$ and the momentum $\mathbf{k}$ should be identified with the superlattice momentum $\tilde{\mathbf{k}}$.

As mentioned previously, CDMFT neglects the phase factors $e^{i\tilde{k}\tilde{x}}$. Correspondingly, the CDMFT approximation to the Laue function reads

$$\Delta^{L}_{\text{CDMFT}} = \sum_{\mathbf{X}} e^{i\mathbf{X}(\mathbf{K}_1+\tilde{\mathbf{k}}_1+\mathbf{K}_2+\tilde{\mathbf{k}}_2-\mathbf{K}_3-\tilde{\mathbf{k}}_3-\mathbf{K}_4-\tilde{\mathbf{k}}_4)} \,, \tag{2.76}$$

which shows that cluster momentum $\mathbf{K}$ is not conserved due to the presence of the phase factors $e^{i\mathbf{kX}}$. Hence CDMFT violates translational invariance with respect to the cluster sites and the cluster sites are not equivalent. This is obvious for clusters with $L_c \geq 3$, where bulk and surface sites of a cluster may be distinguished, but also applies for $L_c = 2$. CDMFT calculations are carried out in the cluster real-space representation (i.e. all quantities are matrices in the cluster sites), since there is no benefit in changing to the cluster $\mathbf{k}$-space representation, which is not diagonal.

Since translational invariance is broken, the lattice quantities are functions of two independent momenta $\mathbf{k}$ and $\mathbf{k}'$. The latter can differ by a reciprocal lattice vector $\mathbf{Q}$, where $Q_i = 0, \ldots, (L_c - 1)2\pi/L_c$. For example, the self-energy is expressed in terms of the cluster self-energy as

$$\Sigma(\mathbf{k}, \mathbf{k}', i\omega) = \frac{1}{N_c} \sum_{\mathbf{Q}} \sum_{\mathbf{X},\mathbf{X}'} e^{i\mathbf{kX}} \Sigma_c(\mathbf{X}, \mathbf{X}', i\omega) e^{-i\mathbf{k}'\mathbf{X}'} \delta(\mathbf{k} - \mathbf{k}' - \mathbf{Q}) \,, \tag{2.77}$$

where the dependence on cluster sites is written explicitly. A translationally invariant solution is obtained by approximating the lattice quantities by the (homogeneous) $\mathbf{Q} = 0$ contribution,

$$\Sigma(\mathbf{k}, i\omega) = \frac{1}{N_c} \sum_{\mathbf{X},\mathbf{X}'} = e^{i\mathbf{k}(\mathbf{X}-\mathbf{X}')} \Sigma_c(\mathbf{X}, \mathbf{X}', i\omega) \,. \tag{2.78}$$

Transforming back to real space shows that the lattice quantities for a given distance $\mathbf{x} - \mathbf{x}'$ are obtained as an average over the cluster quantities for the same distance,

$$\Sigma(\mathbf{x} - \mathbf{x}', i\omega) = \frac{1}{N_c} \sum_{\mathbf{X},\mathbf{X}'} \Sigma_c(\mathbf{X}, \mathbf{X}', i\omega) \, \delta_{\mathbf{X}-\mathbf{X}',\mathbf{x}-\mathbf{x}'} \,. \tag{2.79}$$

Spatial correlations are hence included up to a length determined by the extension of the cluster. Note that (2.79) underestimates the nonlocal contributions, in particular for small clusters. Using the shorthand notation $\Sigma_{\mathbf{X},\mathbf{X}'} = \Sigma(\mathbf{X}, \mathbf{X}')$, one sees that the local self-energy is averaged correctly, $\Sigma(x = 0) = (\Sigma_{c\,00} + \Sigma_{c\,11})/2$, while the nearest-neighbor self-energy contribution according to (2.79) would read $\Sigma(x = 1) = (1/2)\Sigma_{c\,10}$, since $\Sigma_{c\,01}$ contributes to $\Sigma(x = -1)$. It was therefore suggested to reweigh the terms in the sum [65]. For the above example, $\Sigma(x = 1) = \Sigma_{c\,10}$.

When translational invariance is recovered in this way, the solution of the lattice problem may be viewed as shown in Fig. 2.5: The lattice is replaced by a lattice of clusters all of which are embedded in a self-consistent bath. The self-energy on a cluster

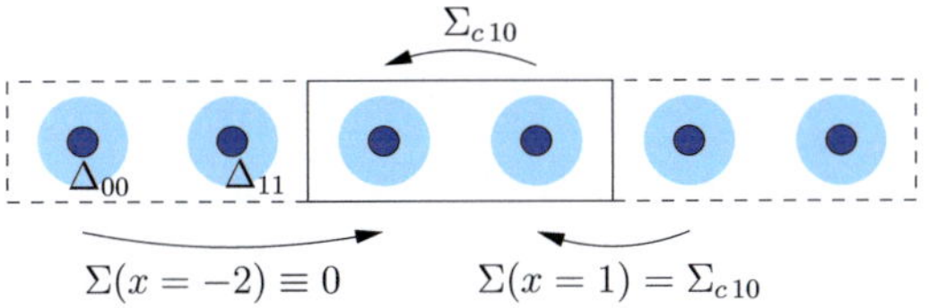


*Figure 2.5:* Illustration of the CDMFT lattice self-energy. The original lattice is replaced by a collection of clusters embedded in a self-consistent bath. For distances not exceeding the maximal distance between cluster sites, the intercluster self-energy $\Sigma(x = 1)$ is approximated by the intracluster self-energy $\Sigma_{c\,10}$ and zero otherwise.

is obtained from the self-consistent solution of the local problem and the intercluster self-energy between sites on neighboring clusters at a distance $\mathbf{x} - \mathbf{x}'$ is artificially set equal to the average of the intracluster self-energy for the same distance. The self-energies for distances exceeding the maximum distance between sites within the cluster are zero. The cluster dual fermion approach presented in chapter 8 is concerned with reintroducing the intercluster self-energy between clusters at all distances perturbatively.

The idea of the DCA is to restore momentum conservation within the cluster by a different choice of the intracluster Fourier transform. In CDMFT, the intracluster transform of the dispersion reads

$$[\mathbf{t}(\tilde{\mathbf{k}})]_{\mathbf{X},\mathbf{X}'} = \frac{1}{N_c} \sum_{\mathbf{K}} e^{i(\mathbf{K}+\tilde{\mathbf{k}})(\mathbf{X}-\mathbf{X}')} \epsilon_{\mathbf{K}+\tilde{\mathbf{k}}} \,, \tag{2.80}$$

while in the DCA, the phase factors $e^{i\tilde{\mathbf{k}}\mathbf{X}}$ are excluded by using the transform

$$[\mathbf{t}_{\mathrm{DCA}}(\tilde{\mathbf{k}})]_{\mathbf{X},\mathbf{X}'} = [\mathbf{t}(\tilde{\mathbf{k}})]_{\mathbf{X},\mathbf{X}'} e^{-i\tilde{\mathbf{k}}(\mathbf{X}-\mathbf{X}')} = \frac{1}{N_c} \sum_{\mathbf{K}} e^{i\mathbf{K}(\mathbf{X}-\mathbf{X}')} \epsilon_{\mathbf{K}+\tilde{\mathbf{k}}} \,. \tag{2.81}$$

Correspondingly, the Laue function reads

$$\Delta_{\mathrm{DCA}}^{\mathrm{L}} = \sum_{\mathbf{X}} e^{i\mathbf{X}(\mathbf{K}_1+\mathbf{K}_2-\mathbf{K}_3-\mathbf{K}_4)} = N_c \delta_{\mathbf{K}_1+\mathbf{K}_2,\mathbf{K}_1'+\mathbf{K}_2'} \,, \tag{2.82}$$

which shows that cluster momentum is conserved. The intracluster hopping in DCA is therefore given by the intracluster Fourier transform of the dispersion, which is obvious by coarse-graining Eq. 2.81. Denoting the coarse-grained hopping as $\bar{\epsilon}_{\mathbf{K}}$, one has $\delta t(\mathbf{K} + \tilde{\mathbf{k}}) = \epsilon_{\mathbf{K}+\tilde{\mathbf{k}}} - \bar{\epsilon}_{\mathbf{K}}$. The analog of (2.70) is hence diagonal in cluster Fourier space:

$$G(\mathbf{K} + \tilde{\mathbf{k}}, i\omega) = g_c(\mathbf{K}, i\omega) + g_c(\mathbf{K}, i\omega)\delta t(\mathbf{K} + \tilde{\mathbf{k}})G(\mathbf{K} + \tilde{\mathbf{k}}, i\omega)$$

$$= \frac{1}{1/g_c(\mathbf{K}, i\omega) - \delta t(\mathbf{K} + \tilde{\mathbf{k}})} \,. \tag{2.83}$$

By viewing these quantities as diagonal matrices with respect to cluster momenta, the equations 2.72-2.75 also apply for the DCA. The self-energy becomes a piecewise constant function in k-space [35].

The variational cluster approach (VCA) is based on the self-energy functional theory (SFT) [66]. In SFT, a functional $\Omega[\Sigma]$ of the self-energy is constructed, which can be shown to be stationary at the physical self-energy. This functional is in general unknown. A numerically solvable reference system is introduced, which shares the interaction part with the original system. The crucial point is that in this case the functional of the original system can be evaluated exactly at the trial self-energies of the reference system and the stationary point can be found on the restricted subset of trial self-energies. The approximation lies in the choice of the reference system. The DMFT and CDMFT can be shown to be special cases of this unifying approach [34]. VCA results are shown in chapter 8.

## 2.8   Broken Symmetry Phases

The DMFT equations are rather straightforwardly generalized to be applicable to commensurate long-range ordered phases. The equations for ferromagnetic long-range order are readily inferred from the previous derivation by separating out the spin from the superindex $\alpha = \{m\sigma\}$ ($m$ labels orbitals). The Green functions, self-energy, etc. are diagonal in spin space, $G_{\sigma\sigma'} = G_\sigma \delta_{\sigma\sigma'}$. Assuming the hopping to be spin independent, the self-consistency condition for the Bethe lattice takes the form

$$\mathcal{G}_\sigma^{-1}(i\omega_n) = (i\omega_n + \mu)\mathbb{1} - t_{\text{loc}} - tG_\sigma(i\omega_n)t \, . \tag{2.84}$$

In case of antiferromagnetism, one needs to account for the fact the unit cell is doubled by the presence of antiferromagnetic order. In general this is done by dividing the lattice into sublattices $A$ and $B$, writing the Hamiltonian in terms of sublattice operators and introducing matrix Green functions (see Sec. 6.10.1). For the Bethe lattice, the relations are obtained straightforwardly. The Green functions acquire a sublattice index and the presence of antiferromagnetic order is expressed by

$$G_{A\sigma}(i\omega_n) = G_{B-\sigma}(i\omega_n) \, . \tag{2.85}$$

The self-consistency condition is obtained from (2.46) by recalling that the cavity lattice Green function is site-diagonal and that the hopping only connects nearest neighbors. As a consequence, the Weiss field on sublattice $A$ is related to the cavity lattice Green function on sublattice $B$ (and vice versa). Identifying the latter with the lattice Green function as before yields the self-consistency condition

$$\mathcal{G}_\sigma^{-1}(i\omega) = (i\omega + \mu)\mathbb{1} - t_{\text{loc}} - t\,G_{-\sigma}(i\omega)\,t \, . \tag{2.86}$$

This relation is used in chapter 4. Incommensurate order has been considered within a generalization of DMFT by imposing an incommensurate spin-density wave structure in the Hubbard model [67]. This is achieved through a suitable unitary transformation of the Hamiltonian and leads to a $2 \times 2$ matrix formalism which remains on-site diagonal. Incommensurate phases can also be established using susceptibilities (see chapter 10).

The generalization to long-range superconducting order is also straightforward. In the previous derivation of the DMFT equations the anomalous averages of the form $F(\mathbf{k}, \tau - \tau') = -\langle c_{\mathbf{k}\uparrow}(\tau)c_{-\mathbf{k}\downarrow}(\tau')\rangle$ and $-\langle c_{\mathbf{k}\uparrow}^*(\tau)c_{-\mathbf{k}\downarrow}^*(\tau')\rangle$ have been assumed to be zero. In the superconducting state, these take on finite values. The corresponding effective action is derived straightforwardly by evaluating (2.42) and retaining anomalous averages. Consider a single-orbital model for simplicity. The sum over $\alpha, \beta$ in (2.43) is then replaced by a sum over spins and one has

$$\frac{1}{2!}\langle(\Delta S)^2\rangle^{(0)} = \frac{1}{2!}\int_0^\beta d\tau \int_0^\beta d\tau' \sum_{ij}\sum_{\sigma\sigma'} \left[t_{0i}c_{0\sigma}^\dagger(\tau)c_{i\sigma}(\tau) + t_{i0}c_{i\sigma}^\dagger(\tau)c_{0\sigma}(\tau)\right] \times$$
$$\times \left[t_{0j}c_{0\sigma'}^\dagger(\tau')c_{j\sigma'}(\tau') + t_{j0}c_{j\sigma'}^\dagger(\tau')c_{0\sigma'}(\tau')\right] \tag{2.87}$$

Multiplying out the product and performing the sum over spins yields the terms

$$\frac{1}{2!}\langle(\Delta S)^2\rangle^{(0)} = \int_0^\beta d\tau \int_0^\beta d\tau' \sum_{ij} \Big[ -t_{0i}t_{0j}c_{0\uparrow}^*(\tau)\langle c_{i\uparrow}(\tau)c_{j\downarrow}(\tau')\rangle^{(0)}c_{0\downarrow}^*(\tau')$$
$$- t_{0i}t_{j0}c_{0\uparrow}^*(\tau)\langle -c_{i\uparrow}(\tau)c_{j\uparrow}^*(\tau')\rangle^{(0)}c_{0\uparrow}(\tau')$$
$$- t_{i0}t_{0j}c_{0\downarrow}(\tau)\langle -c_{i\downarrow}^*(\tau)c_{j\downarrow}(\tau')\rangle^{(0)}c_{0\downarrow}^*(\tau')$$
$$- t_{i0}t_{j0}c_{0\downarrow}(\tau)\langle c_{i\downarrow}^*(\tau)c_{j\uparrow}^*(\tau')\rangle^{(0)}c_{0\uparrow}(\tau')\Big], \tag{2.88}$$

where it has been assumed that the cavity lattice Green functions are spindiagonal, while the anomalous Green functions only contribute for opposite spins. The factor $1/2!$ cancels by performing the sum over spins (the system is assumed to be paramagnetic). The equations can be written in a closed form by introducing Nambu spinors

$$\Psi_{\mathbf{k}}^* = (c_{\mathbf{k}\uparrow}^*, c_{-\mathbf{k}\downarrow}). \tag{2.89}$$

With this notation, the Green function can be written in the form

$$\bar{G}(\mathbf{k}, \tau - \tau') = -\langle \Psi_{\mathbf{k}}(\tau)\Psi_{\mathbf{k}}^*(\tau')\rangle = \begin{pmatrix} G(\mathbf{k}, \tau - \tau') & F(\mathbf{k}, \tau - \tau') \\ F^*(\mathbf{k}, \tau - \tau') & -G(-\mathbf{k}, \tau' - \tau) \end{pmatrix}. \tag{2.90}$$

As before take the $d = \infty$ Bethe lattice with nearest-neighbor hopping as an example. Using that the Green function is site-diagonal and $\langle -c_{i\uparrow}^*(\tau)c_{j\uparrow}(\tau')\rangle^{(0)} = -G_{ji}^{(0)}(\tau' - \tau)$, $\langle -c_{i\uparrow}(\tau)c_{j\downarrow}(\tau')\rangle^{(0)} = -F_{ji}^{(0)}(\tau - \tau')$, it is easy to see that the term in angular brackets in (2.88) can be written in terms of local spinors $\Psi^* = (c_{0\uparrow}^*, c_{0\downarrow})$ in the form

$-t^2 \Psi^*(\tau) \sigma_z \bar{G}^{(0)}(\tau - \tau') \sigma_z \Psi(\tau')$, where $\sigma_z = \mathrm{diag}[1, -1]$ is the Pauli matrix. The effective action then takes the form

$$S_{\mathrm{eff}} = -\sum_{\alpha\delta} \int_0^\beta d\tau \int_0^\beta d\tau' \Psi^*(\tau) \mathcal{G}^{-1}(\tau - \tau') \Psi(\tau') + \int_0^\beta d\tau \tilde{S}_{\mathrm{loc}}[c^*(\tau), c(\tau)] , \quad (2.91)$$

where the Weiss field is given by

$$\mathcal{G}^{-1}(i\omega) = i\omega + \mu\sigma_z - t^2 \sigma_z \bar{G}(i\omega) \sigma_z . \tag{2.92}$$

In this form, the equations are restricted to superconducting order that preserves the symmetry of the lattice. The treatment of d-wave superconductivity is accomplished by introducing a $2 \times 2$ cluster in real space and using a superspinor formalism. It is then possible to account for d-wave superconducting and antiferromagnetic order simultaneously [32].


## 2.9   LDA+DMFT

The combination of the local density approximation with DMFT, the LDA+DMFT approach, allows to capture the effects of local correlations within a description that maintains the material specific aspects of the problem. LDA+DMFT is by now an established approach and extensive reviews are available on the subject (see e.g. Ref. [12]). Only the aspects relevant for the discussion of the combination of the LDA with the dual fermion approach as outlined in Sec. 6.11 are discussed here.

In LDA+DMFT, the many-body Hamiltonian is written in the form

$$H = H_{\mathrm{LDA}} + H_{\mathrm{int}} - H_{\mathrm{DC}} , \tag{2.93}$$

where $H_{\mathrm{LDA}} = \sum_{\mathbf{k}\sigma} \sum_{mm'} h^\sigma_{mm'}(\mathbf{k}) c^\dagger_{m\mathbf{k}\sigma} c_{m'\mathbf{k}\sigma}$ is the LDA part of the Hamiltonian, $H_{\mathrm{int}}$ contains the interaction of the electrons in the correlated orbitals and the double-counting correction $H_{\mathrm{DC}}$ is required to correct for the correlations which are already included in the LDA in an averaged way. This correction is not well defined and consequently different schemes which derive from different physical motivations are in use. For more details, see e.g. Ref. [68].

From a physical point of view, the correlations for e.g. transition-metal oxides and rare earth or actinide compounds are expected to be dominant in the partially filled d- or f-shells, respectively. In practice, however, the ambiguity arises of how to choose the orbitals for which correlations are included. These basis functions will span the "correlated subspace" of the full Hilbert space, but need not be a subset of the full basis which is used to solve the Kohn-Sham equations. The results will depend on the choice of the basis functions of the correlated subspace and e.g. the degree of localization [68].

The impurity problem is solved in the correlated subspace. After the self-energy is found, it has to be upfolded to the lattice, i.e. projected onto the full basis. The lattice Green function is constructed with the upfolded self-energy. In this formulation, the DMFT self-consistency condition amounts to requiring that the Green function of the impurity problem should be equal to the projection of the lattice Green function onto the correlated subspace. The only required changes of the self-consistency loops of Sec. 2.6.2 are the upfolding of the impurity quantities to construct the lattice Green function and the subsequent projection of the latter onto the correlated subspace.

In LDA+DMFT, The interaction part of the Hamiltonian $H_{\text{int}} = \sum_i H_{\text{loc}}[c_i^\dagger c_i]$ (the sum is over all sites) is determined by the local Hamiltonian

$$H_{\text{loc}} = \frac{1}{2} \sum_{m m' m'' m'''} \sum_{\sigma\sigma'} U_{m m' m'' m'''} c_{i m\sigma}^\dagger c_{i m'\sigma'}^\dagger c_{i m''' \sigma'} c_{i m'' \sigma} \ . \tag{2.94}$$

In principle, notwithstanding the sign problem, the continuous-time impurity solvers described in chapter 3 are able to handle the interaction in this general form. The matrix elements

$$U_{m m' m'' m'''} = \langle m, m' | V_{\text{ee}} | m'', m''' \rangle \tag{2.95}$$

are commonly taken with respect to the *screened* Coulomb interaction $V_{\text{ee}} = 1/\epsilon \, |\mathbf{x} - \mathbf{x}'|$. The screening can be determined within constrained RPA and the matrix elements may then be evaluated in terms of Wannier functions [69].

A simpler form of the interaction is obtained by assuming that within the solid, the interactions largely retain their atomic nature. Within a spherical approximation, the matrix elements are expressed in terms of effective (i.e. screened) Slater integrals $F^k$ [70],

$$\langle mm' | V_{\text{ee}} | m'' m''' \rangle = \sum_k a_k(m, m', m'', m''') F^k \ . \tag{2.96}$$

For d-electrons, the non-zero elements are $F^0$, $F^2$ and $F^4$. Defining the average Coulomb and Stoner parameters $U$ and $J$, these are linked to the Slater integrals through (2.96) as $U = F^0$ and $J = (F^2 + F^4)/14$ and can be computed within constrained LDA calculations. The three Slater integrals and hence the matrix elements are unambiguously determined by requiring that the ratio $F^2/F^4$ be equal to its atomic value $\sim 0.625$ [70].

For nearly degenerate bands, the $U$-matrix is often approximated by a parameterization restricted to the three parameters

$$\begin{aligned}
U &= \langle m, m | V_{\text{ee}} | m, m \rangle, \\
U' &= \langle m, m' | V_{\text{ee}} | m, m' \rangle, \\
J &= \langle m, m' | V_{\text{ee}} | m', m \rangle,
\end{aligned} \tag{2.97}$$

with the direct and exchange Coulomb matrix elements. In this case, the fully rotational invariant Hamiltonian (in spin and orbital space) then takes the form

$$
H_{\text{loc}} = \frac{1}{2} \sum_{mm'} \sum_{\sigma} [U - 2J(1 - \delta_{mm'})] n_{m\sigma} n_{m'-\sigma} + \frac{1}{2} \sum_{m \neq m'} \sum_{\sigma} (U' - J) n_{m\sigma} n_{m'\sigma}
$$
$$
+ J \frac{1}{2} \sum_{m \neq m'} \sum_{\sigma} \left( c^{\dagger}_{m\sigma} c^{\dagger}_{m'\bar{\sigma}} c_{m\bar{\sigma}} c_{m'\sigma} + c^{\dagger}_{m\sigma} c^{\dagger}_{m\bar{\sigma}} c_{m'\sigma} c_{m'\bar{\sigma}} \right) , \tag{2.98}
$$

where $U' = U - 2J$ is required by symmetry. The interactions in the first line are of density-density type, while the two terms in the second line correspond to spin-flip and pair-hopping terms, respectively. Often the latter two are neglected, which is not always done for physical reasons (the Hund's rule coupling $J$ is not necessarily small compared to $U$), but in order to reduce the computational complexity of the problem. For example, these terms are known to cause a serious sign-problem for the Hirsch-Fye quantum Monte Carlo solver described in Sec. 3.2. For the strong-coupling solver discussed in Sec. 3.4, the computational effort is strongly reduced if only density-density interactions are considered.

# Chapter 3

# Continuous-Time Quantum Monte Carlo

Monte Carlo simulations are used in many areas of science, not necessarily restricted only to physics [71]. They are used for hypothesis testing, tasks in quantum chemistry, and virtually all areas of physics. Among these are problems that would otherwise defy solution, such as the evaluation of integrals of large dimensionality. The term quantum Monte Carlo generally applies to Monte Carlo simulations that are employed to solve quantum problems. These methods exist in different flavors. For example, the diffusion and variational quantum Monte Carlo algorithms are concerned with finding the ground state wave function and are used in physics and quantum chemistry. The basic idea of path integral quantum Monte Carlo is to write the density matrix as an integral over paths in the spirit of the Feynman path integral formulation and sample these stochastically. There are further variants such as diagrammatic Monte Carlo. The computation of fermionic field-theoretical models became possible with the advent of determinantal quantum Monte Carlo algorithms, due to Blankenbecler, Sugar, Scalapino [72, 73], and Hirsch and Fye [74]. In these algorithms, the partition function is expressed in terms of fermionic determinants by integrating out the fermion fields. Hirsch and Fye later published a method suitable for quantum impurity models [75].

The problems encountered throughout this thesis require the solution of the Anderson impurity problem. For the most part, the weak-coupling continuous-time quantum Monte Carlo impurity solver will be employed. This algorithm belongs to a new class of fermionic continuous-time quantum Monte Carlo algorithms that have been developed recently [56, 76, 77, 78, 79, 80]. Before discussing these algorithms, the foundations of importance sampling Monte Carlo simulations are briefly reviewed. The Hirsch-Fye algorithm is introduced afterwards, before discussing the two complementary continuous-time approaches, the so-called weak- and strong-coupling algorithms. Both methods share the fact that classes of diagrams are collected into fermionic determinants. This is essential to alleviate the sign problem. Because of this, and the favorable scaling of the matrix size with the model parameters, the range of applicability of these solvers is enhanced compared to the Hirsch-Fye method. The continuous-time solvers are presented in some detail, owing to the lack of introductory material on this subject. These methods are steadily gaining importance in the field.

## 3.1　General Principles

In many problems in physics not limited to condensed matter physics, one encounters the problem of computing integrals or sums of large dimensionality. A well-known example is the evaluation of integrals over the $6N$-dimensional phase space of an ensemble of $N$ particles in three spatial dimensions, a problem frequently encountered in classical statistical mechanics. An evaluation using numerical integration is unfeasible for such problems. The reason is that the use of quadrature rules to evaluate multidimensional integrals becomes impractical even for moderately high spatial dimensions. Consider, for example, the composite Simpson's rule for numerical integration, which integrates polynomials of degree three or less exactly. The cumulative error for a one-dimensional integral with the interval discretized into $N$ slices of width $h \sim 1/N$ is $O(h^4)$. This corresponds to a local error of $O(h^4/N) = O(h^5)$. It is easy to see that the local error on a $d$-dimensional cube decreases as $O(h^{4+d})$, while the number of slices to maintain such a discretization scales as $N^d$. The unfavorable exponential scaling to maintain a fixed overall precision of $N^d O(h^{4+d}) = O(h^4)$ renders an accurate evaluation of high-dimensional integrals unfeasible.

The problem when dealing with classical or quantum mechanical systems is that they can assume an exponentially large number of states. Take, for example, the two-dimensional (2D) Ising model

$$H = J \sum_{\langle ij \rangle} S_i S_j , \tag{3.1}$$

where the sum is over all bonds. With only two states $S_i = \{+1, -1\}$ per site, the total number of states for a moderate lattice size of $10 \times 10$ sites is $2^{100} \approx 10^{30}$. Assuming that a single configuration can be enumerated within a single floating point operation (FLOP), the time needed for a modern processor with a performance of order of gigaFLOP/$s$ to simply enumerate all configurations exceeds the age of the universe ($\sim 4.3 \cdot 10^{17} s \approx 13.7 \cdot 10^9 y$). The Ising model analogy is suitable to illustrate some concepts relevant to the continuous-time Monte Carlo algorithms, which may appear obscure owing to the abstract nature of the configuration space.

The abovementioned difficulties can be tackled using Monte Carlo techniques. This is possible because the error in Monte Carlo integration is independent of the dimensionality. The problem of visiting an exponentially large number of states is circumvented by the concept of importance sampling. The idea is to only visit and collect contributions from the states that are important in the sense that their contribution to the desired quantity is large. Physical intuition helps to understand why these seemingly unsolvable problems can be handled by such techniques: the contribution of a given state will not depend on all details of the state, so that large cancellations can be expected to occur when calculating expectation values.

The basic idea of Monte Carlo sampling of multidimensional integrals is to rewrite the integral as

$$I = \int f(X)\,dX = \int g(X)p(X)dX =: \langle g \rangle \, . \tag{3.2}$$

Here $X$ denotes a configuration of the system, which can be either a vector in a $d$-dimensional space or a set of discrete values characterizing the state. The notation $\int dX$ correspondingly denotes $d$-dimensional integration or summation over all states. The function $p(X)$ is chosen such that it can be interpreted as a probability distribution, i.e $p(X)$ is non-negative and normalized, $p(X) \geq 0$, $\int p(X)dX = 1$. $\langle g \rangle_p := \langle g \rangle$ denotes the expectation value of $g$ with respect to $p(X)$. Notwithstanding these conditions, there is an arbitrary number of ways to decompose $f$ in this way and hence $p(X)$ should be chosen in an optimal way. An approximation to the integral is obtained by a sample of $N$ independent values $X_i$ distributed according to $p(X)$, denoted by $X_i \in p(X)$:

$$\int f(X)dX \approx \frac{1}{N} \sum_{\substack{i=1 \\ X_i \in p(X)}}^{N} g(X_i) =: \langle g \rangle_p \tag{3.3}$$

For different, sufficiently large samples, the mean of $g$ sampled according to $p(X)$, by the central limit theorem, is normally distributed around the integral $I$ with standard deviation $[\langle g^2 \rangle_p - \langle g \rangle_p^2]^{1/2}/\sqrt{N}$. This result is obviously independent of the dimensionality of the configuration space. By virtue of Eq. 3.2, $g = f/p$ and it follows that the variance $\langle g^2 \rangle_p - \langle g \rangle_p^2 \approx \langle (f/p - \langle g \rangle)^2 \rangle_p$ is minimized by choosing $p = g/\langle g \rangle$. The probability distribution should thus be chosen similar to the integrand, or in other words, configurations that give large contributions to the integral should be sampled with a higher probability, corresponding to a large overlap between both functions. Sampling configurations according to an optimized probability distribution is referred to as importance sampling. The advantages of this strategy become particularly significant in high spatial dimensions.

In practice, expectation values often naturally decompose according to (3.2), for example into a state variable $g(X)$ and the Boltzmann weight $p(X) = \exp[-\beta E(X)]/\mathcal{Z}$ in the canonical ensemble. In general, for a given decomposition of the integrand, the occurring distribution $\tilde{p}$ need not be normalized, or even positive and cannot be used for Monte Carlo sampling. The freedom to choose the ensemble over which the system is averaged over may be exploited to rewrite the integral as

$$\langle g \rangle = \frac{\int g(X)\tilde{p}(X)dX}{\int \tilde{p}(X)dX} = \frac{\int g(X)\frac{\tilde{p}(X)}{p(X)}p(X)dX}{\int \frac{\tilde{p}(X)}{p(X)}p(X)dX} = \frac{\int g(X)\frac{\tilde{p}(X)}{p(X)}p(X)dX\big/\int p(X)dX}{\int \frac{\tilde{p}(X)}{p(X)}p(X)dX\big/\int p(X)dX} \, . \tag{3.4}$$

Choosing $p(X) = |\tilde{p}(X)|$, $\tilde{p}(X)/p(X) = \text{sgn}[\tilde{p}(X)]$ and assuming for simplicity that $p$ is normalized, one finds

$$\langle g \rangle = \frac{\int g(X)\,\text{sgn}[\tilde{p}(X)]p(X)dX}{\int \text{sgn}[\tilde{p}(X)]p(X)dX} = \frac{\langle g\,\text{sgn}(\tilde{p})\rangle}{\langle \text{sgn}(\tilde{p})\rangle}. \tag{3.5}$$

An estimate of the desired quantity $\langle g \rangle$ can then be obtained from a finite sample of configurations distributed according to $p$: $\langle g \rangle = \langle g \rangle_p = \langle g\,\text{sgn}(\tilde{p})\rangle_p / \langle \text{sgn}(\tilde{p})\rangle_p$. The problem that arises when $\tilde{p}$ becomes negative is referred to as the minus sign problem. It is conceivable that a strongly fluctuating sign will cause large cancellations and error amplification. It can be shown that in such a case the problem becomes exponentially hard [81]. Another application of Eq. 3.4 is reweighing of configurations, in case the probability distribution $\tilde{p}$ is not well-suited to sample $g$. It is desirable to enhance the weight (and acceptance probability) of configurations where $g$ is large. To this end one would choose $p(X) = E(X)\tilde{p}(X)$ (with, e.g. $E \sim g$) so that

$$\langle g \rangle = \frac{\int \frac{g(X)}{E(X)}p(X)dX \big/ \int p(X)dX}{\int \frac{1}{E(X)}p(X)dX \big/ \int p(X)dX} = \frac{\langle g/E \rangle}{\langle 1/E \rangle}. \tag{3.6}$$

### 3.1.1 Markov Process

The probability distribution $p(X)$ often cannot be sampled from directly. In that case, sampling of configurations according to $p(X)$ can be realized using Markov chains. A Markov chain is a sequence of random variables

$$\bar{X}_1, \bar{X}_2, \ldots, \bar{X}_k, \bar{X}_{k+1}, \ldots, \bar{X}_N \,, \tag{3.7}$$

generated by a stochastic process referred to as the Markov process. Let $p(A|B)$ denote the conditional probability for a transition to the state $A$ given the system is in state $B$. The Markov process has the property that

$$p(\bar{X}_{k+1} = X_{k+1}|\bar{X}_1 = X_1, \ldots \bar{X}_k = X_k) = p(\bar{X}_{k+1} = X_{k+1}|\bar{X}_k = X_k) \,, \tag{3.8}$$

i.e. the probability of the system at time $k$ to make a transition to the state $X_{k+1}$ solely depends on the present state $X_k$ and not on previous ones. As a result, the dynamics of the Markov process is determined by the transition probability to a configuration $X'$ from $X$ denoted by $\mathcal{P}(X \rightarrow X')$. In order for the distribution of the elements generated by the Markov process to converge to a given distribution $p(X)$, two criteria have to be met:

- The transition probability fulfills $p(X)\mathcal{P}(X \rightarrow X') = p(X')\mathcal{P}(X' \rightarrow X)$ (detailed balance)

- $\mathcal{P}(X \to X')$ has to guarantee the Markov process to eventually have access to all possible configurations of the system (ergodicity)

The first condition, also referred to as the microreversibility condition, ensures that $p(X)$ is an equilibrium distribution of the Markov process, i.e. if $X_k$ is distributed according to $p(X_k)$, so is $X_{k+1}$. Ergodicity is required to ensure that the distribution of states generated by the Markov process converges to the equilibrium distribution. The number of steps required to attain equilibrium may vary for different systems. In general the first elements in the chain will not be distributed according to $p(X)$. Hence the system should be given some time to equilibrate, before measurements are performed. This time is referred to as thermalization time. There is no general recipe on how to choose it and the thermalization time has to be found empirically. An error due to sampling an initially unequilibrated distribution will decay as $1/N$ and eventually become smaller than the statistical error which decays as $1/\sqrt{N}$. This may, however, require a sizeable $N$ if the initial error is large.

### 3.1.2  Metropolis Algorithm

Originally, the Metropolis algorithm [82] was applied to a liquid described within classical statistics, where the probability of a configuration with energy $E$ of atoms in the liquid is proportional to the Boltzmann factor $\exp(-\beta E)$. Instead of generating all configurations with equal probability and multiplying the state variable by the Boltzmann weight, the algorithm was designed to directly sample configurations according to $\exp(-\beta E)$, hence avoiding the generation of a multitude of samples with very low weight.

For the Ising model (3.1) with $N$ spins on a square lattice, the algorithm considers flipping a spin with probability $1/N$. From the detailed balance condition it follows that

$$\frac{\mathcal{P}(X \to X')}{\mathcal{P}(X' \to X)} = \frac{p(X')}{p(X)} = e^{-\beta[E(X')-E(X)]} \,. \tag{3.9}$$

Hence one sees that the acceptance criterion has to satisfy

$$\mathcal{P}(X \to X') = e^{-\beta[E(X')-E(X)]}\mathcal{P}(X' \to X) \,. \tag{3.10}$$

This is readily seen to be accomplished by the acceptance criterion

$$\mathcal{P}(X \to X') = \min\left(1, e^{-\beta[E(X')-E(X)]}\right): \tag{3.11}$$

the move is always accepted if it reduces the total energy of the system and with a probability proportional to $\exp(-\beta \Delta E)$ if the energy were to increase by $\Delta E$. Assuming $E(X') < E(X)$ (otherwise exchange $X$ and $X'$) implies $\mathcal{P}(X \to X') = 1$ and the probability of the inverse move is $\mathcal{P}(X' \to X) = e^{-\beta[E(X)-E(X')]}$ by virtue of Eq. 3.10. Hence

(3.11) has the desired property. This condition is simply implemented by accepting the move if $\eta < e^{-\beta[E(X')-E(X)]}$, with $\eta$ randomly distributed in the interval $[0, 1]$, and rejecting it otherwise. Note that the Metropolis algorithm only requires knowledge of probability ratios. This is significant, since the absolute probability $\exp(-\beta E)/\mathcal{Z}$ involves the partition function which is in general unknown. The construction of the transition probability of course implicitly assumes that $p(X)$ is normalized. No extra normalization is required for the evaluation of expectation values.

In the above form, the algorithm is not directly applicable to the continuous-time methods. Hastings generalized it to the case where the probability density to propose a move $q(X \to X')$ and its inverse are not equal [83]. The transition probability $\mathcal{P}(X \to X')$ is split into two parts,

$$\mathcal{P}(X \to X') = p(X \to X')q(X \to X') , \tag{3.12}$$

where in case the probability to propose a move from $X$ to $X'$ without loss of generality is larger than its inverse move, $q(X \to X')$ will violate microreversibility:

$$p(X)q(X \to X') > p(X')q(X' \to X) \tag{3.13}$$

(for the reverse inequality, exchange $X$ and $X'$). This can be compensated by choosing $p(X \to X') = 1$. Then detailed balance requires that

$$p(X)q(X \to X')p(X \to X') = p(X')q(X' \to X')p(X' \to X) = p(X')q(X' \to X) , \tag{3.14}$$

which determines $p(X \to X')$. By exchanging the role of $X$ and $X'$, the probability to accept a move is found to be

$$p(X \to X') = \min\left(1, \frac{p(X')q(X' \to X)}{p(X)q(X \to X')}\right) . \tag{3.15}$$

### 3.1.3  Error Estimation

The expectation value of a quantity denoted by $\langle O \rangle$, where $O$ is representative for any observable, can be estimated from the arithmetic mean of $N$ samples,

$$\overline{O} = \frac{1}{N} \sum_{\substack{j=1 \\ j \in p}}^{N} O_j , \tag{3.16}$$

where the shorthand notation $O_j \equiv O(X_j)$ and $\overline{O} \equiv \langle O \rangle_p$ has been introduced for brevity. An important quantity to characterize the errorbar is the variance of the mean, $\sigma_{\overline{O}}^2 = \langle (\overline{O} - \langle \overline{O} \rangle)^2 \rangle = \langle \overline{O}^2 \rangle - \langle \overline{O} \rangle^2$. Inserting the expression (3.16) for the mean yields

$$\sigma_{\overline{O}}^2 = \frac{1}{N}\sigma_{O_i}^2 + \frac{1}{N^2} \sum_{\substack{i \neq j}}^{N} (\langle O_i O_j \rangle - \langle O_i \rangle \langle O_j \rangle) , \tag{3.17}$$

where the variance of the individual measurement $\sigma^2_{O_i}$ has been assumed to be independent of the point in time when it is measured. This assumption is valid provided the Markov process is thermalized, i.e. the equilibrium distribution is being sampled. The second term vanishes given that individual measurements are entirely uncorrelated. In this case the central limit theorem ensures that in the limit of large samples the distribution of mean values is Gaussian with standard deviation $\sigma = \sigma_{O_i}/\sqrt{N}$. The error bar is commonly given as $\overline{O} \pm n\sigma$, with the usual confidence intervals (68% for n=1). The assumption of uncorrelated measurements does not hold for a Markov chain Monte Carlo process. Again assuming time-translation invariance in equilibrium, the variance Eq. 3.17 can be expressed in terms of the normalized autocorrelation function

$$A_O(k) := \frac{\langle O_1 O_{1+k}\rangle - \langle O_1\rangle\langle O_{1+k}\rangle}{\sigma^2_{O_i}} \tag{3.18}$$

as

$$\sigma^2_{\overline{O}} = \frac{1}{N}\sigma^2_{O_i}\left[1 + 2\sum_{k=1}^{N} A(k)\left(1 - \frac{k}{N}\right)\right]. \tag{3.19}$$

At large time differences, the correlations are expected to decay exponentially, $A_O(k) \sim e^{-k/\tau^{ac}_O}$ as $k \to \infty$, with $\tau^{ac}_O$ being the autocorrelation time. The correction $k/N$ is due to the finite sample size and is typically neglected as in any simulation with proper statistics $k \gg \tau^{ac}_O$ and this term is exponentially small when $k$ is comparable to $N$. The term in angular brackets in (3.19) is set equal to $2\tau^{int}_O$, defining the integrated autocorrelation time of the observable $O$. This allows to write the statistical error in the form

$$\sigma_{\overline{O}} = \frac{\sigma_{O_i}}{\sqrt{N}}\sqrt{2\tau^{int}_O} = \frac{\sigma_{O_i}}{\sqrt{N_{\mathrm{eff}}}}, \tag{3.20}$$

which explicitly shows that the effective sample size $N_{\mathrm{eff}} = N/(2\tau^{int}_O)$ can be appreciably reduced by large autocorrelation times. Some care needs to be taken in estimating the autocorrelation, since the relative variance of the estimator of $A_O(k)$ obtained by replacing $\langle O_i O_{i+k}\rangle$ by its mean diverges with $k$. This can be overcome, e.g., by a binning analysis. Note that the above error analysis assumes independent observables. For mutually correlated measurements, the covariance matrix needs to be used. For a more detailed introduction into the statistical analysis of Markov chain Monte Carlo simulations confer e.g. Refs. [84, 85].

## 3.2   Hirsch-Fye Quantum Monte Carlo

In this section, the Hirsch-Fye quantum Monte Carlo algorithm for the impurity problem is described [75]. This allows to put the continuous-time Monte Carlo solvers into perspective and to outline the similarities and main differences between these methods. The algorithm is introduced in the Hamiltonian formulation of the Anderson impurity model for a single impurity orbital and Hubbard interaction. The model is described by the Hamiltonian (cf. Eq. 2.25)

$$
H = \sum_{k\sigma} \epsilon_k f^\dagger_{k\sigma} f_{k\sigma} + \sum_{k\sigma} V_k (c^\dagger_\sigma f_{k\sigma} + f^\dagger_{k\sigma} c_\sigma) - \tilde{\mu} \sum_\sigma n_\sigma + U \left[ n_\uparrow n_\downarrow - \frac{1}{2}(n_\uparrow + n_\downarrow) \right] . \tag{3.21}
$$

The chemical potential $\tilde{\mu} = \mu - U/2$ has been shifted such that $\tilde{\mu} = 0$ corresponds to half-filling. The first part of the Hamiltonian ($H_0$) is Gaussian except for the interaction part in angular brackets (denoted later as $H_{\text{int}}$), which contains a two-fermion interaction which is quadratic in the impurity densities $n_\sigma = c^\dagger_\sigma c_\sigma$. Here $k = 1, \ldots n_s$ labels the $n_s$ conduction electron states or bath sites. It is convenient to denote the impurity state by $k = 0$, so that formally $c = f_0$. The partition function of the system is

$$
\mathcal{Z} = \text{Tr}\, e^{-\beta(H_0 + H_{\text{int}})} . \tag{3.22}
$$

The derivation of the algorithm is based on the Trotter-Suzuki transformation for operators $A$, $B$, namely

$$
e^{(A+B)} = \lim_{L\to\infty} \left( e^{A/L} e^{B/L} \right)^L . \tag{3.23}
$$

This implies that in the limit of small $\Delta\tau$, $\exp(-\Delta\tau(A + B)) = \exp(-\Delta\tau A)\exp(-\Delta\tau B) + O(\Delta\tau^2)$. Hence the exponential of the Hamiltonian in (3.22) is approximately factorized into Gaussian and interacting parts up to an error of order $O(\Delta\tau^2)$ by discretizing the imaginary time interval into $L$ slices of width $\Delta\tau = \beta/L$:

$$
\mathcal{Z} = \text{Tr} \prod_{l=1}^{L} e^{-\Delta\tau H_0} e^{-\Delta\tau H_l^{\text{int}}} + O(\Delta\tau^2) . \tag{3.24}
$$

The Green function is defined as a propagator acting between slices and is approximated by applying the above factorization to the imaginary time evolution operator :

$$
\tilde{g}^{\Delta\tau}_{kk'\,\sigma}(\tau_{l_1}, \tau_{l_2}) = \langle f_{k\sigma}(\tau_{l_1}) f^\dagger_{k'\,\sigma}(\tau_{l_2}) \rangle^{\Delta\tau} = \frac{\text{Tr}[U^{\Delta\tau}_{L-l_1} f_{k\sigma}(\tau_{l_1}) U^{\Delta\tau}_{l_1-l_2} f^\dagger_{k'\,\sigma}(\tau_{l_2}) U^{\Delta\tau}_{l_2}]}{\text{Tr}[U^{\Delta\tau}_L]} . \tag{3.25}
$$

This is the full Green function of the problem, involving the impurity as well as bath states. The partition function is further evaluated by transforming the interacting problem into a noninteracting one. This happens at the cost of introducing auxiliary degrees of freedom and is facilitated by a discrete Hubbard-Stratonovich transformation [86], applied on each of the slices:

$$e^{-\Delta\tau H_I^{\text{int}}} = \frac{1}{2} \sum_{s_l=\pm 1} e^{\lambda s_l(n_\uparrow - n_\downarrow)} \quad \text{with} \quad \cosh \lambda = \exp(\Delta\tau U/2) . \tag{3.26}$$

The interpretation of the auxiliary variables $s_l$ is that of Ising spins. They couple with different signs to the densities with opposite spin. The Hubbard-Stratonovich transformation hence decouples the two-fermion interaction $U n_\uparrow n_\downarrow$ and the interacting fermions are replaced by free fermions subject to an external fluctuating field.

Inserting this into Eq. 3.24, it can be shown that the discretized version of the partition function can be written in the form [74, 15, 12]

$$\mathcal{Z}^{\Delta\tau} = \frac{1}{2^L} \sum_{s_1,\dots,s_L=\pm 1} \prod_{\sigma=\pm 1} \text{Tr}\, e^{-\Delta\tau H_0} e^{[\sum_{kk'=0}^{n_s} f_k^\dagger V_{kk'}^\sigma (s_l) f_{k'}]} = \sum_{s_1,\dots,s_L=\pm 1} \det O^\uparrow(\{S\}) \det O^\downarrow(\{S\}) . \tag{3.27}$$

The $n_s \times n_s$ matrices $V_{kk'}^\sigma(s_l) = \text{diag}[e^{\lambda\sigma s_l}, 1, \dots, 1]$ are diagonal in bath sites and contain the coupling to the Ising spins only in the first row and column corresponding to the impurity state. They should not be confused with the hybridizations in (3.21). The matrices $O^\sigma(\{S\})$ have dimensions $n_s L \times n_s L$ and depend on the particular configuration of the Ising spins denoted by $\{S\}$. The reason that the trace can be expressed in terms of determinants is ultimately a consequence of Wick's theorem, which is applicable since all operators are bilinear after decoupling. The configuration-dependent Green functions are related to these matrices as $g^\sigma(\{S\}) = (O^\sigma)^{-1}(\{S\})$ [74]. These matrices are large and the need to manipulate them explicitly would render the algorithm unfeasible, already for a small number of time slices and bath states. It becomes feasible through the insight of Hirsch and Fye, that the configuration-dependent Green functions for two different configurations are related through a Dyson equation:

$$\tilde{g}'^\sigma(\{S'\}) = \tilde{g}^\sigma(\{S\}) + (\tilde{g}^\sigma(\{S\}) - 1)[e^{V'^\sigma - V^\sigma} - 1]\tilde{g}'^\sigma(\{S'\}) . \tag{3.28}$$

The quantities $\tilde{g}^\sigma$ and $V_{kk'}^\sigma(s_l)$ are matrices of linear dimension $n_s L$ and the latter are diagonal with respect to time and bath labels. From the aforementioned structure of the matrices $V^\sigma$ it is readily seen that the term in angular brackets is a projection operator which projects onto the impurity state. As a consequence, this equation holds in the $k = 0$ subspace and hence for the impurity Green function $g^\sigma := \tilde{g}_{00}^\sigma$ separately. This relates the matrix $g(\{S'\})$ for any configuration $\{S'\}$ to the one in configuration $\{S\}$ by application of an $L \times L$ matrix $A$. Replacing $\tilde{g}$ by $g$ in (3.28) one has $A\, g'^\sigma(\{S'\}) = g^\sigma(\{S\})$ and

$$A = 1 - (g^\sigma(\{S\}) - 1)\left[e^{V'^\sigma - V^\sigma} - 1\right] = e^{V'^\sigma - V^\sigma} - g^\sigma(\{S\})\left[e^{V'^\sigma - V^\sigma} - 1\right] . \tag{3.29}$$

The Green function in the new configuration is obtained by inverting $A$: $g'^\sigma(\{S'\}) = A^{-1}g^\sigma(\{S\})$. The key point of the implementation is that a change of a configuration is attempted via local updates, i.e. single spin flips. The above equation is formulated in the impurity orbital subspace, so that $e^{V^\sigma} = \mathrm{diag}[e^{\lambda\sigma s_1}, \ldots, e^{\lambda\sigma s_1}]$. By flipping a single spin, say, $s_l$, the matrix $e^{V'^\sigma - V^\sigma}$ is nonzero only for the $l, l$-element. From (3.29) it follows that $A$ has the structure $A = 1 + u \otimes e_l$, where $e_l$ is the unit vector in direction $l$ and the direct product (defined by $(u \otimes v)_{ij} = u_i v_j$) adds the column vector $u$ to column $l$ of $A$. The inverse of $A$ can then be evaluated as a special case of the Sherman-Morrison formula [87]. This formula states that a change in a matrix $A$ of the form

$$A \to A' = A + u \otimes v, \tag{3.30}$$

induces a change of the inverse $M = A^{-1}$ of $A$ according to

$$M \to M' = M - \frac{(Mu) \otimes (vM)}{1 + vMu}. \tag{3.31}$$

The determinant ratio of the new and old matrices reads

$$\frac{\det M}{\det M'} = 1 + vMu. \tag{3.32}$$

In this case, $A_{ij} = M_{ij} = \delta_{ij}$, $v_i = (e_l)_i = \delta_{li}$ and $u_i = A_{il} - \delta_{il}$. It follows that $M'_{ij} = \delta_{ij} - u_i\delta_{lj}/(1 + u_l) = \delta_{ij} - u_i\delta_{jl}/A_{ll}$. This is an identity matrix with the column vector $u_i/A_{ll}$ subtracted from column $l$. In column $l$ (for $j = l$) this gives $M'_{ij} = -A_{il}/A_{ll}$ for $i \neq l$ and $1/A_{ll}$ for $i = l$. Inserting the expression for $A$, Eq. 3.29 into $g'^\sigma(\{S'\}) = Mg^\sigma(\{S\})$ hence yields the fast update formula in the form derived by Hirsch and Fye [75],

$$g'^\sigma_{l_1,l_2} = g^\sigma_{l_1,l_2} - [g^\sigma_{l_1,l} - \delta_{l_1,l}][e^{V'^\sigma - V^\sigma} - 1]_{ll}\, M^\sigma_{ll}\, g^\sigma_{l,l_2} \tag{3.33}$$

(no summation over $l$ is implied). This allows to update the Green function in $O(L^2)$ arithmetic operations, while a regular inversion of the matrix $A$ would require $O(L^3)$ operations.

In principle, the Green function can be evaluated using exact numeration of all Ising spin configurations. Gray-code enumeration of configurations is used so that successive configurations differ only by one entry and fast updates can be performed. This has been done for the nonequilibrium Anderson model [57], where the real time evolution of the observables using Monte Carlo is difficult due to the numerical sign problem (see below). Because of the exponential growth of the number of configurations, the range of applicability of the algorithm is significantly enhanced by performing a Markov chain Monte Carlo sampling of configurations. This yields the actual Hirsch-Fye algorithm. The weight of a configuration is proportional to $\left|\det O^\uparrow(\{S\})\det O^\downarrow(\{S\})\right|$. The Monte Carlo sampling can be done using the Metropolis or heat bath algorithm.

Due to the presence of the projection operator in the Dyson equation (3.28) and by using $\tilde{g}^\sigma = (O^\sigma)^{-1}$, the determinant ratio in the Metropolis acceptance criterion is expressed in terms of the measurements of the impurity Green function $g^\sigma$ only:

$$\prod_\sigma \frac{\det O^\sigma(\{S'\})}{\det O^\sigma(\{S\})} = \prod_\sigma \frac{\det g^\sigma(\{S\})}{\det g^\sigma(\{S'\})} = \prod_\sigma \det A^\sigma . \qquad (3.34)$$

Hence the actual algorithm only involves quantities on the impurity site, which is a consequence of the fact that a Gaussian bath can be integrated out exactly. The bare Green function of the problem enters only at the start of the simulation, where the matrix $A$ for an initial configuration $\{S'\}$ is related to $\mathcal{G}_0$ by formally setting $s_1 = \ldots = s_L = 0$, which yields $g(\{S\} = 0) = \mathcal{G}_0$. The matrix $A$ is inverted explicitly for a starting configuration $\{S'\}$. The Monte Carlo average of an observable $O$ is measured using

$$\langle O \rangle = \frac{\sum\limits_{\{S\}} O(\{S\}) \prod\limits_\sigma \mathrm{sgn}\, O^\sigma \left| \prod\limits_\sigma \det O^\sigma(\{S\}) \right|}{\sum\limits_{\{S\}} \prod\limits_\sigma \mathrm{sgn}\, O^\sigma \left| \prod\limits_\sigma \det O^\sigma(\{S\}) \right|}, \qquad (3.35)$$

where $O(\{S\}) = O^{-1}$ for Green's function. The measurements for higher-order correlation functions, such as the two-particle Green function, are constructed from measurements for the Green function. This essentially corresponds to an application of Wick's theorem, which applies since the action is Gaussian. The construction of higher-order correlators is analogous to that in the weak-coupling continuous-time algorithm as detailed in Sec. 3.3.9.

Despite the fact that the algorithm is still widely used, it has practical limitations. While the above update formulas relating the Green function for different configurations are exact, a systematic error of order $O(\Delta\tau^2)$ is committed due to the time discretization. It can be shown that the error of the Green function is of the same order [12]. For obtaining accurate results, an extrapolation of the limit $\Delta\tau \to 0$ has to be performed, which is rather involved since it requires multiple runs with different discretization and in the regime where $\Delta\tau$ is sufficiently small so that the error behaves like $\Delta\tau^2$. This limits the practical applicability of the algorithm in particular in the context of DMFT (although the overall error of self-consistent quantities is empirically found to scale as $\Delta\tau^2$). A so-called multigrid version of the Hirsch-Fye algorithm has recently been proposed to overcome this problem [88, 89]. Simulations for different discretization are performed simultaneously on a parallel computer and the extrapolation is performed between successive runs. This is computationally demanding, but provides a numerically exact result in the sense that systematic errors can be eliminated. Results can, in principle, be obtained to any desired accuracy.

The finite time discretization also makes it difficult to accurately resolve the behavior of the Green function at low temperatures and strong interactions, where it behaves essentially as $\sim e^{-U\tau/2}$. In contrast to the common rule $L \sim \beta U$, it has been proposed that a resolution of $L \sim 5\beta U$ should be used to ensure an accurate representation of $g(\tau)$ on a uniform grid and to yield reasonably accurate results [78]. In addition, the Fourier transform to Matsubara frequencies needed for the DMFT self-consistency loop becomes notoriously difficult on a sparse grid. Note that the computation time approximately scales as $O(L^3)$ and therefore increasing the number of time slices strongly affects the performance of the algorithm.

Another problem arises far away from half-filling when the bare Green function is very steep near $\tau = 0$ or $\beta$ and is not accurately sampled on a discrete grid. This problem is overcome by a static shift of the Green function and a corresponding change of the interaction. A similar technique is discussed in Sec. 3.3.4 for the continuous-time algorithm.

Further complications arise in the treatment of multiorbital systems, relevant to quantum simulations of real materials. Although there is no sign problem in the particle-hole symmetric case for the single-orbital model (the determinants for different spins are equal) and even away from half-filling [90], a severe sign problem arises in the multi-orbital case at low temperatures, in particular for spin-flip and pair-hopping terms. The treatment of general interactions requires to formulate decouplings in analogy to (3.26). Exact decouplings are known for density-density interactions, while general types may require approximate decouplings (at least accurate up to $\Delta\tau^3$ to maintain the $\Delta\tau^2$-scaling of the overall error). A straightforward decoupling of the pair-hopping term leads to a severe sign problem. A solution to this problem has been proposed recently [91], but a decoupling of the fully rotationally invariant Hamiltonian (Sec. 2.9), e.g. relevant for correlated real materials with open d- or f-shells requires a huge number of auxiliary Hubbard-Stratonovich fields. As a consequence, hardly any attempts have been made to incorporate the complete rotationally invariant Hamiltonian into calculations of real materials. The left-out terms may however be of crucial importance. The Hund's rule coupling was found to be responsible for the orbital selective Mott transition in the two-band Hubbard model [51].

In some cases (for not too low temperatures) the algorithm yields qualitatively correct results even for relatively coarse resolutions, and smooth Green functions are obtained for a small number of Monte Carlo sweeps (see, e.g., Fig. 3.6). This advantage is overruled at low temperatures and if numerically exact results are desired. In these cases, one should pass to the continuous-time methods described in the following sections.

## 3.3 Weak-Coupling Expansion

The weak-coupling continuous-time quantum Monte Carlo algorithm for fermions was introduced by Rubtsov *et al.* [56, 76, 77], following the work of Prokof'ev and coworkers [92] who devised a continuous-time scheme to sample the infinite series generated by the path integral representation of the partition function for bosons.

The algorithm is introduced in the path integral formulation for the single-orbital Anderson impurity problem with $H_{\mathrm{loc}} = U n_\uparrow n_\downarrow$. The generalization to the multiorbital case is straightforward and is briefly discussed in Sec. 3.3.3. It has been given including the case of nonlocal-in-time interactions in Ref. [56]. The action for the Anderson impurity model is divided into a Gaussian part $S_0$ and an interaction part as follows:

$$S_0 = \int_0^\beta d\tau \sum_\sigma c_\sigma^*(\tau)[\partial_\tau - \mu + U\alpha_{-\sigma}(\tau)]c_\sigma(\tau) + \int_0^\beta d\tau \int_0^\beta d\tau' \sum_\sigma c_\sigma^*(\tau)\Delta(\tau - \tau')c_\sigma(\tau') \,, \tag{3.36}$$

$$S_U = U \int_0^\beta d\tau \, [c_\uparrow^*(\tau)c_\uparrow(\tau) - \alpha_\uparrow(\tau)][c_\downarrow^*(\tau)c_\downarrow(\tau) - \alpha_\downarrow(\tau)] \,. \tag{3.37}$$

Here the fields $\alpha$, or so-called $\alpha$-parameters have been introduced such that the impurity action $S = S_0 + S_U$ is only changed up to an irrelevant additive constant. They are necessary to control the sign problem, as discussed below. The partition function

$$\mathcal{Z} = \int e^{-S[c^*,c]} \mathcal{D}[c^*,c] \tag{3.38}$$

is written as a functional integral over Grassmann fields. A formal series expansion is obtained by expanding the exponential in the interaction term,

$$\mathcal{Z} = \int e^{-S_0[c^*,c]} \sum_{k=0}^\infty \frac{(-1)^k}{k!} U^k \int_0^\beta d\tau_1 \ldots \int_0^\beta d\tau_k \, [c_\uparrow^*(\tau_1)c_\uparrow(\tau_1) - \alpha_\uparrow(\tau_1)] \times$$
$$\times [c_\downarrow^*(\tau_1)c_\downarrow(\tau_1) - \alpha_\downarrow(\tau_1)] \ldots [c_\uparrow^*(\tau_k)c_\uparrow(\tau_k) - \alpha_\uparrow(\tau_k)][c_\downarrow^*(\tau_k)c_\downarrow(\tau_k) - \alpha_\downarrow(\tau_k)] \mathcal{D}[c^*,c] \,. \tag{3.39}$$

Using the definition of the average over the noninteracting system, i.e. $\langle ... \rangle_0 = (1/\mathcal{Z}_0) \int \mathcal{D}[c^*,c] \ldots \exp(-S_0)$, the partition function can be expressed in the form

$$\mathcal{Z} = \mathcal{Z}_0 \sum_{k=0}^\infty \int_0^\beta d\tau_1 \int_{\tau_1}^\beta d\tau_2 \ldots \int_{\tau_{k-1}}^\beta d\tau_k \, \mathrm{sgn}(\Omega_k) \, |\Omega_k| \,, \tag{3.40}$$

where the integrand is given by

$$\Omega_k = (-1)^k U^k \langle [c_\uparrow^*(\tau_1)c_\uparrow(\tau_1) - \alpha_\uparrow(\tau_1)][c_\downarrow^*(\tau_1)c_\downarrow(\tau_1) - \alpha_\downarrow(\tau_1)] \ldots$$
$$\ldots [c_\uparrow^*(\tau_k)c_\uparrow(\tau_k) - \alpha_\uparrow(\tau_k)][c_\downarrow^*(\tau_k)c_\downarrow(\tau_k) - \alpha_\downarrow(\tau_k)] \rangle_0 \,. \tag{3.41}$$

Here the range of time integration has been changed such that time ordering is explicit: $\tau_k \geq \tau_{k-1} \geq \ldots \geq \tau_1$. For a given set of times all $k!$ permutations of this sequence contribute to Eq. 3.39. These can be brought into the standard sequence by permuting quadruples of Grassmann numbers, and hence without gaining an additional sign. Since all terms are subject to time-ordering, their contribution to the integral is identical, so that the factor $1/k!$ in (3.39) cancels. A configuration is hence fully characterized by specifying a perturbation order $k$ and an (unnumbered) set of $k$ times: $C_k = \{\tau_1, \ldots, \tau_k\}$. The entire configuration space comprises the configurations $C_k$ up to all orders:

$$C = \left\{ C_0 = \{\}, C_1 = \{\tau_1\}, \ldots, C_k = \{\tau_1 \ldots \tau_k\}, \ldots \right\} . \tag{3.42}$$

The algorithm performs importance sampling over the configuration space. The weight of a configuration is taken to be equal to the modulus of the integrand in Eq. 3.40.

Since $S_0$ is Gaussian, the average over the noninteracting system can be evaluated using Wick's theorem. Hence the weight of a configuration is essentially given by a fermionic determinant of a matrix containing the bare Green functions:

$$\Omega_k = (-1)^k U^k \prod_\sigma \det \hat{g}^\sigma, \qquad (\hat{g}^\sigma)_{ij} = g_0^\sigma(\tau_i - \tau_j) - \alpha_\sigma(\tau_i)\delta_{ij} . \tag{3.43}$$

The determinants for different spin orientations factorize since the Green function is diagonal in spin-space. Each configuration to the partition function can be visualized as a collection of Feynman diagrams. The determinant contains all possible contractions of the sequence of Grassmann numbers and corresponds to connecting the Green functions to the vertices in all possible ways. As an example, the four contributions to the partition function for a configuration with $k = 2$ are depicted in Fig. 3.1. Note that here and in the remainder of this thesis, the following definition and diagrammatic representation of the Matsubara Green function (in accordance with Ref. [93]) are used:

$$g_{\alpha\beta}^\sigma(\tau_1 - \tau_2) = -\langle c_{\alpha\sigma}(\tau_1)c_{\beta\sigma}^*(\tau_2)\rangle . \tag{3.44}$$

(The arrow points into the direction of the hole propagation.) With this definition the particle number is obtained as

$$\langle n_{\alpha\sigma}\rangle = \lim_{\tau\to 0^-} g_{\alpha\alpha}^\sigma(\tau) = -g_{\alpha\alpha}^\sigma(\beta) . \tag{3.45}$$

Setting $\alpha_\uparrow = \alpha_\downarrow = 0$ for simplicity, the contribution of this configuration is

$$\begin{aligned}
\Omega_2 &= U^2 \langle c_\uparrow^*(\tau_1)c_\uparrow(\tau_1)c_\downarrow^*(\tau_1)c_\downarrow(\tau_1)c_\uparrow^*(\tau_2)c_\uparrow(\tau_2)c_\downarrow^*(\tau_2)c_\downarrow(\tau_2)\rangle_0 \\
&= U^2 \langle c_\uparrow^*(\tau_1)c_\uparrow(\tau_1)c_\uparrow^*(\tau_2)c_\uparrow(\tau_2)\rangle_0 \langle c_\downarrow^*(\tau_1)c_\downarrow(\tau_1)c_\downarrow^*(\tau_2)c_\downarrow(\tau_2)\rangle_0 .
\end{aligned} \tag{3.46}$$

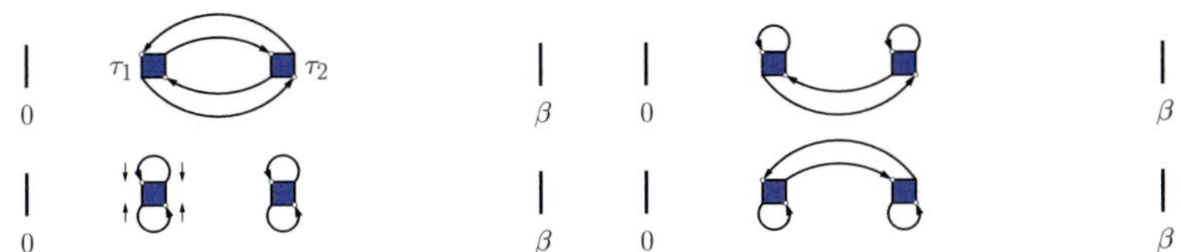

*Figure 3.1:* The four contributions to the partition function for $k = 2$. The interaction vertices are depicted by squares. Bare Green functions are shown as lines. Vertical arrows indicate the spin direction. Connecting the vertices by Green functions in all possible ways is the interpretation of the determinant. Here only four out of 4! terms contribute since the Green function is diagonal in spin-space.

The configuration can be written in the form $\Omega_2 = U^2 \det \hat{g}_\uparrow \det \hat{g}_\downarrow$, where $\hat{g}$ denotes the matrix of Green functions

$$\hat{g}_\sigma = \begin{pmatrix} -\langle c_\sigma(\tau_1)c_\sigma^*(\tau_1)\rangle_0 & -\langle c_\sigma(\tau_1)c_\sigma^*(\tau_2)\rangle_0 \\ -\langle c_\sigma(\tau_2)c_\sigma^*(\tau_1)\rangle_0 & -\langle c_\sigma(\tau_2)c_\sigma^*(\tau_2)\rangle_0 \end{pmatrix}. \tag{3.47}$$

Different columns correspond to different creator times, while the rows correspond to the annihilator times in the particular configuration. Note that the equal time averages defined via the path integral in Sec. 2.3 are to be interpreted as $-\langle c_\sigma(\tau_1 - 0)c^*(\tau_1)\rangle_0 = \langle c_\sigma^*(\tau_1)c(\tau_1 - 0)\rangle_0 = G_\sigma(0^-) = \langle n_\sigma\rangle_0$ (time-ordering is implicit), since the creator acts on a time-slice an instant later than the annihilator. This corresponds to the conventional time-ordering with the largest times appearing to the left. These terms yield the noninteracting density. Multiplying out the determinants gives the $2 \cdot 2! = 4$ contributions of Fig. 3.1. All other out of the 4! possible configurations vanish due to spin conservation.

## 3.3.1 Monte Carlo Sampling

The algorithm samples configurations by performing a Markovian random walk in the configuration space with the weight of a configuration determined by $|\Omega_k|$. Two kinds of Monte Carlo steps are sufficient to ensure ergodicity (other, e.g. global moves can be implemented to increase the sampling efficiency). A vertex is either inserted or removed from the configuration. The probability to insert a vertex (corresponding to adding a row and column to both $\hat{g}_\uparrow$ and $\hat{g}_\downarrow$) into a domain $d\tau$ is taken to be proportional to the fraction of of the domain to the full interval from 0 to $\beta$,

$$q^{\text{add}}(k \to k + 1) = \frac{d\tau}{\beta}. \tag{3.48}$$

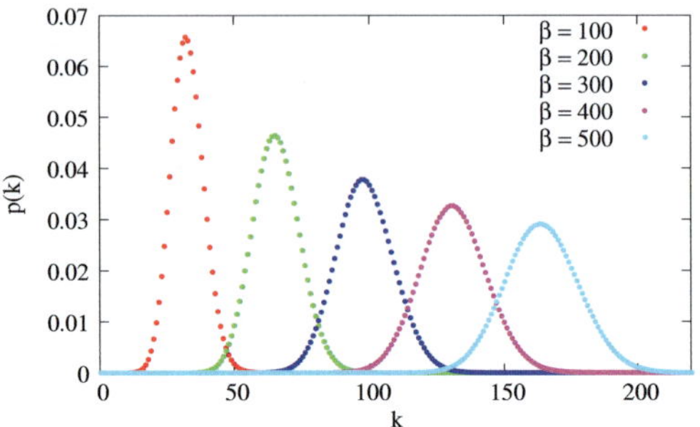

*Figure 3.2:* Probability distribution of the perturbation order $k$ for the single-orbital Anderson impurity model with a wide featureless band (corresponding to $\Delta(i\omega) = (-i/2)\,\mathrm{sgn}(\omega)$) and Hubbard interaction $U = 1$. The histogram shows a shift of the perturbation order to larger values and a broadening of the distribution with increasing inverse temperature $\beta$.

For the inverse move, one needs to remove one out of $k + 1$ vertices, each with the same probability:

$$q^{\mathrm{rem}}(k + 1 \rightarrow k) = \frac{1}{k + 1} \, . \tag{3.49}$$

The probabilities for the old and new configurations are proportional to

$$p^{\mathrm{new}}(k + 1) \sim |U|^{k+1} \, \Pi_\sigma \left|\det \hat{g}_\sigma^{(k+1)}\right| d\tau, \quad p^{\mathrm{old}}(k) \sim |U|^k \, \Pi_\sigma \left|\det \hat{g}_\sigma^{(k)}\right| \tag{3.50}$$

and the Metropolis-Hastings acceptance criterion for this step according to (3.15) is given by

$$p^{\mathrm{acc}} = \min\left(1, \frac{p^{\mathrm{new}}(k + 1)\, q^{\mathrm{rem}}(k + 1 \rightarrow k)}{p^{\mathrm{old}}(k)\, q^{\mathrm{add}}(k \rightarrow k + 1)}\right) = \min\left(1, \frac{\beta\,|U|}{k + 1}\Pi_\sigma \left|\frac{\det \hat{g}_\sigma^{(k+1)}}{\det \hat{g}_\sigma^{(k)}}\right|\right) \, . \tag{3.51}$$

The acceptance criterion for removing a vertex is given by the inverse of (3.51). In the simulation not the matrix $\hat{G}$, but its inverse $M$ is stored and manipulated. The reason is that $M$ gives faster and more direct access to determinant ratios (see appendix D.1), which are needed to compute both acceptance probabilities and Green function measurements. From time to time the $M$-matrix is rebuilt from scratch according to the present configuration either by explicit matrix inversion or sequential updates to avoid accumulation of numerical errors. Either method requires $O(k^3)$ operations. The algorithm performs an importance sampling of the configurations of the system by a Markovian random walk with probability proportional to $|\Omega_k|$. It samples diagrams for finite orders of $k$. Typically the perturbation order distribution has a Gaussian shape with a mean increasing with the interaction $U$ and inverse temperature, as shown in Fig. 3.2. The Gaussian shape is a consequence of the central limit theorem. Note that the average perturbation order $\langle k \rangle_{\mathrm{MC}}$ is an estimator for the average interaction, as can be seen by forming $U \partial \mathcal{Z} / \partial U$ and separating out $k$ or $U n_\uparrow n_\downarrow$ in (3.39).

## 3.3.2 Fast Update Formulae

The key to an efficient algorithm are the so-called fast-update formulas, which allow a fast update of the matrix $M$ and direct access to determinant ratios. Here the fast update formulae are derived using the Sherman-Morrison formula introduced in Sec. 3.2, which states how the *inverse* of a given matrix changes when a row and column are added to the matrix. An alternative derivation which gives more direct access to multiple simultaneous updates is given in appendix D. Inserting a pair of Grassmann numbers into a configuration at order $k$ corresponds to adding a column and a row to the matrix $\hat{g}^{(k)}$. In order to apply these formulas in a simple way, the matrix $\hat{g}^{(k)}$ is extended to be a $k+1 \times k+1$ matrix with zeros in the $k+1$th row and column, except for the $k+1, k+1$-element, which is 1 and the same for its inverse, $M^{(k)}$. Abbreviate the noninteracting Green function as $g_0(\tau_i - \tau'_j) =: g_{ij}$. Adding the column vector $u_{\mathrm{col}} = (g_{1,k+1}, \ldots, g_{k,k+1} | g_{k+1,k+1} - 1)^T$ is expressed in terms of adding the direct product $u_{\mathrm{col}} \otimes e_{k+1}^T$, where $e_{k+1}^T$ is the row vector with entries $(e_{k+1}^T)_i = \delta_{i,l+1}$ ($-1$ compensates the $k+1, k+1$-element of $\hat{g}^{(k)}$ so that $\hat{g}^{(k+1)}$ has the desired form). In a second step, add the row $v_{\mathrm{row}} = (g_{k+1,1}, \ldots, g_{k+1,k} | 0)$ by adding $e_{k+1} \otimes v_{\mathrm{row}}$. Application of the Sherman-Morrison formula for the first step yields the intermediate matrix

$$\bar{M} = M^{(k)} - \frac{M^{(k)} u_{\mathrm{col}} \otimes e_{k+1}^T M^{(k)}}{1 + e_{k+1}^T M^{(k)} u_{\mathrm{col}}} \,. \tag{3.52}$$

Now define $L_{ij} = \sum_{l=1}^{k} M_{il}^{(k)} g_{lj}$ and $R_{ij} = \sum_{l=1}^{k} g_{il} M_{lj}^{(k)}$. Then use $e_{k+1}^T M^{(k)} = e_{k+1}^T$, which implies that $M^{(k)} u_{\mathrm{col}} = (L_{1,k+1}, \ldots L_{k,k+1} | g_{k+1,k+1} - 1)^T$ divided by $1 + e_{k+1}^T M^{(k)} u_{\mathrm{col}} = g_{k+1,k+1}$ is added to the $k+1$th column of $M^{(k)}$ (the $k+1, k+1$-element is $1 - (g_{k+1,k+1} - 1)/g_{k+1,k+1}$):

$$\bar{M} = \left( \begin{array}{ccc|c} & & & -L_{1,k+1}/g_{k+1,k+1} \\ & M^{(k)} & & \vdots \\ & & & -L_{k,k+1}/g_{k+1,k+1} \\ \hline 0 & \cdots & 0 & 1/g_{k+1,k+1} \end{array} \right) . \tag{3.53}$$

Likewise, for the second step

$$M^{(k+1)} = \bar{M} - \frac{\bar{M} e_{k+1} \otimes v_{\mathrm{row}} \bar{M}}{1 + v_{\mathrm{row}} \bar{M} e_{k+1}} \,, \tag{3.54}$$

where now $v_{\text{row}}\bar{M} = (R_{k+1,1}, \ldots R_{k+1,k} \vert - g_{k+1,l}L_{l,k+1}/g_{g+1,g+1})$. Concatenating the updates using the formula for the determinant, $\det M^{(k)}/\det \bar{M} = 1 + e_{k+1}^T M^{(k)} u_{\text{col}}$, yields

$$r = \frac{\det \hat{g}^{(k+1)}}{\det \hat{g}^{(k)}} = \frac{\det M^{(k)}}{\det \bar{M}} \frac{\det \bar{M}}{\det M^{(k+1)}} = (1 + e_{k+1}^T M^{(k)} u_{\text{col}})(1 + v_{\text{row}}\bar{M} e_{k+1})$$

$$= g_{k+1,k+1} - \sum_{ij=1}^{k} g_{k+1,i} M_{ij}^{(k)} g_{j,k+1} \,. \qquad (3.55)$$

The evaluation of the determinant ratio hence requires $O(k^2)$ arithmetic operations for adding a vertex (instead of $O(k^3)$ for a straightforward evaluation). From Eq. D.9 it follows that the determinant ratio in (3.55) is equal to $1/M_{k+1,k+1}^{(k+1)}$. Evaluating the acceptance criterion for a removal update hence is of order $O(1)$, as $\det \hat{g}^{(k-1)}/\det \hat{g}^{(k)} = M_{kk}^{(k)}$ is stored in memory. Evaluating the direct product in (3.54) with (3.53), yields an explicit expression for the update of the matrix elements of $M^{(k+1)}$, which coincides with formulas (D.9) and (D.10). Determinant ratios for matrices which differ by two or more rows and columns can be obtained by successive application of the Sherman-Morrison formulas, or directly using block matrix manipulation (see appendix D.2). Note that for the strong-coupling solver discussed in Sec. 3.4, the structure of the equations is the same.

### 3.3.3  Multiorbital Formalism

The generalization of the algorithm is straightforward, even for the case of nonlocal-in-time interactions, which arise, e.g., when integrating out a phonon-bath. For a general (spin and time-independent) interaction, the action $S = S_0 + S_U$ takes the form

$$S_0 = -\sum_{\alpha\beta}\sum_{\sigma}\int_0^\beta d\tau_1 \int_0^\beta d\tau_1' \left\{ \left[ \mathcal{G}_0^{-1}(\tau_1 - \tau_1') \right]_{\alpha\beta} - \frac{1}{2}\sum_{\gamma\delta}\sum_{\sigma'}\int_0^\beta d\tau_2 \int_0^\beta d\tau_2' \times \right.$$

$$\left. \times \left( U_{\alpha\beta\gamma\delta} + U_{\gamma\delta\alpha\beta} \right) \alpha_{\gamma\delta}^{\sigma'}(\tau_2)\delta(\tau_2 - \tau_2') \right\} c_{\alpha\sigma}^*(\tau_1) c_{\beta\sigma}(\tau_1') \,, \qquad (3.56)$$

$$S_U = \frac{1}{2}\sum_{\alpha\beta\gamma\delta}\sum_{\sigma\sigma'}\int_0^\beta d\tau_1 \int_0^\beta d\tau_1' \int_0^\beta d\tau_2 \int_0^\beta d\tau_2' U_{\alpha\beta\gamma\delta} \times$$

$$\times \left[ c_{\alpha\sigma}^*(\tau_1) c_{\beta\sigma}(\tau_1') - \alpha_{\alpha\beta}^{\sigma}(\tau_1)\delta(\tau_1 - \tau_1') \right]\left[ c_{\gamma\sigma'}^T(\tau_2) c_{\delta\sigma'}(\tau_2') - \alpha_{\gamma\delta}^{\sigma'}(\tau_2)\delta(\tau_2 - \tau_2') \right] \,.$$

$$(3.57)$$

Expanding in the interaction yields the partition function in the form

$$
\mathcal{Z} = \sum_{k=0}^{\infty} \frac{(-1)^k}{2^k k!} \sum_{\alpha_1 \beta_1 \sigma_1} \int_0^{\beta} d\tau_1 \dots \sum_{\gamma_k \delta_k \sigma'_k} \int_0^{\beta} d\tau'_{2k} U_{\alpha_1 \beta_1 \gamma_1 \delta_1} \dots U_{\alpha_k \beta_k \gamma_k \delta_k} \times
$$
$$
\times \langle [c^*_{\alpha_1 \sigma_1}(\tau_1) c_{\beta_1 \sigma_1}(\tau'_1) - \alpha^{\sigma_1}_{\alpha_1 \beta_1}(\tau_1) \delta(\tau_1 - \tau'_1)] \dots
$$
$$
\dots [c^*_{\gamma_k \sigma'_k}(\tau_{2k}) c_{\delta_k \sigma'_k}(\tau'_{2k}) - \alpha^{\sigma'_k}_{\gamma_k \delta_k}(\tau_{2k}) \delta(\tau_{2k} - \tau'_{2k})] \rangle_0 . \tag{3.58}
$$

A change of a configuration is attempted by inserting or removing one of the possible vertices $U_{\alpha\beta\gamma\delta}$, which requires to specify the time, spins and orbitals, and the Metropolis acceptance criterion is modified accordingly [56]. The indices of the $M$-matrix correspond to a superindex $i := \{\tau_i, \sigma_i, \alpha_i\}$.

### 3.3.4 Sign Problem and $\alpha$-Parameters

Apparently, the alternating sign in Eq. 3.39 will lead to a severe sign problem (for $U > 0$). For the single-orbital case, this "trivial" sign problem is completely suppressed by a suitable choice of the $\alpha$-parameters. These correspond to an auxiliary dynamical field (they are chosen randomly). At half-filling the sign problem is suppressed for the choice $\alpha_\downarrow = 1 - \alpha_\uparrow = \alpha$ for any $\alpha$ (by exploiting particle-hole symmetry). Away from half-filling, a sign problem occurs only for $0 < \alpha < 1$. In practice $\alpha$ is chosen close to this interval (slightly above 1) to minimize the interaction term. Instead of introducing the $\alpha$-parameters as in (3.37), one may write the interaction term in symmetrized form,

$$
\frac{U}{2}(n_\uparrow + \alpha_1)(n_\downarrow + \alpha_2) + \frac{U}{2}(n_\uparrow + \alpha_2)(n_\downarrow + \alpha_1) = U n_\uparrow n_\downarrow + \frac{U}{2}(\alpha_1 + \alpha_2)(n_\uparrow + n_\downarrow) + \text{const.} \tag{3.59}
$$

The corresponding change of $S_0$ induces a shift of the chemical potential

$$
\mu \quad \rightarrow \quad \tilde{\mu} = \mu - \frac{U}{2}\langle(\alpha_1 + \alpha_2)(n_\uparrow + n_\downarrow)\rangle \tag{3.60}
$$

in the bare Green function $\mathcal{G}(i\omega) = i\omega + \tilde{\mu} - \Delta(i\omega)$. For the choice $\alpha_S := \alpha_1 + \alpha_2 = 1$, where $\alpha_S$ is the sum of the two $\alpha$-parameters, the Hartree term is subtracted and the chemical potential is conveniently shifted such that $\tilde{\mu} = \mu - U/2 = 0$ corresponds to half-filling. The bare Green function corresponds to that of the half-filled system. Without this shift, the chemical potential $\mu$ instead of $\tilde{\mu}$ would enter the bare Green function, which would correspond to a nearly filled band if $U$ is of order of the bandwidth. As a result, the bare Green function $\mathcal{G}(\tau)$ would be close to zero for small times, causing numerical instabilities. Even with this shift this leads to problems if for example a nearly filled band is desired in presence of strong interactions. A correspondingly large chemical potential $\tilde{\mu} > 0$ of order $U$ enters the noninteracting Green function. It is then advantageous to adjust the filling via the $\alpha$-parameters instead. In the above example for the single-band case, the chemical potential is shifted by $\alpha_S U/2$. The disadvantage of this approach is that the perturbation order increases and one may, as a compromise, adjust the filling by a combination of the two approaches, via $\tilde{\mu}$ and $\alpha_S$ simultaneously. In the multiorbital case, this shift is performed by adjusting the diagonal fields, $\alpha_{\alpha\alpha}^{\sigma}+\alpha_{\gamma\gamma}^{\sigma'}$.

While no sign problem exists for the single-orbital problem [90], this is in general not the case for a multiorbital impurity. No general recipe to avoid the sign problem in this case is available. The choice of $\alpha$-parameters for the multiorbital impurity problem is discussed in detail in Ref. [94].

### 3.3.5 Importance Sampling

The choice of the weight for a configuration to be $|\Omega_k|$ is an obvious one, but it is not a priori clear that this is the optimal choice. Instead of sampling diagrams for Green's function, the algorithm samples diagrams of the partition function. As a consequence, it is not guaranteed that importance sampling is efficient. This requires a large overlap between the partition function and all observables, in the sense that configurations with significant contributions to the partition function also contribute significantly to the respective observables. Experience shows that this prerequisite seems reasonably fulfilled: Enhancing the weight of configurations that give a large contribution to Green's function, e.g. on the first Matsubara frequency by choosing $E(C_k) = |\tilde{g}(\omega_n = \pi/\beta)|$ in Eq. 3.6 did not result in an appreciable improvement of the statistics. This is also observed for the worm-like algorithm proposed by Gull [95]. In this algorithm two additional operators are inserted into a configuration, which make up the head and tail of a worm and correspond to the creation and annihilation operator of the Green function. This way, diagrams of the Green function can directly be sampled. Four-point functions are more strongly fluctuating than single-particle functions and the overlap with the partition function could be appreciably smaller. However, reweighing in this case is impractical.

### 3.3.6 Parallelization

Markov chain Monte Carlo simulations are well suited for parallelization. The simulation is splitted into several Markov processes. At the end of the simulation, the observables are averaged over the threads, which may be assumed independent. Individual Markov chains must be allowed to have sufficient length in order to be thermalized before measurements are taken and to ensure proper statistics. Since typically no or little communication between the threads is needed during the simulation, the parallelization is efficiently performed using MPI (Message passing interface, a communication standard in distributed computing). For complicated systems at low temperature, the perturbation order becomes large. A low-level parallelization can be used to speed up the matrix manipulations and measurements locally on a node. This can be done using the openMP (open Multi-Processing) application programming interface.

### 3.3.7 Global Updates

For the most part of the simulation, local updates of a configuration consisting of insertion or removal of a single vertex (or two, as required for systems with certain symmetries) are performed. These updates are computationally efficient by virtue of the fast update formulas.

Sometimes it is however necessary to perform global updates. This is the case if the configuration space decomposes into regions which are separated by what may be viewed as a tunneling barrier. This occurs for example when the impurity is close to being magnetized, in the sense that two regions of the configuration space contribute to opposite magnetization directions and are separated by unlikely states. The algorithm can become trapped in one of the regions. This problem is related to the critical slowing down at a second-order phase transition. Here it leads to enhanced autocorrelation times and large Monte Carlo errors. One may then erroneously obtain a solution with a finite magnetization that breaks the symmetry. For small polarization, averaging over the spin components is sufficient. Otherwise global updates have to be performed. In a cluster update, the configuration is changed by adding and removing several vertices at a time. In practice this is achieved by concatenating a couple of local updates, with the acceptance probability evaluated for the move as a whole. This way the algorithm may tunnel through the barrier, by avoiding going through unlikely intermediate states. A different global move consists in exchanging two quantum numbers. For example, one may attempt to flip all spins (or two orbitals) within a configuration. An example where we found global updates to be essential is a system of two sites exerted to a thermal bath and an external magnetic field (the magnetization and spin of this system can be approximated analytically for large $U$, see Sec. 3.3.10). Flipping all spins of a configuration did not suffice, but flipping spins of the sites independently. It is desirable to perform global updates only as often as required, because of the need to perform an explicit matrix inversion. It takes $O(N^3)$ operations instead of $O(N^2)$ for the fast updates. An alternative to the above methods is the Wang-Landau algorithm [96, 97], also known as flat histogram sampling. This type of sampling is very efficient at first-order phase transitions. For the continuous-time algorithm, the basic idea is to choose $E(k) = 1/\tilde{p}(k)$ in (3.6), so that the sampling probability $p(k) = E(k)\tilde{p}(k)$ is constant. The Wang-Landau weights $E(k)$ are unknown and are determined up to a finite order $k$ (e.g. up to the average perturbation order). This is done at the start of the simulation, before observables are measured. The algorithm will also visit states with very low a priori weight, but can explore the whole phase space. In particular, the algorithm can return to the $k = 0$ (empty timeline) configuration. This way it need not tunnel e.g. between regions contributing to opposite magnetization directions. Since the same effect is essentially obtained by exploiting the symmetries of the system through properly chosen global updates, results presented in this thesis were obtained using global and cluster updates only.

### 3.3.8   Measurement of Green's Function

An expansion similar to that for the partition function can be written for Green's function. In the multiorbital formulation, it reads

$$g_{\alpha\beta}^{\sigma}(\tau - \tau') = \frac{Z_0}{Z} \sum_{k=0}^{\infty} \int_0^{\beta} d\tau_1 \ldots \int_{\tau_{2k-1}}^{\beta} d\tau_{2k}' \, \tilde{g}_{\alpha\beta}^{\sigma}(\tau, \tau'; \tau_1, \ldots, \tau_{2k}') \, \mathrm{sgn}(\Omega_k) \, |\Omega_k| \, , \quad (3.61)$$

where $\tilde{g}$ denotes a contribution to Green's function which can be expressed as a ratio of fermionic determinants (time labels are suppressed for brevity):

$$\tilde{g}_{\alpha\beta}^{\sigma}(\tau, \tau'; C_k) = \frac{\langle c_{\beta\sigma}^*(\tau)c_{\alpha\sigma}(\tau')(c_{\alpha_1\sigma_1}^* c_{\beta_1\sigma_1} - \alpha_{\alpha_1\beta_1}^{\sigma_1}) \ldots (c_{\gamma_k\sigma_k'}^* c_{\delta_k\sigma_k'} - \alpha_{\gamma_k\delta_k}^{\sigma_k'}) \rangle_0}{\langle (c_{\alpha_1\sigma_1}^* c_{\beta_1\sigma_1} - \alpha_{\alpha_1\beta_1}^{\sigma_1}) \ldots (c_{\gamma_k\sigma_k'}^* c_{\delta_k\sigma_k'} - \alpha_{\gamma_k\delta_k}^{\sigma_k'}) \rangle_0} \, . \quad (3.62)$$

Diagrammatically, the measurement of the Green function corresponds to the removal of all lines connecting a given creator and annihilator from the diagrams of the contribution. This implies the removal of the corresponding row and column of the matrix $\hat{g}$. In order to measure a contribution to, say, $-\langle c_{\uparrow}(\tau_1)c_{\uparrow}^*(\tau_2)\rangle$ in the configuration depicted in Fig. 3.1, remove row 1 and column 2 from the matrix

$$
\begin{array}{cc}
 & \begin{array}{cc} c_{\uparrow}^*(\tau_1) & \hspace{4em} c_{\uparrow}^*(\tau_2) \end{array} \\
\begin{array}{c} c_{\uparrow}(\tau_1) \\ c_{\uparrow}(\tau_2) \end{array} &
\left( \begin{array}{cc} -\langle c_{\uparrow}(\tau_1)c_{\uparrow}^*(\tau_1)\rangle_0 & -\langle c_{\uparrow}(\tau_1)c_{\uparrow}^*(\tau_2)\rangle_0 \\ -\langle c_{\uparrow}(\tau_2)c_{\uparrow}^*(\tau_1)\rangle_0 & -\langle c_{\uparrow}(\tau_2)c_{\uparrow}^*(\tau_2)\rangle_0 \end{array} \right)
\end{array} \, . \quad (3.63)
$$

In doing so, one is left with $-\langle c_{\uparrow}(\tau_2)c_{\uparrow}^*(\tau_1)\rangle_0$ and the full matrix for $\sigma = \downarrow$, which gives two possibilities to connect the Green functions to the vertices. This corresponds to the diagrams in the first row of Fig. 3.3. These are the diagrams with exactly two unconnected endpoints.

As outlined in Sec. 3.3.2 and appendix D.1, the determinant ratio in (3.62) for the measurement of Green's function in imaginary time evaluates to

$$\tilde{g}_{\alpha\beta}^{\sigma}(\tau, \tau'; C_k) = g_{0\,\alpha\beta}^{\sigma}(\tau - \tau') - \sum_{ij=1}^{k} g_{0\alpha\alpha_i}^{\sigma}(\tau - \tau_i) \, M_{ij}^{\sigma} \, g_{0\beta_j\beta}^{\sigma}(\tau_j - \tau') \, , \quad (3.64)$$

where $M$ is the inverse of the matrix $\hat{g}$ of noninteracting Green functions evaluated at the times corresponding to the particular configuration. Note that the Green function is measured as a correction to a known function ($g_0$). This is a consequence of the

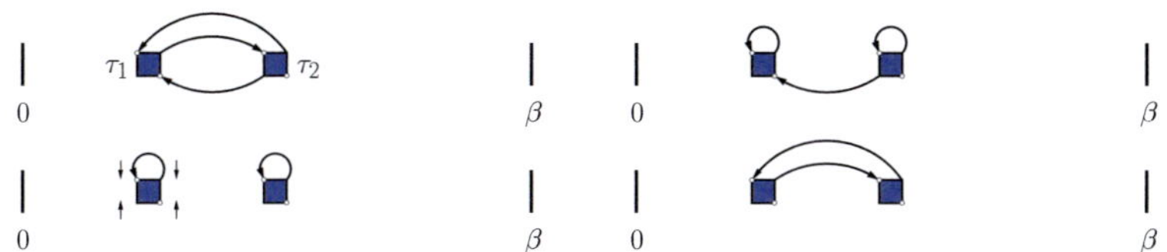

*Figure 3.3:* The four contributions to $\Omega_2$ after removing all Green functions connected to $c_\uparrow(\tau_1)$ and $c_\uparrow^*(\tau_2)$. The diagrams in the first row have only two unconnected endpoints and contribute to the Green function $g_\uparrow(\tau_1 - \tau_2) = -\langle c_\uparrow(\tau_1)c_\uparrow^*(\tau_2)\rangle$.

expansion in the interaction. For the strong-coupling solver this is not the case (see Sec. 3.4.1). Individual configurations are not time-translation invariant and the quantities $\tilde{g}(\tau, \tau', C_k)$ depend on both times separately. The Monte Carlo averaging eventually restores time-translational invariance and the Green function may be measured as a function of the time-difference only:

$$g_{\alpha\beta}^\sigma(\tau - \tau') = \sum_{\{k,C_k\}} \tilde{g}_{\alpha\beta}^\sigma(\tau, \tau'; C_k)\,\mathrm{sgn}(\Omega_k) \Big/ \sum_{\{k,C_k\}} \mathrm{sgn}(\Omega_k) =: \langle \tilde{g}_{\alpha\beta}^\sigma(\tau, \tau'; C_k)\rangle_{\mathrm{MC}}. \qquad (3.65)$$

The Fourier transform of (3.64) is given by

$$\tilde{g}_{\alpha\beta}^\sigma(\omega, \omega'; C_k) = g_{0\,\alpha\beta}^\sigma(\omega)\delta_{\omega,\omega'} - \frac{1}{\beta}\sum_{ij=1}^{k} e^{i\omega\tau_i} g_{0\,\alpha\alpha_i}^\sigma(\omega)\, M_{ij}^\sigma\, g_{0\,\beta_j\beta}^\sigma(\omega')e^{-i\omega'\tau_j}\,, \qquad (3.66)$$

which allows to directly measure the Green function on Matsubara frequencies. This has the advantage that a time discretization for a measurement in imaginary time is avoided, so that the algorithm remains truly continuous in time. In addition, the DMFT self-consistency condition is implemented in the diagonal Matsubara representation, so that the Matsubara measurement saves one explicit Fourier transform. The actual Green function is obtained by averaging the diagonal elements: $g_{\alpha\beta}^\sigma(\omega) = \langle \tilde{g}_{\alpha\beta}^\sigma(\omega, \omega'; C_k)\delta_{\omega,\omega'}\rangle_{\mathrm{MC}}$.

The measurement for a single frequency takes $O(k^2)$ operations (summing over all elements of the $M$-matrix). This can be recast into a linear ($O(k)$) operation by inserting (D.10) into (D.5):

$$\tilde{g}_{\alpha\beta}^\sigma(\omega, \omega'; C_k) = \tilde{g}_{\alpha\beta}^{\sigma\,\mathrm{old}}(\omega, \omega'; C_k) - \frac{1}{\beta M_{kk}^\sigma}\Big(\sum_{i=1}^{k} g_{0\,\alpha\alpha_i}^\sigma(\omega)\, M_{ik}^\sigma e^{i\omega\tau_i}\Big)\Big(\sum_{j=1}^{k} M_{kj}^\sigma\, g_{0\,\beta_j\beta}^\sigma(\omega')e^{-i\omega'\tau_j}\Big),$$

$$(3.67)$$

where $\tilde{g}^{\mathrm{old}}$ is the estimator for Green's function obtained in the previous step. Hence this measurement has to be performed at every step and the measurements will be correlated.

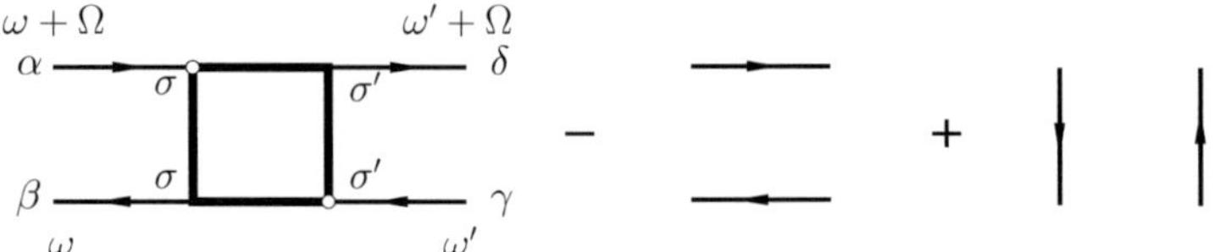

*Figure 3.4:* Diagrammatic representation of the local two-particle Green function in terms of scattering of particle-hole pairs, $\chi^{\sigma\sigma'}_{\alpha\beta\gamma\delta}(\omega_1, \omega_2, \omega_3, \omega_4) := \langle c_{\alpha\sigma}(\omega_1)c^*_{\beta\sigma}(\omega_2)c_{\gamma\sigma'}(\omega_3)c^*_{\delta\sigma'}(\omega_4)\rangle$. Defining the (bosonic) transferred frequency as $\Omega := \omega_1 - \omega_2$, which is conserved in scattering processes, it can be written $\chi^{\sigma\sigma'}_{\alpha\beta\gamma\delta}(\omega, \omega', \Omega) := \chi^{\sigma\sigma'}_{\alpha\beta\gamma\delta}(\omega + \Omega, \omega, \omega', \omega' + \Omega)$. The first term contains the vertex part (see chapter 10) depicted by the square. Lines are fully dressed propagators. The direction of the arrows is in accordance with (3.44). Frequencies and orbital degrees of freedom are labeled counterclockwise, starting at the top left. The two terms on the right will be referred to as the trivial part, which describes renormalized propagation without scattering. Only these terms contribute in absence of interaction.

In addition, formula (3.67) is susceptible to accumulation of roundoff errors. On the contrary, a measurement of the Green function according to (3.66) is performed only after the new configuration is independent of the previous one. This can be assumed after all elements of a configuration have been changed on average, i.e. after a number of steps corresponding to the average perturbation order. Hence the linear measurement performed every step has no advantage compared to an $O(k^2)$ measurement performed every $k$-th step.

### 3.3.9   Measurement of $2n$-Point Correlation Functions

Two- and three-particle Green functions are extensively used in this thesis. As shown in appendix D.2, the determinant ratio for $2n$-point correlation functions can be represented as the determinant of an $n \times n$ matrix containing the measurements of single-particle (2-point) Green functions. This property is a consequence of Wick's theorem and provides a mnemonic rule to construct measurements for higher-order moments of the impurity: One (symbolically) enumerates all complete contractions of the Grassmann numbers or operators and replaces each contraction by the corresponding measurement (determinant ratio) of the single particle Green function.

For example, consider the two-particle imaginary time correlation function $\chi^{\sigma\sigma'}_{\alpha\beta\gamma\delta}(\tau) := \langle c_{\alpha\sigma}(\tau)c^*_{\beta\sigma}(\tau)c_{\gamma\sigma'}(0)c^*_{\delta\sigma'}(0)\rangle$, which arises in the calculation of the double occupancy $d = \langle n_\uparrow n_\downarrow \rangle$, or the spin- and charge susceptibilities. Greek indices label additional degrees of freedom such as orbitals or sites in cluster calculations. The mea-

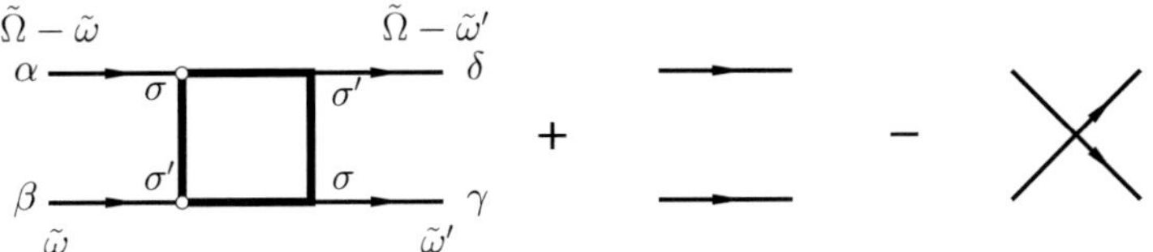

*Figure 3.5:* Two-particle Green function in terms of scattering of particle-particle pairs. The contributions of the trivial part are recognized as direct and exchange terms. This representation is related to the one in Fig. 3.4 by exchanging the bottom endpoints using the antisymmetry property of $\chi$: $\chi^{pp\,\sigma\sigma'}_{\alpha\beta\gamma\delta}(\omega_1,\omega_2,\omega_3,\omega_4) := \langle c_{\alpha\sigma}(\omega_1)c_{\beta\sigma'}(\omega_2)c^*_{\gamma\sigma}(\omega_3)c^*_{\delta\sigma'}(\omega_4)\rangle = -\langle c_{\alpha\sigma}(\omega_1)c^*_{\gamma\sigma}(\omega_3)c_{\beta\sigma'}(\omega_2)c^*_{\delta\sigma'}(\omega_4)\rangle$. When written in this form, energy conservation requires that $\tilde\omega_1 + \tilde\omega_2 = \tilde\omega_3 + \tilde\omega_4$ and the bosonic frequency is defined as $\tilde\Omega := \tilde\omega_1 + \tilde\omega_2 = \tilde\omega_3 + \tilde\omega_4$.

surement for this function is constructed in the following way:

$$\langle c_{\alpha\sigma}(\tau)c^*_{\beta\sigma}(\tau)c_{\gamma\sigma'}(0)c^*_{\delta\sigma'}(0)\rangle \longrightarrow c_{\alpha\sigma}(\tau)c^*_{\beta\sigma}(\tau)\ c_{\gamma\sigma'}(0)c^*_{\delta\sigma'}(0)$$

$$- c_{\alpha\sigma}(\tau)c^*_{\delta\sigma'}(0)\ c_{\gamma\sigma'}(0)c^*_{\beta\sigma}(\tau)$$

$$\longrightarrow \tilde{g}^{\sigma}_{\alpha\beta}(\tau,\tau)\ \tilde{g}^{\sigma'}_{\gamma\delta}(0,0) - \delta_{\sigma\sigma'}\tilde{g}^{\sigma}_{\alpha\delta}(\tau,0)\ \tilde{g}^{\sigma}_{\gamma\beta}(0,\tau)\ . \quad (3.68)$$

By Fourier transform (cf. appendix C), it is easy to see that this rule applies equally well to the corresponding quantities in Matsubara representation. For example, the determinant ratio for the general two-particle Green function $\chi^{\sigma\sigma'}_{\alpha\beta\gamma\delta}(\omega_1,\omega_2,\omega_3,\omega_4) :=$ $\langle c_{\alpha\sigma}(\omega_1)c^*_{\beta\sigma}(\omega_2)c_{\gamma\sigma'}(\omega_3)c^*_{\delta\sigma'}(\omega_4)\rangle$ is obtained as

$$\langle c_{\alpha\sigma}(\omega_1)c^*_{\beta\sigma}(\omega_2)c_{\gamma\sigma'}(\omega_3)c^*_{\delta\sigma'}(\omega_4)\rangle \longrightarrow c_{\alpha\sigma}(\omega_1)c^*_{\beta\sigma}(\omega_2)\ c_{\gamma\sigma'}(\omega_3)c^*_{\delta\sigma'}(\omega_4)$$

$$- c_{\alpha\sigma}(\omega_1)c^*_{\delta\sigma'}(\omega_4)\ c_{\gamma\sigma'}(\omega_3)c^*_{\beta\sigma}(\omega_2)$$

$$\longrightarrow \tilde{g}^{\sigma}_{\alpha\beta}(\omega_1,\omega_2)\ \tilde{g}^{\sigma'}_{\gamma\delta}(\omega_3,\omega_4)$$

$$- \delta_{\sigma\sigma'}\tilde{g}^{\sigma}_{\alpha\delta}(\omega_1,\omega_4)\ \tilde{g}^{\sigma}_{\gamma\beta}(\omega_3,\omega_2)\ . \quad (3.69)$$

Energy conservation requires $\omega_1 - \omega_2 + \omega_3 - \omega_4 = 0$. The two-particle Green function hence depends on three independent frequencies. The different sign on the frequencies follows from the definition of the Fourier transform. It is convenient to introduce a (bosonic) transferred frequency associated with the energy of the scattered particle-hole pair, which is conserved in scattering processes. Here it is defined as

$$\Omega := \omega_1 - \omega_2 = \omega_4 - \omega_3\ . \quad (3.70)$$

With this definition, the two-particle Green function reads $\chi^{\sigma\sigma'}_{\alpha\beta\gamma\delta}(\omega, \omega', \Omega) := \chi^{\sigma\sigma'}_{\alpha\beta\gamma\delta}(\omega + \Omega, \omega, \omega', \omega' + \Omega)$. $\chi$ is unsymmetric with respect to the fermionic frequencies, which can be seen in Fig. 3.10. It is hence measured for three independent, positive and negative fermionic frequencies $\omega_n = (2n + 1)\pi/\beta$, i.e. $n = -N_\omega, \ldots, N_\omega - 1$. Measuring one bosonic and two fermionic frequencies is equivalent. Using the symmetry $\chi(\omega, \omega', -\Omega) = \chi^*(-\omega, -\omega', \Omega)$ (C.5) associated with reversing the sign of $\Omega$, only positive bosonic frequencies $\Omega_m = 2m\pi/\beta$ need to measured, with $m = 0, \ldots, N_\Omega = 2N_\omega - 1$. In any case, ignoring further symmetries, the total number of observables for a consistent cut-off is $(2N_\omega)^3$ times the number of spin- and orbital configurations. It is therefore useful to exploit symmetries, in order to reduce the computational cost and memory requirements.

It is straightforward to exploit spin symmetry. Due to conservation of the $z$-component of the spin only 6 configurations are nonzero: $\chi^{\uparrow\uparrow\uparrow\uparrow}, \chi^{\uparrow\uparrow\downarrow\downarrow}, \chi^{\uparrow\downarrow\downarrow\uparrow}$ and three more by reversing all spins. The last one is related to the second by exchanging the creators and making use of the antisymmetry property of $\chi^{(4)}$. Hence it is sufficient to consider spin configurations of the form $\chi^{\sigma\sigma\sigma'\sigma'}$ and to use the shorthand notation $\chi^{\sigma\sigma'} := \chi^{\sigma\sigma\sigma'\sigma'}$. In the paramagnetic state, $\chi$ is invariant under exchange of all spins. In that case, only two independent configurations, $\chi^{\uparrow\uparrow}$ and $\chi^{\uparrow\downarrow}$, exist. In practice, it is however beneficial to measure $\chi^{\uparrow\uparrow}$ and $\chi^{\downarrow\downarrow}$ and average the two, to reduce the noise due to spurious polarization effects. The same applies for $\chi^{\uparrow\downarrow}$. The number of observables can be reduced using $\chi^{\uparrow\uparrow}(\omega_1, \omega_2, \omega_3, \omega_4) = \chi^{\uparrow\uparrow}(\omega_3, \omega_4, \omega_1, \omega_2)$ and $\chi^{\uparrow\downarrow}(\omega_1, \omega_2, \omega_3, \omega_4) = \chi^{\downarrow\uparrow}(\omega_3, \omega_4, \omega_1, \omega_2)$. These are exact symmetries in the sense that they are preserved by the Monte Carlo measurements as obvious from (3.69). Nothing is gained by measuring the left- and right-hand sides simultaneously. The symmetries are in general reduced in presence of orbital degrees of freedom. The two-particle Green function is depicted diagrammatically in Figs. 3.4 and 3.5.

The full (reducible) two-particle impurity vertex function is extracted from the two-particle Green function. Diagrammatically, it corresponds to the square in Figs. 3.4 and 3.5. It is hence obtained by subtracting the Green functions disconnected from the vertex to obtain the connected (nontrivial) part of the two-particle Green function,

$$\tilde{\chi}^{\sigma\sigma'}_{\alpha\beta\gamma\delta}(\omega_1, \omega_2, \omega_3, \omega_4) := \chi^{\sigma\sigma'}_{\alpha\beta\gamma\delta}(\omega_1, \omega_2, \omega_3, \omega_4) - g^\sigma_{\alpha\beta}(\omega_1)g^{\sigma'}_{\gamma\delta}(\omega_3)\delta_{\omega_1,\omega_2}\delta_{\omega_3,\omega_4}$$
$$+ \delta_{\sigma\sigma'}g^\sigma_{\alpha'\delta'}(\omega_1)g^\sigma_{\gamma'\beta'}(\omega_3)\delta_{\omega_1,\omega_4}\delta_{\omega_3,\omega_2}$$

$$(3.71)$$

and removing the four Green function legs (see also Eqs. A.63, A.64):

$$\Gamma^{\sigma\sigma'}_{\alpha\beta\gamma\delta}(\omega_1, \omega_2, \omega_3, \omega_4) := g^{\sigma\,-1}_{\alpha\alpha'}(\omega_1)g^{\sigma'\,-1}_{\gamma\gamma'}(\omega_3)\tilde{\chi}^{\sigma\sigma'}_{\alpha'\beta'\gamma'\delta'}(\omega_1, \omega_2, \omega_3, \omega_4)g^{\sigma\,-1}_{\beta'\beta}(\omega_2)g^{\sigma'\,-1}_{\delta'\delta}(\omega_4) \, .$$

$$(3.72)$$

In the dual fermion approach (see chapter 6), it is also required to compute the three-particle Green function. The measurement for this quantity is readily evaluated to be

$$
-\chi^{(6)\,\sigma_1\sigma_2\sigma_3}_{\alpha\beta\gamma\delta\epsilon\zeta}(\omega_1,\omega_2,\omega_3,\omega_4,\omega_5,\omega_6) :=
$$

$$
\langle c_{\alpha\sigma_1}(\omega_1)c^*_{\beta\sigma_1}(\omega_2)c_{\gamma\sigma_2}(\omega_3)c^*_{\delta\sigma_2}(\omega_4)c_{\epsilon\sigma_3}(\omega_5)c^*_{\zeta\sigma_3}(\omega_6)\rangle
$$

$$
\longrightarrow\ c_{\alpha\sigma_1}(\omega_1)c^*_{\beta\sigma_1}(\omega_2)c_{\gamma\sigma_2}(\omega_3)c^*_{\delta\sigma_2}(\omega_4)c_{\epsilon\sigma_3}(\omega_5)c^*_{\zeta\sigma_3}(\omega_6)
$$

$$
+\ c_{\alpha\sigma_1}(\omega_1)c^*_{\beta\sigma_1}(\omega_2)c_{\gamma\sigma_2}(\omega_3)c^*_{\delta\sigma_2}(\omega_4)c_{\epsilon\sigma_3}(\omega_5)c^*_{\zeta\sigma_3}(\omega_6)
$$

$$
+\ c_{\alpha\sigma_1}(\omega_1)c^*_{\beta\sigma_1}(\omega_2)c_{\gamma\sigma_2}(\omega_3)c^*_{\delta\sigma_2}(\omega_4)c_{\epsilon\sigma_3}(\omega_5)c^*_{\zeta\sigma_3}(\omega_6)
$$

$$
+\ c_{\alpha\sigma_1}(\omega_1)c^*_{\beta\sigma_1}(\omega_2)c_{\gamma\sigma_2}(\omega_3)c^*_{\delta\sigma_2}(\omega_4)c_{\epsilon\sigma_3}(\omega_5)c^*_{\zeta\sigma_3}(\omega_6)
$$

$$
+\ c_{\alpha\sigma_1}(\omega_1)c^*_{\beta\sigma_1}(\omega_2)c_{\gamma\sigma_2}(\omega_3)c^*_{\delta\sigma_2}(\omega_4)c_{\epsilon\sigma_3}(\omega_5)c^*_{\zeta\sigma_3}(\omega_6)
$$

$$
+\ c_{\alpha\sigma_1}(\omega_1)c^*_{\beta\sigma_1}(\omega_2)c_{\gamma\sigma_2}(\omega_3)c^*_{\delta\sigma_2}(\omega_4)c_{\epsilon\sigma_3}(\omega_5)c^*_{\zeta\sigma_3}(\omega_6)
$$

$$
\longrightarrow\ \tilde{g}^{\sigma_1}_{\alpha\beta}(\omega_1,\omega_2)\ \tilde{g}^{\sigma_2}_{\gamma\delta}(\omega_3,\omega_4)\ \tilde{g}^{\sigma_3}_{\epsilon\zeta}(\omega_5,\omega_6)
$$

$$
-\delta_{\sigma_2,\sigma_3}\ \tilde{g}^{\sigma_1}_{\alpha\beta}(\omega_1,\omega_2)\ \tilde{g}^{\sigma_2}_{\gamma\zeta}(\omega_3,\omega_6)\ \tilde{g}^{\sigma_3}_{\epsilon\zeta}(\omega_5,\omega_4)
$$

$$
-\delta_{\sigma_1,\sigma_2}\ \tilde{g}^{\sigma_1}_{\alpha\delta}(\omega_1,\omega_4)\ \tilde{g}^{\sigma_2}_{\gamma\beta}(\omega_3,\omega_2)\ \tilde{g}^{\sigma_3}_{\epsilon\zeta}(\omega_5,\omega_6)
$$

$$
-\delta_{\sigma_1,\sigma_3}\ \tilde{g}^{\sigma_1}_{\alpha\zeta}(\omega_1,\omega_6)\ \tilde{g}^{\sigma_2}_{\gamma\delta}(\omega_3,\omega_4)\ \tilde{g}^{\sigma_3}_{\epsilon\beta}(\omega_5,\omega_2)
$$

$$
+\delta_{\sigma_1,\sigma_2}\,\delta_{\sigma_1,\sigma_3}\ \tilde{g}^{\sigma_1}_{\alpha\zeta}(\omega_1,\omega_6)\ \tilde{g}^{\sigma_2}_{\gamma\beta}(\omega_3,\omega_2)\ \tilde{g}^{\sigma_3}_{\epsilon\delta}(\omega_5,\omega_4)
$$

$$
+\delta_{\sigma_1,\sigma_2}\,\delta_{\sigma_1,\sigma_3}\ \tilde{g}^{\sigma_1}_{\alpha\delta}(\omega_1,\omega_4)\ \tilde{g}^{\sigma_2}_{\gamma\zeta}(\omega_3,\omega_6)\ \tilde{g}^{\sigma_3}_{\epsilon\beta}(\omega_5,\omega_2)\,. \tag{3.73}
$$

In this case, energy conservation requires that $\omega_1 + \omega_3 + \omega_5 = \omega_2 + \omega_4 + \omega_6$ and a transferred frequency may be defined in analogy to Eq. 3.70. The three-particle Green function depends on five independent Matsubara frequencies so that the number of observables is $20 \times (2N_\omega)^5$ for the twenty nonzero spin configurations. It is vital to use the symmetry properties in analogy to those of the two-particle Green function. For example only the independent spin configurations $\chi^{\uparrow\uparrow\uparrow\uparrow\uparrow\uparrow}$, $\chi^{\uparrow\uparrow\uparrow\uparrow\downarrow\downarrow}$ and those with all spins reversed need to be measured. The orbital indices in (3.73) are given for completeness. At present a multiorbital calculation of the three-particle Green function is computationally not feasible due to the large number of observables and memory requirements. The three-particle vertex function is defined in Eqs. A.74, A.75.

Estimation of the Monte Carlo error for the single-particle quantities as outlined in Sec. 3.1.3, becomes impractical for higher-order correlation functions. This would entirely dominate the simulation. Typically, the higher-order correlators are used in subsequent computations, such as the summation of diagrams. These calculations reduce the information to a single or a few numbers by operations which are too complex to allow for an error propagation analysis. In such a case, at least a consistency check of the data and the convergence of the Monte Carlo simulation can be obtained by monitoring the error of a few selected observables and by performing backups of the $2n$-point functions at a few points of the simulation, which are separated by a sufficiently large number of sweeps. The convergence can then be judged on the basis of the fluctuations of the derived quantities.

### 3.3.10  Imaginary Time Correlation Functions

The usefulness of two-particle correlation functions is based on the Kubo formula, which establishes a relation to observable quantities. For example, the dynamical susceptibility describes the response (e.g. the magnetization) of a system to a time-dependent external field. The magnetic susceptibility is defined as

$$\chi_{zz} = \frac{\partial M_z}{\partial h_z} . \tag{3.74}$$

For simplicity, here only the $zz$-component of the susceptibility tensor is considered and the index $z$ is dropped in the following. If the external field (the perturbation) is small, the susceptibility is independent of the field (linear response). The field couples to the magnetic moment $m$ of the electron through the perturbation $V(t) = -mh(t)$, where $m = g\mu_B S_z$ and $g$ and $\mu_B$ are the $g$-factor and the Bohr magneton, respectively. Within linear response theory, the magnetization is given by

$$M(t) = -i \int_{-\infty}^t dt' \langle [m(t), V(t')] \rangle = i \int_{-\infty}^t dt' \langle [m(t), m(t')] \rangle h(t') , \tag{3.75}$$

which can be expressed in terms of the following retarded response function

$$\chi^{\text{ret}}(t - t') = -i\theta(t - t') \langle [S_z(t - t'), S_z(0)] \rangle \tag{3.76}$$

as

$$M(t) = -(g\mu_B)^2 \int_{-\infty}^\infty dt' \chi^{\text{ret}}(t - t') h(t') . \tag{3.77}$$

Hence the Fourier transform of the retarded function is related to the dynamical susceptibility through $\chi(\omega) = -\chi^{\text{ret}}(\omega)$ (in units of $(g\mu_B)^2$), which contains information about the response of the system to a time-dependent field (or on the stability of the system itself, see chapter 10). In the linear response regime, the response to an arbitrary signal is obtained by superposition of individual Fourier components. The (negative) retarded function is obtained from the corresponding (time-ordered) Matsubara correlation function

$$\chi(i\Omega_m) = \int_0^\beta d\tau e^{i\Omega_m \tau} \langle S_z(\tau) S_z(0) \rangle \tag{3.78}$$

by analytical continuation, i.e. by letting $i\Omega_m \to \omega + i\delta$ [98]. $\Omega_m = 2\pi m/\beta$ are bosonic Matsubara frequencies. The static susceptibility is directly obtained by setting $\Omega_m$ to zero. It is sometimes written as

$$\chi(i\Omega_m = 0) = \int_0^\beta d\tau \langle S_z(\tau) S_z(0)\rangle e^{i\Omega_m \tau}\Big|_{\Omega_m=0} = \int_0^\beta d\tau \langle S_z(\tau) S_z(0)\rangle \, . \tag{3.79}$$

The full frequency dependence can be either obtained by numerical Fourier transform of the imaginary time response function (see Fig. 3.10), or alternatively from the two-particle Green function in Matsubara representation[1] (see Appendix C):

$$\chi(i\Omega) = \frac{1}{2} \int_0^\beta d\tau \langle n_\uparrow(\tau) n_\uparrow(0) - n_\uparrow(\tau) n_\downarrow(0)\rangle e^{i\Omega \tau}$$

$$= \frac{1}{2}\frac{1}{\beta^2} \sum_{\omega,\omega'} \left[\chi^{\uparrow\uparrow}(\omega+\Omega, \omega, \omega'\omega'+\Omega) - \chi^{\uparrow\downarrow}(\omega+\Omega, \omega, \omega'\omega'+\Omega)\right] \, . \tag{3.80}$$

The sum in the last line is over all Matsubara frequencies from $n = -\infty$ to $n = +\infty$. The two-particle Green function is measured for a finite number of Matsubara frequencies and the susceptibility has to be approximated by properly including the asymptotics as described in [94]. Depending on the method of numerical analytical continuation, the input data is either given as a function of bosonic frequencies or imaginary time (see Sec. 3.5).

As a test for the measurement of equal time correlation functions, results for the local moment given in Ref. [99] for the symmetric AIM ($E_d = -U/2$) as a function of temperature are reproduced in Fig. 3.6. The local moment is given by the expectation value

$$\chi(\tau = 0^+) = \lim_{\tau \to 0^+} \langle S_z(\tau) S_z(0)\rangle = \chi(\beta) \, . \tag{3.81}$$

The last equality holds since for this case $\chi$ is symmetric, $\chi(\tau) = \chi(\beta - \tau)$. As for the original data, $\langle \sigma_z^2\rangle = 4\langle S_z^2\rangle$ is plotted. All energies are in units of the impurity resonance (see chapter 5). The data shows the formation of a local moment below a temperature $T \sim \Delta = 1/2$. In the high temperature limit, the impurity occupation is $1/2$ on average, giving $\langle S_z^2\rangle \overset{T\to\infty}{=} 1/2 \cdot 1/4$ or $\langle \sigma_z^2\rangle = 1/2$. For temperatures small compared to $\Delta$, $\langle \sigma_z^2\rangle$ approximately tends to unity as $U \to \infty$.

The data from the original publication (not shown) has been retrieved by using an image digitizer tool on the figure [100]. The overall agreement is good. The local moment in the original data is slightly overestimated (by $\sim 2\%$ for $U = 4$ and low temperature $T = 0.1$; the estimated CTQMC error is $\sim 0.1\%$). This is quite remarkable, since a time discretization down to $\Delta\tau = \beta/L = 0.25$ and a number of typically 5000 sweeps for the calculation in this reference were used[2].

---

[1] The following relation holds in the paramagnetic state.

[2] Computational resources were limited, the year of publication is 1987.

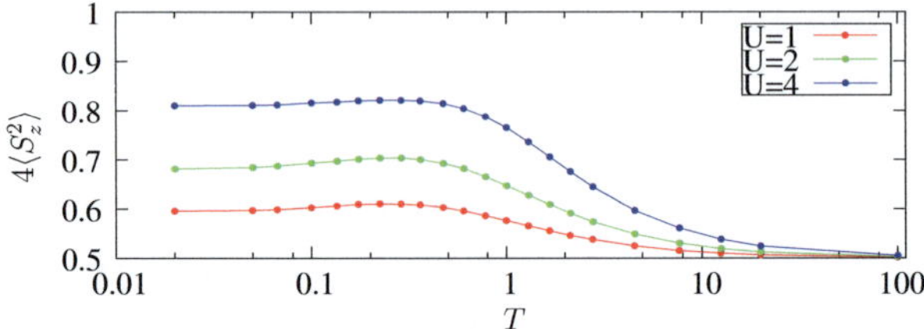

*Figure 3.6:* Local moment of the Anderson impurity model as a function of temperature and different values of Hubbard repulsion $U$, obtained by measurement of the equal-time correlation function (3.81) in CTQMC. The hybridization $\Delta(i\omega) = (-i/2)\,\mathrm{sgn}(\omega)$ is that of a wide featureless band (see chapter 5). The Monte Carlo error is smaller than the symbol size (of the order of the linewidth).

Further tests (including cluster calculations) have been performed by C. Herkt-Januschek during the course of his diploma thesis [101]. A comparison with analytical results is possible e.g. by considering the one-dimensional Hubbard model at half-filling. The sites are connected by hopping $t$ and exerted to a thermal bath at inverse temperature $\beta$ described by the Green function $G^{-1}(i\omega) = i\omega + \mu$. In the limit of large repulsion $U$, i.e. $U \gg 1/\beta$, $U \gg t$ and $U \gg h$, using second order degenerate perturbation theory, the half-filled system can be shown to approximately map to a Heisenberg model (see, e.g. Ref. [102]), with Hamiltonian

$$H = J \sum_i \mathbf{S}_i \, \mathbf{S}_{i+1} - h \sum_i S_i^z \tag{3.82}$$

and an antiferromagnetic effective exchange $J_{\mathrm{eff}} = 4t^2/U > 0$. The temperature dependence of the total spin or other correlation functions is readily obtained by exact diagonalization of the Heisenberg Hamiltonian. For example, the thermal expectation value for the total spin of a two-site chain reads

$$\langle \mathbf{S}^2 \rangle = \frac{1}{\mathcal{Z}} \mathrm{Tr}\, \mathbf{S}^2 e^{-\beta H} = 2 \frac{1 + \cosh(\beta h)}{1 + \cosh(\beta h) + e^{\beta J}} \, , \tag{3.83}$$

where $\mathcal{Z} = \mathrm{Tr}\, e^{-\beta H}$ is the partition function. In the validity range of the Heisenberg approximation, convincing agreement was found for single- and two-particle correlation functions for various values of the parameters $\beta$, $t$ and $h$ [101]. Applications of these concepts, including results for the imaginary time-dependence of spin-correlators and their analytical continuation, are shown in chapter 4.

## 3.4 Strong-Coupling Expansion

The strong-coupling algorithm was initially introduced by P. Werner *et al.* [78] and has been generalized to multiorbital systems with general interactions [79, 95, 80]. Here the algorithm is discussed in the segment representation, which exploits the possibility of a very fast computation of the trace for density-density type of interactions. The action is regrouped into the atomic part

$$S_{\text{at}} = \int_0^\beta d\tau' \sum_\sigma c_\sigma^*(\tau)[\partial_\tau - \mu]c_\sigma(\tau) + U \int_0^\beta d\tau c_\uparrow^*(\tau)c_\uparrow(\tau)c_\downarrow^*(\tau)c_\downarrow(\tau) \tag{3.84}$$

and the part of the action $S_\Delta$ which contains the hybridization term:

$$S_\Delta = -\sum_\sigma \int_0^\beta d\tau' \int_0^\beta d\tau c_\sigma(\tau)\Delta(\tau - \tau')c_\sigma^*(\tau'). \tag{3.85}$$

Here the sign is taken out by reversing the original order of $c$ and $c^*$ to avoid an alternating sign in the expansion. To simplify the notation, consider first the spinless fermion model, which is obtained by disregarding the spin sums and interaction in (3.84,3.85). The series expansion for the partition function is generated by expanding in the hybridization term:

$$\mathcal{Z} = \int e^{-S_{\text{at}}} \sum_k \frac{1}{k!} \int_0^\beta d\tau_1' \int_0^\beta d\tau_1 \ldots \int_0^\beta d\tau_k' \int_0^\beta d\tau_k \times$$

$$\times c(\tau_k)c^*(\tau_k') \ldots c(\tau_1)c^*(\tau_1')\Delta(\tau_1 - \tau_1') \ldots \Delta(\tau_k - \tau_k')\,\mathcal{D}[c^*, c]. \tag{3.86}$$

The important observation now is that, at any order, the diagrams can be collected into a determinant of hybridization functions. In order to see this, one can use a similar reasoning as in the weak-coupling case. The range of integration in (3.86) is changed such that $\tau_k \geq \tau_k' \geq \ldots \geq \tau_1 \geq \tau_1'$. Note that time ordering is not explicitly indicated since it is implicit in the construction of the path integral (Sec. 2.3). For any given time-ordered sequence of $2k$ times $\tau_k, \tau_k', \ldots \tau_1, \tau_1'$, the integration in (3.86) generates exactly $2k!$ terms with a different order of times for which the Grassmann numbers can be brought into the same order. These are the $k!$ permutations with times $\tau_i$ permuted among themselves and correspondingly for the times $\tau_i'$. The Grassmann numbers can always be commuted by permuting pairs of Grassmann numbers in a first step, which does not yield an additional sign and in the second step permuting the annihilators among themselves. The latter operation is associated with an eventual sign depending on the number of permutation required. Specifically, for $k = 2$, one has the $2k! = 4$ terms

$$c(\tau_2)c^*(\tau_2')c(\tau_1)c^*(\tau_1')\Delta(\tau_1 - \tau_1')\Delta(\tau_2 - \tau_2'),$$
$$c(\tau_1)c^*(\tau_1')c(\tau_2)c^*(\tau_2')\Delta(\tau_2 - \tau_2')\Delta(\tau_1 - \tau_1'),$$
$$c(\tau_2)c^*(\tau_1')c(\tau_1)c^*(\tau_2')\Delta(\tau_2 - \tau_1')\Delta(\tau_1 - \tau_2'),$$
$$c(\tau_1)c^*(\tau_2')c(\tau_2)c^*(\tau_1')\Delta(\tau_2 - \tau_1')\Delta(\tau_1 - \tau_2'). \tag{3.87}$$

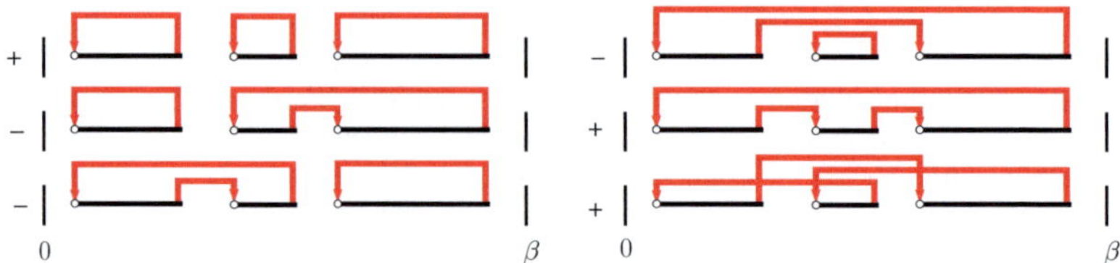

*Figure 3.7:* Diagrammatic representation of the six contributions to the partition function for spinless fermions at $k = 3$. An electron is inserted at the start of a segment (marked by an open circle) and removed at the segment endpoint. The hybridization function lines $\Delta(\tau_i - \tau'_j)$ (shown in red) are connected to the segments in all possible ways. The sign of each diagram is given on the left. The diagrams collect into a determinant. Reproduced from Ref. [78].

Upon time ordering, the Grassmann numbers in the last three lines are brought into the same order as in the first line. It is easy to check that only the last two lines acquire a minus sign. Therefore, these terms can be collected to give

$$c(\tau_2)c^*(\tau'_2)c(\tau_1)c^*(\tau'_1)\,2\,\det\begin{pmatrix} \Delta(\tau_1 - \tau'_1) & \Delta(\tau_1 - \tau'_2) \\ \Delta(\tau_2 - \tau'_1) & \Delta(\tau_2 - \tau'_2) \end{pmatrix}. \tag{3.88}$$

There are always $k!$ terms which can be brought into the same order by permuting pairs of Grassmann numbers and hence the factor $1/k!$ in (3.86) cancels. The partition function then takes the form

$$\mathcal{Z} = \mathcal{Z}_{\text{at}} \sum_k \int_0^\beta d\tau'_1 \int_{\tau'_1}^\beta d\tau_1 \dots \int_{\tau_{k-1}}^\beta d\tau'_k \int_{\tau'_k}^{\circ\tau'_k} d\tau_k \times$$
$$\times \langle c(\tau_k)c^*(\tau'_k)\dots c(\tau_1)c^*(\tau'_1)\rangle_{\text{at}} \, \det \hat{\Delta}^{(k)}, \tag{3.89}$$

where the average is over the states of the atomic problem described by $S_{\text{at}}$. Here $\det \hat{\Delta}^{(k)}$ denotes the determinant of the matrix of hybridizations $\hat{\Delta}_{ij} = \Delta(\tau_i - \tau'_j)$. The diagrams contributing to the partition function for $k = 3$ are shown in Fig. 3.7. A diagram is depicted by a collection of segments, where a segment is symbolic for the time interval where the impurity is occupied. The collection of diagrams obtained by connecting the hybridization lines in all possible ways corresponds to the determinant. Collecting the diagrams into a determinant is essential to alleviate or completely suppress the sign problem. Note that the imaginary time interval in (3.89) is viewed as a circle denoted by $\circ\tau'_k$. The trajectories in the path integral are subject to antiperiodic boundary conditions which is accommodated by an additional sign if a segment winds around the circle.

It is straightforward to generalize this result to general density-density interactions, for which Grassmann numbers of a given flavor (spin, orbitals) appear in alternating

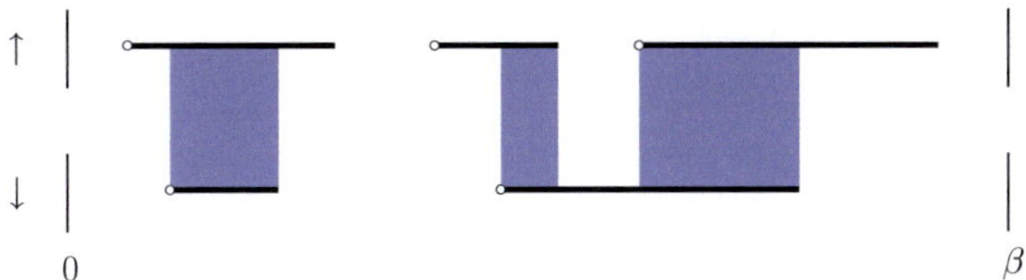

*Figure 3.8:* Typical configuration for the simulation of a single-impurity Anderson model with Hubbard interaction. A separate timeline is drawn for each spin. The shaded regions correspond to the intervals where the impurity is doubly occupied. The segments on each timeline are connected by hybridization lines in all possible ways as in Fig. 3.7 (not shown). The hybridization does not connect segments across timelines, because it is spin-diagonal.

order. For general interactions the segment picture has to be abandoned and the trace over a number of matrices that scales with the perturbation order has to be computed explicitly. For d- or f-systems, the number of states becomes very large, ($2^{10}$ and $2^{14}$, respectively) and different optimizations for the efficient computation of the trace have been proposed, such as storing the operators in a binary tree [95], an adjustable base [80], or a Krylov implementation [103].

The diagrams of the partition function are sampled by randomly inserting or removing segments of varying length (and further moves to increase sampling efficiency). Empty and fully occupied states also have to be sampled. The weight of a configuration is proportional to

$$w^{(k)} \sim \left| \langle c(\tau_k)c^*(\tau_k') \dots c(\tau_1)c^*(\tau_1') \rangle_{\text{at}} \det \hat{\Delta}^{(k)} \right| . \tag{3.90}$$

The insertion of $c^*$ is attempted anywhere in the interval from 0 to $\beta$. If the place is already occupied, the move is rejected. Otherwise the $c$ can be inserted at a later time in the interval $[\tau_{k+1}', \tau_{k+1}' + l_{\max})$, where $l_{\max}$ is the distance to the starting point of the next segment. The probability to add a segment is hence given by $q^{\text{add}} = d\tau^2/\beta l_{\max}$. The probability to remove a pair is $q^{\text{rem}} = 1/(k+1)$. The probabilities of the configuration before and after the move are proportional to $p^{\text{new}} \sim w^{(k+1)}d\tau^2/\beta^2$ and $p^{\text{old}} \sim w^{(k)}$. This yields the acceptance criterion

$$p^{\text{add}} = \min\left(1, \frac{1}{k+1} \frac{l_{\max}}{\beta} \frac{w^{(k+1)}}{w^{(k)}}\right) , \tag{3.91}$$

which involves the computation of ratios of determinants and the average over the atomic states. The latter is evaluated using exact diagonalization of the atomic problem.

For the spinless fermion model, this is particularly simple: $H_{\text{at}} = -\int_0^\beta d\tau \mu c^\dagger(\tau) c(\tau)$ is already diagonal in the occupation number basis $\{|0\rangle, |1\rangle\}$ and

$$\text{Tr}[T_\tau e^{\mu \int_0^\beta d\tau n(\tau)} c(\tau_2) c^\dagger(\tau_2') c(\tau_1) c^\dagger(\tau_1')] =$$
$$\text{Tr}[e^{\mu(\beta-\tau_2)n} c(\tau_2) e^{\mu(\tau_2-\tau_2')n} c^\dagger(\tau_2') e^{\mu(\tau_2'-\tau_1)n} c(\tau_1) e^{\mu(\tau_1-\tau_1')n} c^\dagger(\tau_1') e^{\mu(\tau_1)n}] = e^{\mu l} \,. \tag{3.92}$$

where $l$ is the length of the time interval on which the impurity is occupied. Here only a single "path" through alternating states $|0\rangle$ and $|1\rangle$ contributes, since the operators have only a single nonzero matrix element. For a large basis, several paths leading through different matrix elements contribute.

For the single-orbital Anderson impurity model with Hubbard interaction the segment picture still holds and gives a very intuitive picture of the imaginary time dynamics. A configuration is visualized by two separate timelines, one for each spin. The additional sum over spins, $\sum_{\sigma_1...\sigma_k}$, which enters in the first line of Eq. 3.89 generates contributions such as the one shown in Fig. 3.8. The only difference to the spinless fermion model is that in case the impurity is doubly occupied, the energy $U$ has to be paid and the trace is $e^{\mu(l_\uparrow + l_\downarrow)} e^{-U l_d}$, where $l_\sigma$ is the time spent on the impurity for an electron with spin $\sigma$ and $l_d$ is the time the impurity is doubly occupied. The acceptance criterion is modified accordingly.

### 3.4.1 Measurement of Green's Function

In the strong-coupling formalism, the expansion for Green's function is given by

$$g(\tau - \tau') = -\frac{Z_{\text{at}}}{Z} \sum_k \int_0^\beta d\tau_1' \int_{\tau_1'}^\beta d\tau_1 \ldots \int_{\tau_{k-1}}^\beta d\tau_k' \int_{\tau_k'}^\beta d\tau_k \, \tilde{g}(\tau - \tau'; C_k) \times$$
$$\times \langle c(\tau_k) c^*(\tau_k') \ldots c(\tau_1) c^*(\tau_1') \rangle_{\text{at}} \det \hat{\Delta}(\tau_i - \tau_j') \,, \tag{3.93}$$

where now

$$\tilde{g}(\tau - \tau'; C_k) = \frac{\langle c(\tau) c^*(\tau') c(\tau_k) c^*(\tau_k') \ldots c(\tau_1) c^*(\tau_1') \rangle_{\text{at}}}{\langle c(\tau_k) c^*(\tau_k') \ldots c(\tau_1) c^*(\tau_1') \rangle_{\text{at}}} \tag{3.94}$$

is a measurement for Green's function. Hence $\tilde{g}$ is obtained by computing a ratio of traces. Denoting the trace over a product of $k + 1$ pairs of operators in the numerator in (3.94) in the short-hand notation form $\text{Tr}^{(k+1)}$, the integrand for the Green function at some perturbation order $k$ can be written

$$\text{Tr}^{(k+1)} \det \hat{\Delta}^{(k)} = \frac{\text{Tr}^{(k+1)}}{\text{Tr}^{(k)}} \text{Tr}^{(k)} \det \hat{\Delta}^{(k)}$$
$$= \frac{\det \hat{\Delta}^{(k)}}{\det \hat{\Delta}^{(k+1)}} \text{Tr}^{(k+1)} \det \hat{\Delta}^{(k+1)} \,. \tag{3.95}$$

The weight of the partition function appears to the right. Therefore the Green function can either be measured by adding a pair to the trace and compute the Monte Carlo average of the ratio of traces (according to the first line), or remove a column and a row from the matrix $\hat{\Delta}$ and calculate the Monte Carlo average of the ratio of determinants (second line). Diagrammatically, the first possibility corresponds to adding a pair of operators into one of the timelines in Fig. 3.7 (not a segment) and the second to cutting all hybridization lines connecting a given pair. In each case, two unconnected operators $c(\tau)$, $c^\dagger(\tau')$ contribute to $g(\tau - \tau')$. It is clear how to compute the ratio of traces. The ratio of determinants for the strong-coupling case can be obtained using the following identity from linear algebra ($A^{-1}$ is the inverse of the matrix of hybridization functions, i.e. the $M$-matrix):

$$(A^{-1})_{ji} = \frac{1}{\det(A)} C_{ij}, \tag{3.96}$$

where $C_{ji}$ is the $j, i$ cofactor of $A$, i.e. the determinant of the $j, i$ minor of $A$ times the factor $(-1)^{i+j}$, which takes care of the sign acquired in the row/column permutations. Note the order of indices, $j, i$. After cutting the hybridization lines connecting an arbitrary pair of operators, it can be commuted through to the beginning of the trace yielding a factor $(-1)^{i+j}$, so that the right-hand side of (3.96) is the desired determinant ratio. Hence

$$g(\tau) = \left\langle \sum_{i,j=1}^{k} M_{ji} \bar{\delta}(\tau, \tau_i^e - \tau_j^s) \right\rangle_{\mathrm{MC}} \quad \text{where} \quad \bar{\delta}(\tau, \tau') := \left\{ \begin{array}{ll} \delta(\tau - \tau') & \tau' > 0 \\ -\delta(\tau - (\tau' + \beta)) & \tau' < 0 \end{array} \right. . \tag{3.97}$$

Here $\tau_j^s$ corresponds to the start of a segment, i.e. to the creation operator and $\tau_i^e$ to the end or annihilator and the $\bar{\delta}$ function arises from the $\beta$-antiperiodic definition of the Green function. The density can be accurately determined from the average length of the segments. The measurement for the two-particle Green function can be obtained by generalizing (3.96).

The impurity Green function in imaginary time obtained from a converged calculation in strong-coupling CTQMC is shown in Fig. 3.9 in comparison with exact diagonalization (ED) for a single bath site. For the strong-coupling solver, the hybridization is defined as [79]

$$\Delta(\tau) = \sum_{k} |V_k|^2 \frac{e^{-\epsilon_k(\beta - \tau)}}{1 + e^{-\beta \epsilon_k}}. \tag{3.98}$$

By virtue of Eq. 3.85, its Fourier transform $\Delta(i\omega)$ is related to the hybridization function in Sec. 2.4 by reversing the sign of the argument. A comparison to ED is e.g. possible by choosing the hybridization to be constant with $\Delta(\tau) = V^2/2$, corresponding to the ED bath site parameters $V_k = V$, $\epsilon_k = 0$.

In order to compare the imaginary-time and frequency measurements for the strong-coupling solver, the latter was implemented into the strong-coupling CTQMC code by P. Werner [78]. To ensure the results to be unaffected by differences in autocorrelation

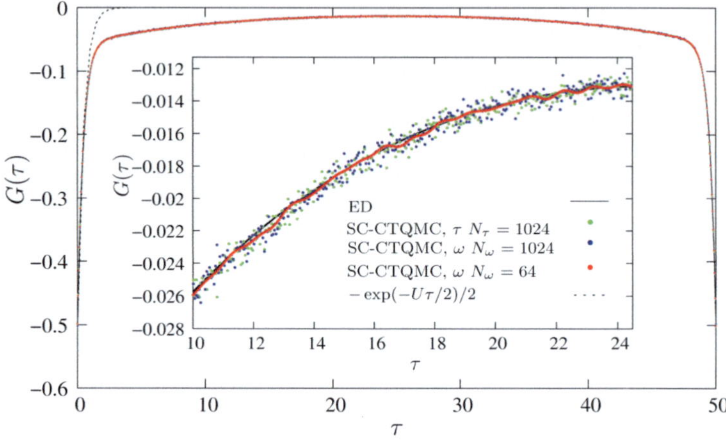


*Figure 3.9:* Imaginary-time Green function $G(\tau)$ obtained within strong-coupling CTQMC by imaginary time and frequency measurements in comparison with ED for a single bath site. The parameters are $U = 4$, $\beta = 50$, $V = 0.25$ and $\epsilon = 0$. Comparison to $-\exp(-U\tau/2)/2$ reveals the steep drop at small and large times, which requires an adequate time discretization. The number of imaginary time bins $N_\tau = 1024$ follows the suggestion of $N_\tau \sim 5\beta U = 10^3$ [78]. The inset reveals a similar noise level for the frequency measurement for $N_\omega = 1024$ frequencies and subsequent Fourier transform including high-energy asymptotics for frequencies above the cutoff. The high-frequency part contributes significantly to the noise, as seen by comparison with $G(\tau)$ obtained by Fourier transform using a reduced cutoff $N_\omega = 64$. Oscillations with a characteristic period $\Delta\tau \sim 2\beta/(2N_\omega + 1)$ due to the small cutoff can be seen to the right.

times, the frequency and time measurement have been performed simultaneously within the same simulation.

The comparison of Green functions is shown in imaginary time mainly for visualization purposes. In addition, the Fourier transform from the frequency to the time domain does not introduce significant systematic errors. As seen in the inset of Fig. 3.9, the noise level of Green functions is similar for imaginary-time and frequency measurements, reflecting the fact that both contain the same information. By taking a small cutoff frequency $N_\omega = 64$ for the Fourier transform, the oscillations are damped considerably, showing that the high-frequency part converges slowly and therefore contributes significantly to the noise. The function is still accurate, but now contains spurious oscillations with a period corresponding to the cutoff frequency. When measuring in the frequency domain, the cutoff should therefore be chosen not too large, but sufficiently large to ensure that at frequencies above the cutoff, the Green function is well represented by a truncated high-frequency expansion. The latter is expressed in terms of Green function moments and can be computed from the knowledge of a few equal-time correlation functions which are measured during the simulation [80].

Here the tail was represented using the parameterization

$$g(i\omega_n) = \frac{1}{2}\left(\frac{1}{i\omega - \mu_1} + \frac{1}{i\omega - \mu_2}\right), \tag{3.99}$$

by adjusting the two parameters $\mu_{1,2}$ to capture the high-frequency behavior. This is difficult if the system is far from half-filling, so that a more complex parameterization or the calculation using moments should be preferred. The Fourier transform to $g(\tau)$ is performed numerically up to the cutoff frequency $N_\omega$ on the measured Green function with the tail subtracted. The Fourier transform of the tail, which can be evaluated analytically, is added back afterwards.

The fact that the error does not decay with frequency is contrary to the weak coupling case. In the weak coupling approach, the Green function is measured as a correction to $g_0$, so that the correction and the error decay with frequency. A measurement in the frequency domain is also more costly than in imaginary time. For a measurement of the Green function at a given frequency, a sum over the $k^2$ elements of the $M$ matrix has to be performed. More severely, the computation of the exponentials $\exp(\pm i\omega_n\tau)$ is computationally highly demanding and it has been suggested to store them for all frequencies on a fine imaginary time grid [95]. In the imaginary time measurement, the time interval has to be discretized into bins. A given configuration produces measurements for $k^2$ times at perturbation order $k$, which are distributed among all imaginary time bins. The Green function for a given bin is thus not measured every time a measurement is performed. The total number of measurements must therefore be large enough to ensure proper statistics for each individual bin. The Fourier transform from imaginary time to Matsubara frequencies is technically more involved than for the inverse direction. The Green function is interpolated by an, e.g., cubic spline adapted to antiperiodic functions, with the boundary conditions chosen such that the correct high-frequency behavior is obtained (see, e.g. Ref. [95] and references therein). This transform was originally developed for the Hirsch-Fye algorithm. It becomes inherently difficult for a coarse imaginary time grid. Since increasing the number of bins however does not directly affect the performance of the algorithm, it is possible to measure the Green function on a dense grid of imaginary time points, so that a systematic error from the discretization becomes negligible. The problem of Fourier transform from the time to the frequency domain is hence alleviated and there is no major benefit from the measurement in the frequency domain, once the Fourier transform from imaginary time to Matsubara frequencies is accurately implemented.

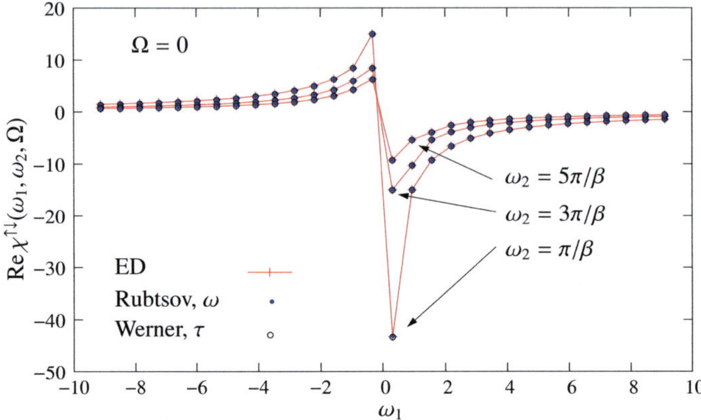



*Figure 3.10:* Two-particle Green function $\chi^{\uparrow\downarrow}(\omega_1, \omega_2, \Omega)$ for fixed bosonic frequency $\Omega = 0$ and three different values of $\omega_2$ as a function of $\omega_1$. It has been obtained for testing purposes by exact diagonalization (ED), within weak-coupling CTQMC (Rubtsov) by measurement on Matsubara frequencies and within strong-coupling CTQMC (Werner) by imaginary-time measurement and subsequent Fourier transform. The system is a single-orbital Anderson impurity model with hybridization (3.98) to a single bath site. Parameters are $U = 1, \beta = 10, V = 0.2, \epsilon = 0$.

### 3.4.2  Measurement of Four-Point Functions

The construction for the measurement of four-point functions proceeds in an analogous fashion as in the previous section. It is possible to insert four operators into the trace and evaluate the ratio of traces, or compute the measurement from the $M$-matrix. Again the first way is straightforward. For the second, the corresponding expression is obtained from the fast-update formulae or by generalization of (3.96). The two-particle Green function is measured as

$$\chi^{\sigma\sigma'}(\tau_1, \tau_2, \tau_3) = \left\langle \sum_{i,j=1}^{k} \sum_{k,l=1}^{k} \left[ M_{j,i}^{\sigma} M_{l,k}^{\sigma'} - \delta_{\sigma\sigma'} M_{li}^{\sigma} M_{jk}^{\sigma} \right] \times \right.$$

$$\left. \times \bar{\delta}(\tau_1, \tau_i^e - \tau_l^s) \bar{\delta}(\tau_2, \tau_j^s - \tau_l^s) \bar{\delta}(\tau_3, \tau_k^e - \tau_l^s) \right\rangle_{\text{MC}} . \qquad (3.100)$$

Note that binning and Fourier transform are less straightforward than for the single-particle Green function. This is because of multiple discontinuities, in particular along the diagonal, which require case differentiation. In principle, one can measure the function at both sides of the discontinuity, where two or more time arguments are equal, through special correlation functions. This is similar to measuring the density instead of the expectation value of $-c(\tau)c^\dagger(\tau + \epsilon)$. However, the accuracy is different than for the other time points and many different correlation functions have to be measured in the multi-orbital case. The Fourier transform is performed after subtracting the disconnected part (cf. Eq. 3.71), which removes the discontinuities. These problems are avoided measuring in the frequency domain.

For illustration, results for $\chi^{\uparrow\downarrow}$ measured in strong- and weak-coupling CTQMC are shown in Fig. 3.10 in comparison with results obtained from ED. The strong-coupling CTQMC data has been obtained by measurement on a grid of $100^3$ imaginary time points and subsequent Fourier transform to Matsubara space, using the implementation by P. Werner. The weak-coupling data was measured directly in the frequency domain and the ED data is obtained from the Lehmann representation of the two-particle Green function (see Sec. 9.3). The results from all approaches are in very good agreement. This illustrates that although based on expansions around different limits, the two continuous-time methods are complementary. Both algorithms converge to the same results, but perform differently depending on the parameter regime. They are numerically exact in the sense that, notwithstanding the sign problem, results can in principle be obtained to any desired accuracy. A detailed performance analysis of the two approaches in comparison with the Hirsch-Fye method has been conducted in Ref. [104].

## 3.5 Analytical Continuation

Quantum Monte Carlo simulations are restricted to the imaginary time domain as the oscillatory behavior of the real time evolution renders the sampling highly inefficient[3]. In order to obtain quantities which are directly accessible to the experiment, such as the density of states, the spectral function $A(\mathbf{k}, \omega)$ or the dynamical susceptibility $\chi(\mathbf{q}, \omega)$, the imaginary time quantities have to be analytically continued to the real axis. Specifically, the task is to find a spectral function $A(\omega)$, such that

$$G(\tau) = \int_{-\infty}^{\infty} d\omega K(\tau, \omega)A(\omega) , \qquad K(\tau, \omega) = \frac{e^{-\omega\tau}}{1 \pm e^{-\beta\omega}} , \qquad (3.101)$$

where the positive (negative) sign is for fermions (bosons). This is a mathematically ill-posed problem: The kernel $K(\tau, \omega)$ is exponentially damped at large frequencies $|\omega|$. When discretized, the resulting matrix is ill-conditioned. Inverting (3.101) to find $A(\omega)$ therefore requires regularization. The problem is further complicated through the fact that Monte Carlo simulations provide noisy and incomplete data (the Green function is known only on a discrete set of times). Due to the exponential damping through the

---

[3]QMC algorithms have nevertheless been employed to study e.g. the real time evolution of the nonequilibrium Anderson model. Accessible times are limited by the sign problem. Accurate measurements of observables can be obtained for an average sign as low as $10^{-3}$ [105].

kernel, the spectral function at large frequencies depends on small and subtle details of the Green function, which may be obscured through statistical errors. In such a situation the solution is not unique and has to be selected by certain criteria. A common strategy is to determine the most probable spectral density using concepts of conditional probabilities (Bayesian inference) [106].

A maximum likelihood estimate can be obtained by minimizing the least squares deviation

$$\chi^2 := \sum_{ij}^{L} (\langle G \rangle_i - G_i)[C^{-1}]_{ij}(\langle G \rangle_j - G_j), \tag{3.102}$$

where $L$ is the number of discrete $\tau$ values and $C$ is the covariance matrix. $\langle G \rangle :=$ $(1/N)\sum_i G^{(i)}$ denotes the Monte Carlo average. In practice, the values $G_i$ are obtained by evaluating a discretized version of (3.101), i.e. $G_i = \sum_{ij} K_{ij}A_j$ with $K_{ij} = K(\tau_i, \omega_j)$ and $A_i = A(\omega_i)\Delta\omega_i$. In particular for a large number of parameters $A_i$, this strategy leads to overfitting of the QMC data and thus to noisy and non-unique results. In the maximum entropy method (MaxEnt), the procedure is regularized by instead *maximizing* the functional

$$F[A] := e^{\alpha S[A] - \frac{1}{2}\chi^2[A]} . \tag{3.103}$$

In addition to minimizing $\chi^2$, the solution ought to simultaneously maximize the entropy $S$, the latter being given through its information-theory based definition,

$$S = - \int d\omega A(\omega) \ln[A(\omega)/m(\omega)] \tag{3.104}$$

(or a suitable discretized version thereof). $S$ takes its maximum $S = 0$ for $A(\omega) = m(\omega)$, where $m(\omega)$ is the so-called default model. For large $\alpha$, the solution will be close to the default model, while a small $\alpha$ leads to overfitting of the QMC data. It is important that in the so-called classic MaxEnt used here, $\alpha$ is not a free adjustable parameter, but is uniquely determined using Bayesian methods. Here the implementation by A. Sandvik was employed, where the default model is taken to be a constant. Only the diagonal of the covariance matrix (the variance) enters the calculation. In general, the maximum entropy procedure tends to overlook and smear out fine or sharp features in the spectrum due to the entropy term. This is particularly relevant for problems in the context of LDA+DMFT, where the method should reproduce the fine structure of the quasiparticle peak (the renormalized LDA bands).

Due to such difficulties, it is valuable to have alternative methods at hand. An approach which is rather different in spirit is based on stochastic optimization. The algorithm developed in Ref. [107] was specifically designed to resolve sharp features in the spectrum, namely by overcoming the limitations of a predetermined discretization of the spectral function and the assumption that the likelihood function can be described by a single Gaussian peak [106]. The idea is to randomly generate independent samples of $A(\omega)$, which optimize a given deviation function, in this case

$d = \int_0^{\tau_{\max}} d\tau \left| G(\tau) - \tilde{G}(\tau) \right| G^{-1}(\tau)$. A configuration $\tilde{A}(\omega)$ is parameterized as a collection of rectangles, for which (3.101) can be evaluated analytically to yield $\tilde{G}(\tau)$. Monte Carlo updates modify the configuration through insertion, or removal, merging or splitting of rectangles or changing their parameters (position, width and height). Updates are performed subject to the normalization constraint and until a near optimal parameterization with nearly minimal deviation is found. To obtain a meaningful solution, the final spectrum is computed as the average over a representative set of statistically independent samples, $A(\omega) = (1/K) \sum_{k=1}^{K} \tilde{A}(\omega)$, which leads to large cancellations of the noise. Regarding practical applicability, this method has the disadvantage that it is computationally demanding. Obtaining smooth spectra requires computation time of up to a day on a single CPU and additional refinement procedures may be necessary. However, for the systems studied here, the gross features of the spectrum were visible after averaging a few spectra and provided a consistency check of the MaxEnt. A third method which has been employed here is the analytical continuation via Padé approximants, i.e. the approximation of a given function by rational functions. For details of the algorithm see Ref. [108].

From the structure of the kernel one could expect that estimates for $A(\omega)$ in the vicinity of the Fermi level or for small frequencies are reliable, independent of the particular method used. Indirect evidence for MaxEnt and Mishchenko's method (for bosonic frequencies) is provided by the results of chapter 4. An explicit demonstration for the Padé method is shown in Fig. 9.15.

Good quality of the data is essential to obtain reliable results. This is confirmed in practice. The results presented in this thesis obtained by MaxEnt were cross-checked against Mishchenko's method. No qualitative differences were found in all cases, except for the density of states shown in Fig. 8.3. For some Monte Carlo runs the MaxEnt did not resolve the sharp peaks at the gap edge (although small features were visible hinting towards their presence). For complicated multiband models, the discrepancy is expected to become more significant. For such problems, the full covariance information should be used. In chapter 9, an alternative approach for the approximate solution of the impurity problem is proposed, which is free of statistical errors. One may therefore expect the analytical continuation to be more stable. This method also allows compute dynamical quantities directly on the real axis, hence circumventing the problem of numerical analytical continuation.

# Chapter 4

# Metal-Insulator Transition by Suppression of Spin Fluctuations

H. Hafermann, M. I. Katsnelson
and A. I. Lichtenstein
EPL **85**, 37006 (2009)

## 4.1  Introduction

Different kinds of Metal-insulator transitions (MITs) are observed in condensed-matter systems. Metal to band insulator transitions take place in weakly or uncorrelated systems and can successfully be described by an independent electron theory, such as density functional theory. An inherently different scenario emerges when electron-electron interactions become important: in such a situation the system can undergo a Mott MIT [14]: the electrons localize due to the Coulomb repulsion. This has, e.g., been observed in d-electron systems such as transition-metal oxides. Examples for a different kind of MIT are the metal to singlet insulator transitions found in the Ruthenium oxide $Hg_2Ru_2O_7$ [109] and in the transition metal spinel $MgTi_2O_4$ [110]. Singlet insulator ground states are also found in quasi one-dimensional (1D) systems such as TiOCl [111] and $CuGeO_3$ [112] due to dimerization via a spin-Peierls mechanism. They are further characterized by the presence of a spin-gap. The formation of dynamical singlets on pairs of Vanadium atoms has been proposed as the mechanism for the MIT in $VO_2$ [7].

A very successful description of the Mott physics is provided through dynamical mean-field theory, which maps the lattice model onto a quantum impurity model subject to a self-consistency condition (see Sec. 2.6). In the Fermi liquid state close to the Mott transition, the spectral function exhibits a characteristic three-peak structure [15]: Two broad incoherent peaks, the Hubbard bands, are present at high energies. They coexist with the coherent quasi-particle peak at the Fermi level, the Abrikosov-Suhl resonance.

Within the dynamical mean-field description, the quasiparticle peak of the metallic state is identified with the Kondo resonance of the impurity model and the Kondo temperature plays the role of a coherence scale on which Fermi liquid behavior sets in [113].

The Kondo effect, originally considered by Jun Kondo to describe the resistance minimum of dilute magnetic alloys [114], is believed to be responsible for heavy-fermion behavior [4] of several intermetallic compounds based on rare-earth and actinide elements. The Kondo effect is a consequence of local quantum spin fluctuations. It is known that in order to have this effect, the spins need to be quantum, i.e. it cannot be obtained either in the limit of zero or infinite (classical) spins; for the latter case the *double*-peak structure without the Kondo resonance was demonstrated near the MIT point [115]. In fact, within a path integral approach, the Kondo effect was explained as a consequence of quantum tunneling between degenerate valleys of a double well potential [116], which explicitly stresses the quantum nature of the phenomenon. For more details on the Kondo effect, see chapter 5. It is expected that dimerization, i.e. the singlet formation between neighboring sites, leads to a quenching of the Kondo effect, since it inhibits the singlet formation between the local moment and the conduction electron spins. Since the Kondo effect provides quasiparticle weight at the Fermi level, tuning the dimerization is a possible mechanism for driving the MIT.

Starting from earlier work on the stacked Hubbard model on the Bethe lattice, which addressed the competition between the RKKY interaction and Kondo physics in the paramagnetic phase [117], more recently the Mott to band insulator transition was studied in the 2D bilayer Hubbard model [118]. Further studies focused on the ground state phase diagram [119], the role of frustration and dimensionality [120] and the magnetic and transport properties for the doped case [121]. Here the MIT is studied in the half-filled model on the Bethe lattice, driven by tuning the coupling between dimer sites. The focus is on the role of the short-range singlet correlations, which drive the MIT. This mechanism is rather different from the interaction driven MIT.

## 4.2   The Model

The $D = \infty$ two-plane Hubbard model on the Bethe lattice is illustrated in Fig. 4.1. Sites within a plane are connected by hopping $t$ and coupled to a neighbor on the other plane by a perpendicular hopping $t_\perp$. Two electrons occupying the same site in either plane have to pay the Coulomb repulsion energy $U$. The model is governed by the Hamiltonian

$$H = -t \sum_{\langle ij \rangle, \sigma} (a_{i\sigma}^\dagger a_{j\sigma} + b_{i\sigma}^\dagger b_{j\sigma}) - t_\perp \sum_{i\sigma} (a_{i\sigma}^\dagger b_{i\sigma} + b_{i\sigma}^\dagger a_{i\sigma})$$

$$+ U \sum_{i\sigma} (n_{ai\uparrow} n_{ai\downarrow} + n_{bi\uparrow} n_{bi\downarrow}) , \tag{4.1}$$

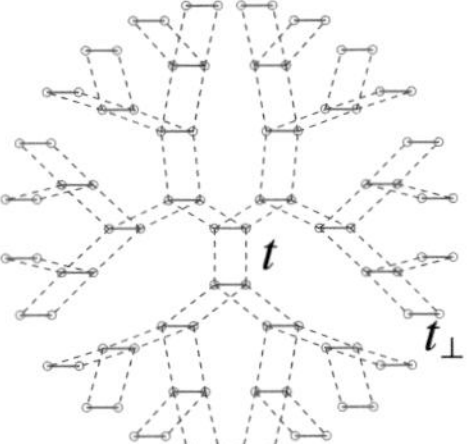



*Figure 4.1:* The two-plane Hubbard model on the Bethe lattice visualized for coordination number $z = 3$. It can be viewed as a lattice of dimers, or equivalently as two planes with opposing sites coupled by a perpendicular hopping $t_\perp$.

where the operators $a^\dagger(a)$ create (annihilate) an electron on plane $a$, $n_{a\sigma} = a^\dagger_\sigma a_\sigma$ and correspondingly for plane $b$. The hopping $t$ has to be rescaled by $1/\sqrt{z}$ as described in Sec. 2.6.1 in order to give a nontrivial model in the limit of infinite coordination number $z$. The effective action for this model has the same form as (2.45), provided the matrix Green function is defined as

$$G_\sigma(\tau - \tau') := \begin{pmatrix} -\langle a_\sigma(\tau)a^*_\sigma(\tau')\rangle & -\langle a_\sigma(\tau)b^*_\sigma(\tau')\rangle \\ -\langle b_\sigma(\tau)a^*_\sigma(\tau')\rangle & -\langle b_\sigma(\tau)b^*_\sigma(\tau')\rangle \end{pmatrix} . \tag{4.2}$$

The self-consistency condition has previously been derived in Ref. [117]. A straightforward way to obtain it is to think of the model as a lattice of dimers. It is then possible to apply the self-consistency condition (2.47). A dimer may be viewed as a two-orbital atom with interorbital hopping matrix $t^{ab}_{\text{loc}} = t^{ba}_{\text{loc}} = t_\perp$ and $t^{aa}_{\text{loc}} = t^{bb}_{\text{loc}} = 0$. Noting further that the intra-plane hopping $t$ only connects neighbors within a plane, so that the hopping matrix $t^{\alpha\beta} = t\delta_{\alpha\beta}$ is diagonal, leads to

$$\mathcal{G}^{-1}_\sigma(i\omega_n) = \begin{pmatrix} i\omega_n + \mu & t_\perp \\ t_\perp & i\omega_n + \mu \end{pmatrix} - t^2 G_{-\sigma}(i\omega_n) . \tag{4.3}$$

This condition relates the Weiss field with spin projection $\sigma$ to the Green function with opposite spin and accounts for the commensurate antiferromagnetic (AF) order within a plane according to (2.86). In order to assure that an AF solution is obtained instead of a metastable paramagnetic one, a small spin- and site- dependent symmetry breaking field has been added to the right-hand side of (4.3) on the first DMFT iteration:

$$\hat{h} = \begin{pmatrix} 0 & h\sigma \\ -h\sigma & 0 \end{pmatrix} , \tag{4.4}$$

where $\sigma = +(-)1$ for spin up (down). In the AF phase, the two planes couple antiferromagnetically. From (4.3) the hybridization function is read off to be $\Delta_\sigma(i\omega) = t^2 G_{-\sigma}(i\omega)$.

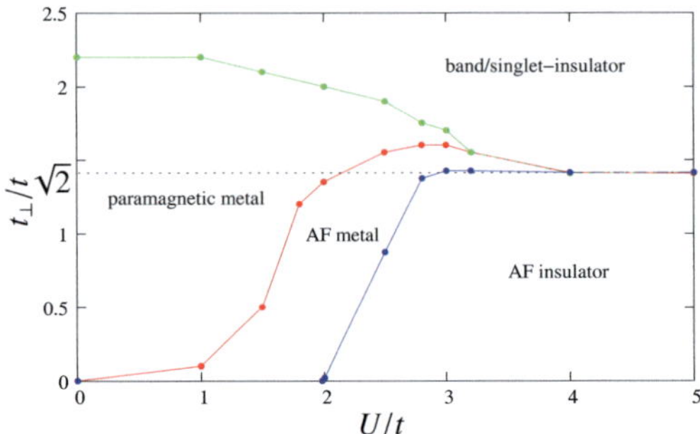


*Figure 4.2:* Finite temperature phase diagram of the two-plane Hubbard model on the Bethe lattice at temperature $T/t = 0.1$. The mean-field value of $t_\perp$ for the AF to singlet insulator transition for large $U$ is marked by a dashed line.

Theoretically, this model is particularly appealing, since it can be solved exactly within DMFT. The continuous-time quantum Monte Carlo (CTQMC) algorithm described in chapter 3 is employed for the solution of the associated local Anderson impurity problem. This method is numerically exact in the sense that it does not involve any additional approximations. The finite temperature results presented here may therefore be viewed as essentially exact within the Monte Carlo (MC) error.

## 4.3   Finite Temperature Phasediagram

Before considering the finite temperature phase diagram in presence of interaction, it is instructive to consider the noninteracting case first. For $U = 0$, the model is paramagnetic and $G_{11} = G_{22}$, $G_{12} = G_{21}$ by symmetry. The solution may readily be obtained by changing to a diagonal basis. Viewing the dimer as a molecule, the symmetric and antisymmetric combinations of the wavefunctions correspond to bonding and antibonding orbitals. The self-consistency condition in this basis reads $(\mathcal{G}^{b/a})^{-1} = i\omega + \mu \mp t_\perp - t^2 G^{b/a}$ (for the half-filled case considered here, $\mu = 0$). The noninteracting Green function $G_0$ for this model is obtained by identifying the Weiss field and Green function in (4.3) with $G_0$ and solving the resulting equation for $G_0$. The solution to this quadratic equation is

$$G_0^{b/a} = \frac{i\omega + \mu \pm t_\perp - i\sqrt{4t^2 - (i\omega + \mu \pm t_\perp)^2}}{2t^2} \; . \tag{4.5}$$

The sign in front of the square root is determined by the condition that the density of states (DOS) be positive. In the paramagnetic phase, the DMFT equations may be

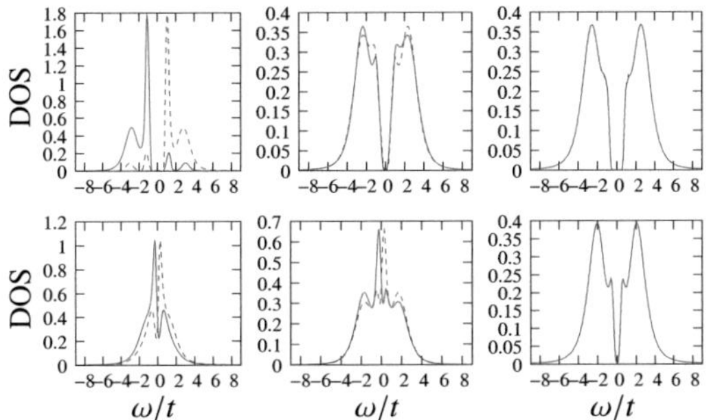

*Figure 4.3:* Characteristic local density of states in the different phases of the two-plane Hubbard model. Upper panel: $U/t = 4.0$ and $t_\perp/t = 0.5, 1.41$ and $2.0$ (from left to right). Lower panel: $U/t = 2.0$, $t_\perp/t = 0.5, 1.2$ and $2.0$.

solved in the diagonal basis. All quantities can then be obtained by transforming back to the original basis, $G_{11} = (G^b + G^a)/2$ and $G_{12} = (G^b - G^a)/2$. The DOS obtained from (4.5) for the bonding and antibonding bands is given by

$$D^{b/a}(E) = -\frac{1}{\pi} \operatorname{Im} G_0^{b/a}(E) = \frac{1}{2\pi t^2} \sqrt{4t^2 - (E + \mu \mp t_\perp)^2} \tag{4.6}$$

for $|E + \mu \mp t_\perp| < 2t$ and $0$ otherwise. For $t_\perp = 0$ this is the familiar semielliptical density of states of bandwidth $W = 4t$. As $t_\perp$ is increased, the bonding and antibonding bands split and fully separate at $t_\perp = 2t$ (at zero temperature, $T = 0$) and hence the model undergoes a transition to a band insulating state.

The finite temperature phase diagram in the $t_\perp$-$U$-plane obtained from the CTQMC calculations is depicted in Fig. 4.2. The calculations were performed at inverse temperature $\beta = 20$ in units of the half bandwidth, corresponding to $T/t = 0.1$. The model exhibits four different phases: A paramagnetic and an antiferromagnetic (AF) metal phase for small $U$ and AF as well as band/singlet insulator phases for large $U$. For $U = 0$ the transition to the band insulating state is found at a somewhat larger value as expected for $T = 0$, due to temperature broadening of the bands. In the weak coupling regime, the system is metallic and magnetic order is suppressed for small values of $t_\perp$. As $U$ is increased, a local moment forms and stabilizes the antiferromagnetism, which is also present in the decoupled lattices. Further increase of $U$ drives the system into the antiferromagnetic insulating phase. For $t_\perp = 0$, the transition occurs at $U_c \sim W/2$. For intermediate $U$, the antiferromagnetic metal phase is stable up to fairly large values of $t_\perp$. The boundary to the paramagnetic phase is determined by the condition that the local magnetization

$$m := \langle S^z \rangle = \frac{\langle n_\uparrow \rangle - \langle n_\downarrow \rangle}{2} \tag{4.7}$$

becomes zero. The boundary to the insulating state for large values of $t_\perp$ has been determined by the condition that the total spin of the dimer is zero within the MC error (see below). This yields essentially the same results as the condition of a vanishing maximum entropy density of states at the Fermi level. The boundary to the insulating state turns to smaller values of $t_\perp$ as the interaction increases. The phase boundaries finally merge at $U/t \approx 4$ at the mean-field value $t_\perp/t \approx \sqrt{2}$, separating the AF insulating from a singlet insulator phase. The phase diagram is qualitatively similar to those given in Refs. [119, 121] for the two-dimensional model.

Some characteristic local maximum entropy density of states (DOS) for two different values of the on-site repulsion $U$ and different inter-plane hoppings are shown in Fig. 4.3. For an on-site repulsion of $U/t = 4.0$ (upper panel) and small $t_\perp$, the system is clearly an AF insulator. The spin splitting is pronounced and the DOS displays a four peak structure. The outer peaks can be identified as the lower and upper Hubbard bands, while the inner peaks are characteristic to the AF state. At $t_\perp/t \approx \sqrt{2} \approx 1.414$ these peaks have almost disappeared and the magnetization is close to zero. For larger $t_\perp$ the DOS shows no spin polarization and the system remains insulating. For smaller $U$, the situation is different. For $U/t = 2.0$ the system is at the transition to the insulating state for vanishing $t_\perp$. Therefore, when becoming metallic for small perpendicular hopping, it exhibits a pseudogap structure and displays the characteristics of an AF metal. For larger $t_\perp$ the DOS shows the Kondo resonance at the Fermi level, which is slightly spin-split due to the antiferromagnetism. At enhanced perpendicular hopping the spin-splitting disappears as the system crosses the transition line to the paramagnetic state. For even larger $t_\perp$ the gap develops again and finally fully opens at the metal to singlet insulator transition line.

## 4.4   Large $U$

For large $U$ the phase diagram is considerably simpler than for intermediate $U$ and therefore this case is considered first. For all results the MC error was negligible, i.e. considerably smaller than the symbol size, except where shown. In the strongly correlated regime, the model exhibits only a single transition from an AF to a paramagnetic insulating phase. In Fig. 4.4, the magnetization on both sites of a dimer is shown as a function of the interplane hopping $t_\perp$ for $U/t = 4.0$. For small $t_\perp$ the solution is antiferromagnetic with different sign of the magnetization on opposite sides of the planes (and for nearest neighbors). In the limit of zero coupling between dimer sites, the magnetization is slightly smaller than $1/2$ due to intraplane hopping and thermal fluctuations. The magnetization decreases as the perpendicular hopping is increased and finally reaches

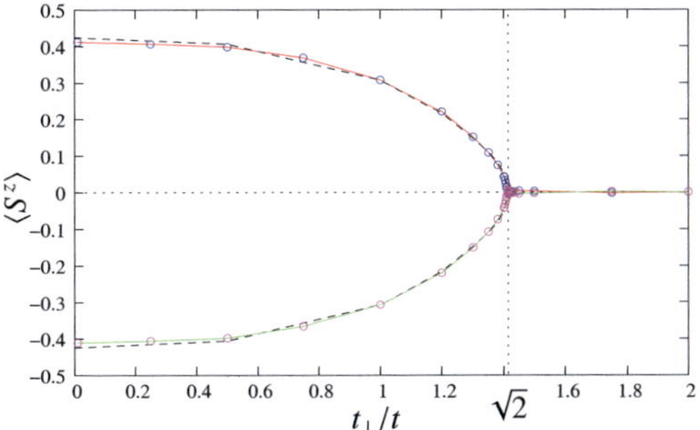

*Figure 4.4:* Magnetization $\langle S_i^z \rangle$ on opposite sides of the two Bethe lattices for $U/t = 4$ and temperature $T/t = 0.1$. The dashed lines show the corresponding result for $T/t = 0.04$.

zero at the value $t_\perp/t \approx \sqrt{2}$. The same behavior is observed for lower temperatures, where the maximal magnetization is somewhat larger, consistent with a larger local moment due to an alleviated effect of thermal fluctuations. The magnetization vanishes continuously. The critical slowing down at this second-order transition manifests itself through a strongly decelerated convergence of the DMFT self-consistency loop. When approaching the transition from below $t_\perp < t_\perp^c$, the number of iterations to obtain a converged solution diverges. In the vicinity of the transition the perpendicular hopping was increased in small steps and the hybridization from the previous DMFT calculation was used as a starting guess for the next. This is necessary to obtain the correct critical value, which otherwise is strongly overestimated.

The observed critical value of $t_\perp/t = \sqrt{2}$ can be understood by noting that for sufficiently large $U$, the model can be approximately mapped onto an effective Heisenberg model (see also Sec. 3.3.10). Neighboring sites within a plane are then coupled by an effective AF Heisenberg exchange. Virtual transitions provide the relevant energy scale $J_\parallel \sim t_\parallel^2/U$. A competing energy scale emerges due to hopping processes between planes. The number of bonds connecting the dimers is twice the number of bonds on a dimer. The mean-field condition for the interplane exchange $J_\perp \sim t_\perp^2/U$ to overcome the exchange due to hopping within the planes hence is $J_\perp = 2J_\parallel$, so that $t_\perp/t = \sqrt{2}$.

The formation of a singlet between neighboring sites on a dimer is thus expected as the origin of the transition to the non-magnetic state. In order to prove this consideration, the spin correlations are examined. To this end, the averages of spin operators are decomposed into four-point correlation functions. For example, the total spin of a dimer can be written

$$\langle \mathbf{S}^2 \rangle = \langle (\mathbf{S}_1 + \mathbf{S}_2)^2 \rangle = \sum_{ij} \langle \mathbf{S}_i \cdot \mathbf{S}_j \rangle \,, \tag{4.8}$$

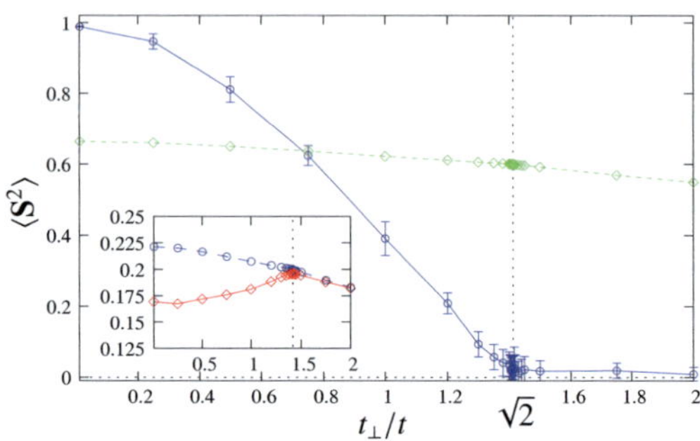

*Figure 4.5:* Spin-correlations $\langle \mathbf{S}_i^2 \rangle$ on opposite sides of the two Bethe lattices (dashed line) and total spin $\langle \mathbf{S}^2 \rangle$ (solid line) for the dimer for $U/t = 4$ and temperature $T/t = 0.1$. The inset compares the correlations $\langle S_i^z S_i^z \rangle$ (upper dashed line) and $-\langle S_i^z S_j^z \rangle$ for $i \neq j$ (solid line). The transition point is marked by the vertical dashed line.

where the sum is over the dimer sites. The product of spin operators can further be decomposed using

$$\mathbf{S}_i \cdot \mathbf{S}_j = S_i^z S_j^z + \frac{1}{2}\left( S_i^+ S_j^- + S_i^- S_j^+ \right) . \tag{4.9}$$

Finally, the operator for the $z$-component of the spin and the spin raising and lowering operators are expressed in terms of $c$-operators as

$$S^z = \frac{n_\uparrow - n_\downarrow}{2}, \qquad S^+ = c_\uparrow^\dagger c_\downarrow, \qquad S^- = c_\downarrow^\dagger c_\uparrow . \tag{4.10}$$

The results for the total spin of a single site and the dimer are shown in Fig. 4.5. The total on-site spin $\langle \mathbf{S}_i^2 \rangle$ is somewhat smaller than $= S(S + 1) = 3/4$, and varies only slightly as a function of the perpendicular hopping. In particular, no remarkable feature occurs at the critical value $t_\perp^c$. The total spin of a dimer on the other hand continuously decreases as the coupling $t_\perp$ between dimer sites is increased. The results are consistent with a total spin of zero at $t_\perp/t \approx \sqrt{2}$. This proves the assumption that a singlet is formed between neighboring spins on opposite sides of the planes. In contrast to the MC error for all other measured quantities, the error of the total spin of the dimer is considerable, reflecting the fact that this quantity is strongly fluctuating in the simulation. The picture of a singlet formation is further underlined by comparing the quantities $\langle S_i^z S_i^z \rangle$ and $-\langle S_i^z S_j^z \rangle$ for $i \neq j$ on a dimer as shown in the inset of Fig. 4.5. The square of the $z$-projection of the spin decreases due to enhanced double occupancy (note that for the Hubbard model $\langle S^z S^z \rangle = (1/2 - \langle n_\uparrow n_\downarrow \rangle)/2$ holds). Correlations $\langle S_i^z S_j^z \rangle$ for $i \neq j$ have negative sign, in accordance with the tendency to AF coupling between the

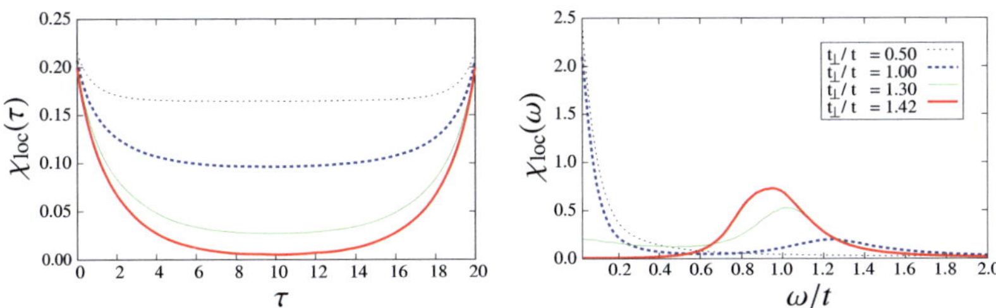

*Figure 4.6:* Imaginary time correlation function $\chi_{\text{loc}}(\tau) := \langle S_z(\tau) S_z(0) \rangle$ (left) and corresponding dynamical susceptibility $\chi_{\text{loc}}(\omega)$ obtained by analytical continuation for different values of the perpendicular hopping at $U/t = 4.0$. A spin gap develops for $t_\perp/t \gtrsim \sqrt{2}$.

spins. For weak coupling, the on-site and inter-site correlations differ. This is expected, since incoherent spin flips caused by the conduction electrons destroy the correlation between sites. For larger values of the coupling however, the magnitude of the intersite correlation approaches the on-site value. This unambiguously confirms the formation of a coherent state: While the local moments on each of the dimer sites flip dynamically so that the total magnetization is zero, they always point in opposite directions. The degree of entanglement can be varied continuously by tuning the perpendicular hopping between the two planes. Due to the absence of spatial correlations within the planes, the transition occurs at the mean-field value $t_\perp^c/t = \sqrt{2}$.

The dynamical susceptibility $\chi_{\text{loc}}(\omega)$ is plotted in Fig. 4.6 for different values of the perpendicular hopping for $U/t = 4.0$. It is obtained by analytical continuation of the Fourier components of the corresponding imaginary time correlation function $\chi(\tau) = \langle S^z(\tau) S^z(0) \rangle$ using the method of Ref. [107] (see Sec. 3.5). For small values of $t_\perp$, a pronounced peak is present at zero energy. This can be attributed to the Goldstone mode present in the symmetry-broken (ordered) state. Therefore the peak diminishes as the border to the disordered state is approached by increasing the perpendicular hopping. At the same time a new feature starts to develop at higher energies at around $\omega/t \sim 1$. This energy scale is the same as the effective (in-plane) exchange $J_\parallel = 4t^2/U$ (for $U/t = 5$ this feature occurs at $\omega/t \sim 0.8$). This excitation cannot be attributed to a singlet-triplet transition, which would be associated with the energy scale of the interplane exchange. The origin is unknown at present. It is notable, however, that the opening of the spin gap occurs at $t_\perp/t \sim \sqrt{2}$, i.e. at the same point at which the spin of the dimer vanishes.

These results bear some resemblance to a bilayer square-lattice Heisenberg model, although the two models are quite different. Two modes exist in the spectrum corresponding to the in-phase and out-of-phase excitations of the two layers [122]. In order to compare this with the present results, one should bear in mind that on the Bethe lat-

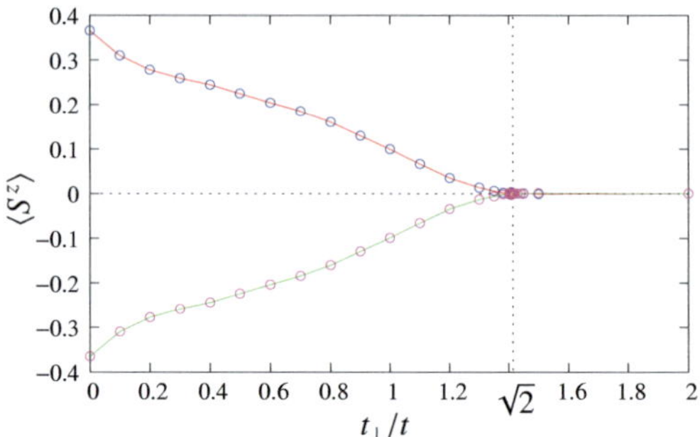

*Figure 4.7:* Magnetization as a function of the perpendicular hopping for $U/t = 2.0$.

tice, there are only two k-points: The $\Gamma$-point ($\mathbf{k} = \mathbf{0}$) and the $M$-point ($\mathbf{k} = (\pi, \ldots, \pi)$).
For the bilayer Heisenberg model the energies of the in-phase excitation at the $\Gamma$-point
and the out-of-phase excitation at the $M$-point (and vice versa) are equal in the ordered
state and similar in the disordered phase. Hence the two modes cannot be distinguished
here. The excitation is gapless in the ordered phase in agreement with Fig. 4.6. The
out-of-phase excitation at the $\Gamma$-point and the in-phase excitation at the $M$-point have
the highest energy in the spectrum. Presumingly the intensity is too low so that this
mode is not detected. This is supported by a calculation of $\chi_{\mathrm{loc}}$ for a dimer using exact diagonalization (not shown). In the disordered phase the in-phase excitation at $\Gamma$ is
gapped, in accordance with the present results.

This similarity is a consequence of the fact that the model approximately maps to the
Heisenberg model in this parameter regime. Note, however, that in the bilayer square
lattice model, the number of intraplane bonds per interplane bond is 4 instead of 2 for
the Bethe lattice. Hence the mean-field value of the critical exchange and the critical
hopping is $J^c_\perp/J_\parallel = 4$ and $t^c_\perp/t = 2$, respectively, as opposed to $J^c_\perp/J_\parallel = 2$ for the
Bethe lattice. In this parameter regime, charge fluctuations are effectively frozen and
the physics is dominated by the spin degrees of freedom. The transition is an order-
disorder transition between the two insulating phases.


## 4.5    Intermediate $U$

The case of intermediate $U$ is more complicated, due to the interplay of spin and charge
degrees of freedom. In Fig. 4.7, the magnetization is shown for $U/t_\perp$ as a function of
perpendicular hopping. Compared to Fig. 4.4, the maximum value for $t_\perp = 0$ is slightly
smaller as double occupancy is more likely. The magnetization vanishes at a smaller

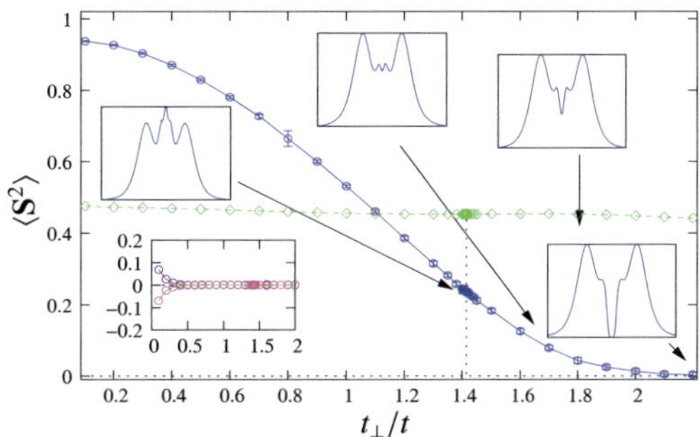

*Figure 4.8:* On-site spin correlation $\langle \mathbf{S}_i^2 \rangle$ (dashed line) and total spin $\langle \mathbf{S}^2 \rangle$ of a dimer (solid line), together with corresponding density of states (small insets) for $U/t = 1.5$. The magnetization $\langle S^z \rangle$ as a function of $t_\perp/t$ is shown in the inset (bottom left).

value of $t_\perp$ compared to the case of large $U$ (for this value of $U$, it is however still close to $t_\perp/t = \sqrt{2}$). The way the magnetization vanishes is qualitatively differently from the case of large $U$. Here the slope of the magnetization as a function of $t_\perp$ approaches zero as $\langle S^z \rangle \to 0$, while it diverges for large $U$, suggesting a different mechanism for the transition to the paramagnetic state. As seen from the phasediagram in Fig. 4.2, the vanishing of the magnetization is not accompanied by a transition to an insulating state. This behavior is also found for smaller values of $U$.

In order to study the interplay between the spin and charge degrees of freedom at the metal-insulator transition, consider an on-site repulsion of $U/t = 1.5$. The spin-correlations are shown in Fig. 4.8. The local moment is considerably smaller than in Fig. 4.5. Notably, the transition to the singlet state is not observed at $t_\perp/t = \sqrt{2}$, since the Heisenberg picture is no longer valid for this value of $U$. Nevertheless, a transition to a singlet state still takes place as the total spin goes to zero at $t_\perp/t \gtrsim 2.1$. For this value of $U$ the magnetization vanishes already for small $t_\perp$ and the transition takes place in the paramagnetic phase, as shown in the inset of Fig. 4.8. Qualitatively the same behavior is observed for the AF metal to singlet insulator transition which takes place in a narrow region of the phase diagram (cf. Fig. 4.2).

In Fig. 4.8, the density of states is plotted at some characteristic points close to where the singlet is formed. The DOS is not spinpolarized as the system is deep in the paramagnetic regime. For small $t_\perp$, the usual three peak structure is found. This structure is visible over a broad range of $t_\perp$ values. By further increasing $t_\perp$, the coherent peak at the Fermi level disappears and a pseudogap starts to develop. The singlet formation between dimer sites leads to a quenching of the Kondo effect as it inhibits the formation of the singlet between the local moment and the conduction electron spins.

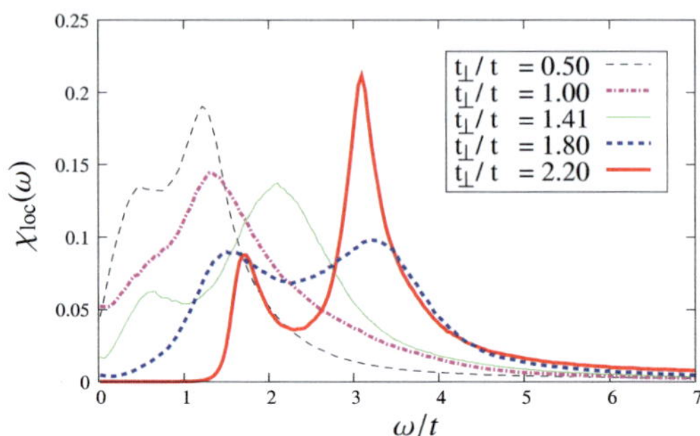

*Figure 4.9:* Dynamical susceptibility $\chi_{\text{loc}}(\omega)$ for different values of the perpendicular hopping at $U/t = 1.5$. Here the spin gap opens for $t_\perp/t \gtrsim 2.1$.

The spectral weight provided by the local spin fluctuations is gradually suppressed and the gap develops continuously as the singlet on the dimer is formed. The singlet formation leads to electron localization. In accordance with the results for large $U$, the gap is fully opened where the total spin reaches zero within the MC error. This observation is found to be robust against the presence of magnetic order, i.e. it is found in the AF metal regime, where magnetism survives up to the point where the singlet is formed. For large interplane hopping the model is thus a singlet insulator for all values of $U$. Note that in contrast to the Mott metal-insulator transition, the density of states close to the transition does not show a preformed gap.

In Fig. 4.9 the dynamical susceptibility is shown for $U/t = 1.5$ and various values of $t_\perp$. The interplay between spin and charge degrees of freedom leads to a more complicated behavior compared to the case of large $U$. Here the Goldstone mode at zero energy is absent since the system is in the disordered (paramagnetic) phase for the values of $t_\perp$ used in this figure. The results have been compared for different values of $U$ and to exact diagonalization of a dimer. It appears however to be difficult to identify the origin of individual peaks, which may further be obscured by artifacts of the analytical continuation, in particular at high energies. However, the opening of the spin gap is found to be always consistent with the opening of a charge gap in the simulation. Here the spectrum is gapped for $t_\perp/t \gtrsim 2.1$, consistent with Fig. 4.8.

## 4.6 Conclusions and Outlook

The application of cluster DMFT to this model allowed to describe the transition between antiferromagnetic and quantum singlet states. The results illustrate that short-range correlations can significantly alter the spectral properties of a system. By explicit calculation of the spin correlations, the local spin fluctuations caused by the conduction electrons are found to be suppressed by the formation of a singlet state on a dimer. The quenching of the Kondo effect leads to a reduction of the spectral weight at the Fermi level. Correspondingly, the quasi-particle peak diminishes and a pseudogap starts to develop. The transition is continuous. This behavior is in contrast to the usual Mott MIT, where a preformed gap exists slightly below the transition and the width of the quasiparticle peak decreases as the transition is approached from the metallic side.

The spin and charge gaps open simultaneously, at the point where the total spin becomes equal to zero within the Monte Carlo error. For large values of $U$, the charge degrees of freedom are effectively frozen and the physics is dominated by the spin degrees of freedom. The DMFT solution approaches the (static) mean-field solution for the two-plane Heisenberg model, for which an order-disorder transition takes place. The conjecture of a transition to a singlet insulating state was found to apply for all finite values of $U$ and is also valid at $U = 0$. For $t_\perp > W$ the bonding and antibonding bands split, so that one band is completely empty and the other completely filled (at $T = 0$). This state is described by a single wave function, which is a Slater determinant. The magnetization and total spin in this state must be zero (in analogy to an atomic closed-shell configuration).

It is interesting to study the two-plane Hubbard model in two dimensions. The spatial correlations within a plane provide a competing energy scale which may change the results even qualitatively. While the model has been studied within CDMFT, which takes short-range correlations into account [119], it has not been possible so far to study the effect of long-range correlations due to limited computational resources. A method which can overcome these limitations is presented in chapter 11. In this approach, correlations between the planes can be accounted for through a cluster calculation, while long-range correlations are treated perturbatively. This approach does not break translational invariance.

The tools developed here may further be applied to nanosystems, such as atoms on a metallic surface, or frustrated quantum magnets. An example of how a nanosystem can be modeled through the Anderson impurity model is given in the following chapter.

# Chapter 5

# Correlations around a Kondo Impurity: The Kondo Screening Cloud

K. R. Patton, H. Hafermann, S. Brener,
A. I. Lichtenstein and M. I. Katsnelson
Phys. Rev. B **80**, 212403 (2009)

The observation of a resistance minimum in the resistivity of gold due to magnetic impurities in the 1930s was first explained by Jun Kondo [114] about thirty years after. The application of Wilson's numerical renormalization group (NRG, for a review, see [123]), later showed that at low temperatures the impurity spin is completely screened by the conduction electron spins to form a true many-body quantum state. This happens below an energy scale set by the Kondo temperature $T_K$. Though known for a long time, the Kondo effect continues to be under active research [124]. For example, it is believed to underly the physics of the heavy fermion compounds [4]. The orbital Kondo resonance observed on a chromium (001) surface due to degenerate surface states [125] and an increase of the conductance in quantum dots are different facets of the same physical phenomenon.

The local aspects of this effect, the screening of the impurity spin and the formation of the Kondo resonance have been investigated both theoretically and experimentally and are well understood. The spatial dependence of spin correlations between the magnetic impurity and the conduction electrons has been studied numerically [99, 126, 127, 128]. Conduction electrons within a certain range around the impurity defined by the Kondo screening length $\xi_K$ participate in the formation of the singlet. However, half a century after the discovery and its explanation, the theoretical prediction regarding the spatial dependence of the spin correlations around a magnetic impurity in the Kondo regime – the Kondo cloud – has not yet been observed experimentally.

Nanosystems provide a possible experimental setup to confirm this expectation. Scanning tunneling microscopy (STM) is a powerful tool to probe a nanosystem with

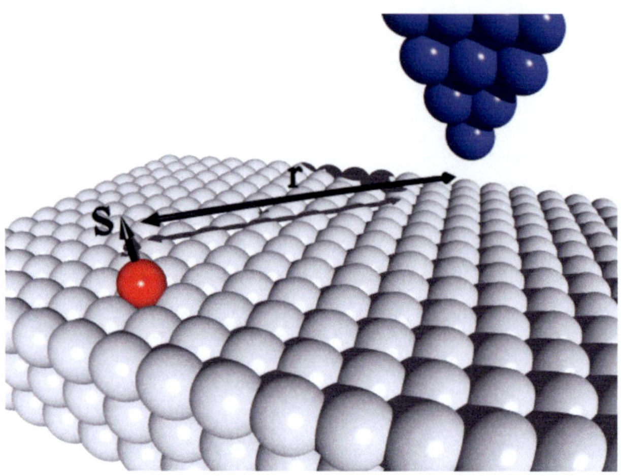

*Figure 5.1:* Illustration of the STM setup. The STM tip (blue) probes the surroundings of a magnetic atom (red) lying on a (semi-)metallic substrate at a distance **r** from the impurity. Courtesy K. R. Patton.

spatial resolution on the atomic scale. In this chapter, numerical results are presented which are a contribution to the paper entitled *Probing the Kondo screening cloud via tunneling-current conductance fluctuations* by K. R. Patton *et al.* [129]. The main result of the paper is that a local probe – a spin-polarized STM – may be used to experimentally extract a signature of the Kondo cloud by measuring conductance fluctuations of the tunneling current. Before turning to the numerical results, in order for this chapter to be self-contained, the derivation of the formulae that relate the spin and charge correlations to the conductance fluctuations is briefly sketched as derived by K. R. Patton. Then the formalism used to establish a connection to computable quantities within CTQMC is introduced. More details can be found in Ref. [129].

## 5.1   Relation to an STM Experiment

The experimental setup is illustrated in Fig. 5.1. The Hamiltonian of the system is modeled as $H = H_{\text{STM}} + H_{\text{sub}} + H_{\text{tun}}$, where $H_{\text{STM}} = \sum_{\mathbf{k}}(\epsilon_{\mathbf{k}\sigma} - \mu - eV)a^{\dagger}_{\mathbf{k}\sigma}a_{\mathbf{k}\sigma}$ is the Hamiltonian of the STM, which is assumed to be noninteracting, $H_{\text{sub}}$ is the Hamiltonian of the substrate which includes the (interacting) magnetic impurity and $H_{\text{tun}} = \sum_{\mathbf{k}\mathbf{k}'\sigma}[\bar{T}a^{\dagger}_{\mathbf{k}\sigma}b_{\mathbf{k}',\sigma} + \text{H.c.}]$ describes the tunneling between tip and sample. An

STM experiment measures the tunneling current $I$ between the STM tip and substrate. Usually, from this, the $\partial I/\partial V$ (conductance) signal is extracted, since (under certain assumptions, see Ref. [98]) it is proportional to the density of states of the substrate. Within the standard weak tunneling formalism [98], the nonequilibrium expectation value of the current can be obtained within linear response theory by treating $H_{\text{tun}}$ as the perturbation and assuming the STM and the substrate to be separately in equilibrium. The conductance operator $G$ is defined such that $\langle G \rangle = \partial I/\partial V$. Now define the spectral density of the conductance fluctuations as

$$S(\mathbf{r}, \omega) = \frac{1}{2} \int dt\, e^{i\omega t} \langle \{\delta G(\mathbf{r}, t), \delta G(\mathbf{r}, 0)\} \rangle \,, \tag{5.1}$$

(where $\delta G = g - \langle G \rangle$). This quantity can be determined experimentally through noise spectroscopy of the conductance signal. With the above assumptions, the homogeneous component of the spectral density of the conductance for an STM experiment can be shown to be given by

$$S(\mathbf{r}, \omega = 0) = \pi^2 e^4 \left|\bar{T}\right|^4 [\rho_{\text{STM}}(eV)]^2 \tilde{\chi}_{\text{sub}}^{\text{ch}}(\mathbf{r}, \omega = 0)$$
$$+ 32\pi^2 e^4 \left|\bar{T}\right|^4 [m_{\text{STM}}(eV)]^2 \tilde{\chi}_{\text{sub}}^{\text{sp}}(\mathbf{r}, \omega = 0) \,. \tag{5.2}$$

Here $\rho_{\text{STM}} = \sum_\sigma \rho_{\text{STM}}^\sigma$ and $m_{\text{STM}} = \rho_{\text{STM}}^\uparrow - \rho_{\text{STM}}^\downarrow$ are the charge and magnetization densities of the STM, respectively, $e$ is the electron charge and $\bar{T}$ is the tunneling amplitude between the STM tip and the substrate. The experimentally accessible conductance fluctuations are hence determined by the charge-charge, and for a spin-polarized STM ($m_{\text{STM}} \neq 0$) by spin-spin *density* correlations. $\tilde{\chi}_{\text{sub}}^{\text{sp/ch}}(\mathbf{r}, t)$ denotes the nontrivial part (see below) of the full correlation functions which are defined as ($\delta n = n - \langle n \rangle$)

$$\chi_{\text{sub}}^{\text{sp}}(\mathbf{r}, t) = \langle s_z(\mathbf{r}, t) s_z(\mathbf{r}, t) \rangle_{\text{sub}} \,, \tag{5.3}$$
$$\chi_{\text{sub}}^{\text{ch}}(\mathbf{r}, t) = \langle \delta n(\mathbf{r}, t) \delta n(\mathbf{r}, t) \rangle_{\text{sub}} \,. \tag{5.4}$$

The average is taken over the Hamiltonian of the substrate. Provided that the local conduction electron spin density correlations contain a signature of the Kondo screening cloud, these are measurable by a spinpolarized STM experiment which *locally* probes the surroundings of an impurity. Previous works focused on the spin-spin correlation between the impurity and conduction electron spins [99, 126, 127, 128]. While this quantity was found to contain a signature of the spin-compensation cloud, it is not accessible via an STM experiment. Therefore the focus will be on the local correlation function of the substrate.

## 5.2 Spatial Correlations on the Substrate

In order to make predictions for an experimental observation using an STM, the system is envisioned as a magnetic impurity lying on a (semi-)metallic substrate. The conduction electrons in the substrate are further assumed to be noninteracting. This is precisely the situation which is described by the Anderson impurity model (AIM, see Sec. 2.4). For the AIM the Hamiltonian is given by Eq. 2.25. It should be noted that the formalism is completely general and not necessarily restricted to a single magnetic impurity. The extension to a system of multiple impurities, which may have internal structure (orbitals) and may interact via, e.g. an RKKY-exchange, is straightforward and is given in Ref. [129]. In contrast to the Kondo model, which was been considered previously in this context and has a fixed spin, the AIM describes the more realistic situation where also charge fluctuations are possible.

In order to simplify the notation, the formalism is introduced for a single impurity relevant for the numerical results shown below. The action for this model is obtained along the lines presented in Sec. 2.4. In Matsubara representation, it is given by

$$
\begin{aligned}
S = & -\sum_{\omega \mathbf{k} \sigma} f^*_{\omega \mathbf{k} \sigma} \left[ i\omega_n - (\epsilon_{\mathbf{k}\sigma} - \mu) \right] f_{\omega \mathbf{k} \sigma} + \sum_{\omega \mathbf{k} \sigma} \left[ c^*_{\omega \sigma} V_{\mathbf{k}\sigma} f_{\omega \mathbf{k} \sigma} + f^*_{\omega \mathbf{k} \sigma} V^*_{\mathbf{k}\sigma} c_{\omega \sigma} \right] \\
& - \sum_{\omega \sigma} c^*_{\omega \sigma} \left[ i\omega_n - e_\sigma \right] c_{\omega \sigma} + S_{\mathrm{loc}}[c^*, c] \,,
\end{aligned}
\tag{5.5}
$$

where the conduction electron (impurity) degrees of freedom are represented by $f$ ($c$). The numerical calculations were performed for the case of a Hubbard interaction

$$
S_{\mathrm{loc}}[c^*, c] = U \int_0^\beta d\tau\, n_\uparrow(\tau) n_\downarrow(\tau) \,,
\tag{5.6}
$$

with $n_\sigma = c^*_\sigma c_\sigma$. The partition function for this system is given by

$$
\mathcal{Z} = \int e^{-S[c^*, c; f^*, f]} \mathcal{D}[c^*, c; f^*, f] \,.
\tag{5.7}
$$

In order to obtain the correlation functions of this model, in particular the conduction electron correlations, sources are coupled to both the impurity and conduction electron degrees of freedom,

$$
\begin{aligned}
\mathcal{Z}_J = \int \exp\Bigg( & - S[c^*, c; f^*, f] + \sum_{\omega \sigma} \left[ J^{c*}_{\omega \sigma} c_{\omega \sigma} + c^*_{\omega \sigma} J^c_{\omega \sigma} \right] + \sum_{\omega \mathbf{k} \sigma} \left[ J^{f*}_{\omega \mathbf{k} \sigma} f_{\omega \mathbf{k} \sigma} + f^*_{\omega \mathbf{k} \sigma} J^f_{\omega \mathbf{k} \sigma} \right] \Bigg) \times \\
& \times \mathcal{D}[c^*, c; f^*, f] \,.
\end{aligned}
\tag{5.8}
$$

The conduction electrons can be integrated out exactly. This is another application of the Gaussian integral (A.9). At the location of the impurity, the effect of the conduction electrons is then described by an electronic bath described by a time-dependent Green function. The resulting action reads

$$
\mathcal{Z}_J = \det\left[i\omega_n - (\epsilon_{\mathbf{k}\sigma} - \mu)\right] \int e^{-S_{\mathrm{eff}}[c^*,c]} \exp\Bigg( \sum_{\omega\sigma} [J^{c*}_{\omega\sigma} c_{\omega\sigma} + c^*_{\omega\sigma} J^c_{\omega\sigma}] - \sum_{\omega\mathbf{k}\sigma} \frac{J^{f*}_{\omega\mathbf{k}\sigma} V^*_{\mathbf{k}\sigma} c_{\omega\sigma}}{i\omega - (\epsilon_{\mathbf{k}\sigma} - \mu)}
$$
$$
- \sum_{\omega\mathbf{k}\sigma} \frac{c^*_{\omega\sigma} V_{\mathbf{k}\sigma} J^f_{\omega\mathbf{k}\sigma}}{i\omega - (\epsilon_{\mathbf{k}\sigma} - \mu)} - \sum_{\omega\mathbf{k}\sigma} \frac{J^{f*}_{\omega\mathbf{k}\sigma} J^f_{\omega\mathbf{k}\sigma}}{i\omega - (\epsilon_{\mathbf{k}\sigma} - \mu)} \Bigg) \mathcal{D}[c^*, c; f^*, f] \, .
$$
$$(5.9)$$

The effective action, bare Green function and the hybridization are given by Eqs. 2.31-2.33, respectively. As usual, arbitrary correlation functions are now obtained by a suitable functional derivative with respect to the sources. For example, taking the following functional derivative of Eq. 5.8 gives the single-particle Green function of the bath:

$$
-\frac{1}{\mathcal{Z}} \frac{\delta^2 \mathcal{Z}_J}{\delta J^f_{\omega'\mathbf{k}'\sigma'} \delta (J^f_{\omega\mathbf{k}\sigma})^*}\Bigg|_{J=0} = -\frac{1}{\mathcal{Z}} \int f_{\omega\mathbf{k}\sigma} f^*_{\omega'\mathbf{k}'\sigma'} e^{-S[c^*,c;f^*,f]} \mathcal{D}[c^*, c, f^*, f]
$$
$$
= G^\sigma_f(\mathbf{k}, \mathbf{k}', i\omega) \delta_{\omega\omega'} \delta_{\sigma\sigma'} \, ,
$$
$$(5.10)$$

(the index '$f$' is omitted in the following). Taking the same derivative, using instead the form (5.9) for the partition function, relates the bath Green function to the corresponding impurity Green function:

$$
G^\sigma(\mathbf{k}, \mathbf{k}', i\omega) = [i\omega - (\epsilon_{\mathbf{k}\sigma} - \mu)]^{-1} \delta_{\mathbf{k}\mathbf{k}'} - \frac{V^*_{\mathbf{k}\sigma}}{i\omega - (\epsilon_{\mathbf{k}\sigma} - \mu)} \langle c_{\omega\sigma} c^*_{\omega\sigma} \rangle_{\mathrm{imp}} \frac{V_{\mathbf{k}'\sigma}}{i\omega - (\epsilon_{\mathbf{k}'\sigma} - \mu)} \, .
$$
$$(5.11)$$

Translational invariance is broken due to the presence of the impurity and the Green function depends on both momenta separately. Here the average is defined with respect to the impurity effective action, $\langle \ldots \rangle_{\mathrm{imp}} := (1/\mathcal{Z}) \int \ldots \exp(-S_{\mathrm{eff}}[c^*, c]) \mathcal{D}[c^*, c]$. The same result can be obtained using the equation of motion technique [4]. This also applies for the two-particle correlation functions obtained below. The noninteracting Green function is abbreviated as

$$
G^\sigma_0(\mathbf{k}, \mathbf{k}', i\omega) := [i\omega - (\epsilon_{\mathbf{k}\sigma} - \mu)]^{-1} \delta_{\mathbf{k}\mathbf{k}'} \, .
$$
$$(5.12)$$

Introducing the combined index $1 \equiv \{\omega\mathbf{k}\sigma\}$ to simplify the notation, the result (5.11) for the Green function reads

$$
G_{12} = G_{0\,12} + G_{0\,1\bar{1}} V^*_{\bar{1}1'} g_{1'2'} V_{2'\bar{2}} G_{0\,\bar{2}2} \, ,
$$
$$(5.13)$$

Here $g$ denotes the impurity Green function $-\langle c_1 c^*_2 \rangle$. This equation has the form of a $T$-matrix equation for the bath propagator [4]. In order to express the two-particle Green function of the bath in terms of the impurity two-particle Green function, consider the following functional derivative:

$$\frac{1}{Z} \frac{\delta^4 Z_J}{\delta J_4^f \delta(J_3^f)^* \delta J_2^f \delta(J_1^f)^*}\bigg|_{J=0} = \frac{1}{Z} \int f_1 f_2^* f_3 f_4^* e^{-S[c^*,c;f^*,f]} \mathcal{D}[c^*,c;f^*,f] =: \chi_{1234} \, . \quad (5.14)$$

Taking the corresponding functional derivative of (5.9) is straightforward, but results in a somewhat lengthy expression when written in terms of the two-particle Green function of the impurity. The result can be written in a simple form by separating off the trivial part of the two-particle Green functions[1]:

$$\tilde{\chi}_{1234} = G_{0\,1\bar{1}} V_{\bar{1}1'}^* G_{0\,3\bar{3}} V_{\bar{3}3'}^* [\tilde{\chi}_{\text{imp}}]_{1'2'3'4'} V_{2'\bar{2}} G_{0\,\bar{2}2} V_{4'\bar{4}} G_{0\,\bar{4}4} \, , \quad (5.15)$$

where the tilde denotes the nontrivial part (see Eq. 3.71). Another quantity of interest is the impurity-bath correlation function

$$\langle c_1 c_2^* f_3 f_4^* \rangle = \frac{1}{Z} \frac{\delta^4 Z_J}{\delta J_4^f \delta(J_3^f)^* \delta J_2^c \delta(J_1^c)^*}\bigg|_{J=0} = g_{12} G_{0\,34} + G_{0\,3\bar{3}} V_{\bar{3}3'}^* \langle c_1 c_2^* c_{3'} c_{4'}^* \rangle V_{4'\bar{4}} G_{0\,\bar{4}4}$$

$$= g_{12} G_{34} - g_{14'} V_{4'\bar{4}} G_{0\,\bar{4}4} G_{0\,3\bar{3}} V_{\bar{3}3'}^* g_{3'2}$$

$$+ G_{0\,3\bar{3}} V_{\bar{3}3'}^* [\tilde{\chi}_{\text{imp}}]_{123'4'} V_{4'\bar{4}} G_{0\,\bar{4}4} \, , \quad (5.16)$$

where in the last line Eq. 5.13 was used.

Isotropic s-wave coupling of the impurity to the surrounding conduction electron bath, $V_{\mathbf{k}} \equiv V$, is assumed for the calculations. When coupled to a bath, the impurity level at energy $E_d$ is not an exact eigenstate and hence acquires a finite lifetime already for $U = 0$. The Lorentzian lineshape centered at $E_d$ has a width $2\Gamma(\omega = 0) = 2\pi |V|^2 N_\sigma(\omega = 0)$ (Fermi's golden rule) and height $\Gamma(0)$, where $N_\sigma(\omega = 0)$ denotes the spin-resolved density of states. The full width of the resonance $2\Gamma(0)$ is taken as the energy unit and units are further chosen such that $\hbar = k_B = 1$. Approximating the density of states as $N^\sigma(\epsilon) = (1/N) \sum_{\mathbf{k}} \delta(\epsilon - \epsilon_{\mathbf{k}\sigma}) \approx N_0^\sigma = \text{const.}$ yields the hybridization function

$$\Delta(i\omega_n) = \sum_{\mathbf{k}} \frac{|V|^2}{i\omega_n + \mu - \epsilon_{\mathbf{k}}} \approx |V|^2 N_0^\sigma \int_{-D}^{D} \frac{d\epsilon}{i\omega_n + \mu - \epsilon} = -\frac{\Gamma(0)}{\pi} \ln\left(\frac{D - i\omega_n - \mu}{-D - i\omega_n - \mu}\right) \, , \quad (5.17)$$

which tends to $-i\Gamma(0)\,\text{sgn}(\omega_n)$ for $D \to \infty$. In the above units, for a wide featureless band, the hybridization is hence given by

$$\Delta(i\omega_n) = -i\frac{1}{2}\,\text{sgn}(\omega_n) \, . \quad (5.18)$$

---

[1]Note that this result formally takes the same form as the relationship between the two-particle Green functions of lattice and auxiliary (dual) fermions, Eq. A.132. The auxiliary fermions are introduced by essentially the same Gaussian identity which here is used to integrate out the bath fermions.

In two dimensions, the density of states is constant and this result is exact. For the three-dimensional case, which is chosen in order to allow comparison to Ref. [99], this is still a good approximation in the vicinity of the Fermi level.

It remains to relate the impurity correlation functions to the experimentally relevant ones appearing in Eqs. 5.3 and 5.4. The spin-spin density correlation function is defined by the expectation value

$$\chi^{\mathrm{sp}}(\mathbf{r}, \mathbf{r}', \tau - \tau') := \langle s_z(\mathbf{r}, \tau) s_z(\mathbf{r}', \tau') \rangle . \tag{5.19}$$

Since the presence of the impurity breaks translational invariance, the correlation function depends on two coordinates separately. Writing the spin density in terms of field operators, $s_z(\mathbf{r}) = 1/2 \sum_{\sigma\sigma'} \psi_\sigma^\dagger(\mathbf{r}) \sigma_{\sigma\sigma'}^z \psi_{\sigma'}(\mathbf{r})$, where $\sigma_{\sigma\sigma'}^z = \langle \sigma | \sigma^z | \sigma' \rangle$ is the Pauli matrix, and expanding the field operators in a single-particle basis, $\psi_\sigma(\mathbf{r}) = \sum_k \phi_k(\mathbf{r}) f_{k\sigma}$, leads to the following expression

$$\chi^{\mathrm{sp}}(\mathbf{r}, \mathbf{r}', t - t') = \frac{1}{2} \sum_{k_1 \ldots k_4} \sum_\sigma \sigma \phi_{k_1}^*(\mathbf{r}) \phi_{k_2}(\mathbf{r}) \phi_{k_3}^*(\mathbf{r}') \phi_{k_4}(\mathbf{r}') \langle f_{k_1\uparrow}^\dagger(t) f_{k_2\uparrow}(t) f_{k_3\sigma}^\dagger(t') f_{k_4\sigma}(t') \rangle ,$$

$$\tag{5.20}$$

with $\sigma = \pm 1$. The dynamical response $\chi^s(\mathbf{r}, \mathbf{r}', \omega)$ is given by Fourier transform of this equation. By similar arguments as given in Sec. 3.3.10, the expectation value in (5.20) is obtained from the corresponding time-ordered Matsubara correlation function. The nontrivial part $\tilde{\chi}_s(\mathbf{r}, \mathbf{r}', \omega = 0)$ of the dynamical response is expressed in terms of $\tilde{\chi}_{k_1 k_2 k_3 k_4}^{\sigma\sigma'}(\omega_1, \omega_2, \omega_3, \omega_4) := \langle f_{k_1\sigma}(\omega_1) f_{k_2\sigma'}^*(\omega_2) f_{k_3\sigma'}(\omega_3) f_{k_4\sigma'}^*(\omega_4) \rangle$ as:

$$\tilde{\chi}^{\mathrm{sp}}(\mathbf{r}, \mathbf{r}', i\Omega_m) = \frac{1}{2} \sum_{k_1 \ldots k_4} \sum_\sigma \sum_{\omega_n \omega_n'} \sigma \phi_{k_1}(\mathbf{r}) \phi_{k_2}^*(\mathbf{r}) \phi_{k_3}(\mathbf{r}') \phi_{k_4}^*(\mathbf{r}') \times$$

$$\times \tilde{\chi}_{k_1 k_2 k_3 k_4}^{\uparrow\sigma}(\omega_n + \Omega_m, \omega_n, \omega_n', \omega_n' + \Omega_m) . \tag{5.21}$$

Inserting plane waves $\phi_{\mathbf{k}}(\mathbf{r}) = 1/(2\pi)^{d/2} e^{i\mathbf{k}\mathbf{r}}$ and the result (5.15) with (5.12) into this expression for a continuous spectrum, leads to the following integrals,

$$G_0^\sigma(\mathbf{r}, i\omega_n) = \frac{1}{(2\pi)^d} \int d^d k \frac{e^{i\mathbf{k}\mathbf{r}}}{i\omega_n - (\epsilon_{\mathbf{k}\sigma} - \epsilon_F)} , \tag{5.22}$$

where the dimension is chosen $d = 3$ in accordance with Ref. [99]. We expect no qualitative changes of the results presented below for $d < 3$ (note however, that the power of the $\mathbf{r}$-dependence will be different). For a parabolic band, $\epsilon_{\mathbf{k}\sigma} = k_F^2 (\mathbf{k}/k_F)^2/2m = \epsilon_F (\mathbf{k}/k_F)^2$ ($k_F$ and $\epsilon_F$ are the Fermi wavevector and Fermi energy, respectively), the integral is straightforwardly evaluated to give

$$G_0^\sigma(\mathbf{r}, i\omega_n) = -\frac{\pi N_\sigma(\epsilon_F)}{k_F |\mathbf{r}|} e^{-\sqrt{-1 - i\omega_n/\epsilon_F} k_F |\mathbf{r}|} , \tag{5.23}$$

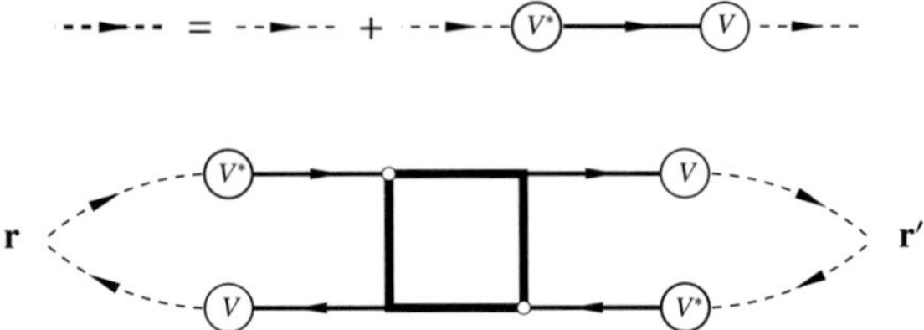

*Figure 5.2:* Top: Diagrammatic representation of the full conduction electron Green function (5.13, thick dashed line). The thin dashed line denotes the bare conduction electron propagator. The second diagram on the right describes bare propagation in the conduction electron bath, subsequent transfer to the impurity through the hybridization $V$, renormalized propagation on the impurity described by the impurity Green function (thick solid line) and transfer back to and propagation in the bath. Bottom: Non-trivial part of the spin/charge correlation function Eq. 5.15 (5.24,5.25). The square denotes the two-particle impurity vertex function $\gamma$.

where $N^\sigma(\epsilon_F) = |\mathbf{k}_F|^3/4\pi^2\epsilon_F$. Considering the static local quantity (i.e. $i\Omega = 0$, $\mathbf{r} = \mathbf{r}'$) and the nontrivial part only, which enters Eq. 5.2, leads to the final result

$$\tilde{\chi}^{\mathrm{sp}}(\mathbf{r},\mathbf{r},i\Omega = 0) = \frac{1}{2}\left[\frac{\pi N_\sigma(\epsilon_F)\Gamma(0)}{k_F^2\,|\mathbf{r}|^2}\right]^2\frac{1}{\beta^2}\sum_{\omega_n\omega'_n}e^{-2\sqrt{-1-i\omega_n/\epsilon_F}\,k_F|\mathbf{r}|}\,e^{-2\sqrt{-1-i\omega'_n/\epsilon_F}\,k_F|\mathbf{r}|}\times$$

$$\times g(i\omega_n)g(i\omega_n)\left[\Gamma^{\uparrow\uparrow}(i\omega_n,i\omega_n,i\omega'_n,i\omega'_n) - \Gamma^{\uparrow\downarrow}(i\omega_n,i\omega_n,i\omega'_n,i\omega'_n)\right]g(i\omega'_n)g(i\omega'_n)\,. \quad (5.24)$$

Likewise, the charge susceptibility (5.4) is obtained by expressing the density in terms of field operators as $n(\mathbf{r}) = \sum_\sigma \psi_\sigma^\dagger(\mathbf{r})\psi_\sigma(\mathbf{r})$. The result is

$$\tilde{\chi}^{\mathrm{ch}}(\mathbf{r},\mathbf{r},i\Omega = 0) = 2\left[\frac{\pi N_\sigma(\epsilon_F)\Gamma(0)}{k_F^2\,|\mathbf{r}|^2}\right]^2\frac{1}{\beta^2}\sum_{\omega_n\omega'_n}e^{-2\sqrt{-1-i\omega_n/\epsilon_F}\,k_F|\mathbf{r}|}\,e^{-2\sqrt{-1-i\omega'_n/\epsilon_F}\,k_F|\mathbf{r}|}\times$$

$$\times g(i\omega_n)g(i\omega_n)\left[\Gamma^{\uparrow\uparrow}(i\omega_n,i\omega_n,i\omega'_n,i\omega'_n) + \Gamma^{\uparrow\downarrow}(i\omega_n,i\omega_n,i\omega'_n,i\omega'_n)\right]g(i\omega'_n)g(i\omega'_n)\,. \quad (5.25)$$

In order to compare with previous results, the non-local impurity-bath spin correlation function between the impurity and a point in the bath at distance $|\mathbf{r}|$ from the impurity is also considered. It is obtained by evaluating 5.16 in place of 5.15. The equal-time correlation function $\chi^s(\mathbf{r},0) := \langle S^z_{\mathrm{imp}}(\mathbf{0},\tau)s^z(\mathbf{r},0)\rangle$ is given by

$$\tilde{\chi}_{\text{nl}}^{\text{sp}}(\mathbf{r}, \mathbf{r}, \tau = 0) = \frac{1}{2} \left[ \frac{\pi N_\sigma(\epsilon_F)\Gamma(0)}{k_F^2 \, |\mathbf{r}|^2} \right] \frac{1}{\beta^3} \sum_{\Omega_m} \sum_{\omega_n \omega_n'} e^{-\sqrt{-1-i(\omega_n+\Omega_m)/\epsilon_F}\, k_F|\mathbf{r}|} e^{-\sqrt{-1-i\omega_n/\epsilon_F}\, k_F|\mathbf{r}|} \times$$
$$\times g(i\omega_n + \Omega_m)g(i\omega_n) \left[ \Gamma^{\uparrow\uparrow}(i\omega_n + \Omega_m, i\omega_n, i\omega_n', i\omega_n' + \Omega_m) - \right.$$
$$\left. - \Gamma^{\uparrow\downarrow}(i\omega_n + \Omega_m, i\omega_n, i\omega_n', i\omega_n' + \Omega_m) \right] g(i\omega_n' + \Omega_m)g(i\omega_n') \, .$$
$$(5.26)$$

The trivial part is needed to ensure comparability with known results. It stems from the second term in the next to last line of Eq. 5.16 (the first term, $G_{12}G_{34}$, does not contribute to the susceptibility) and reads

$$\chi_{\text{nl}}^{0\,\text{sp}}(\mathbf{r}, \mathbf{r}, \tau = 0) = -\frac{1}{2} \left[ \frac{\pi N_\sigma(\epsilon_F)\Gamma(0)}{k_F^2 \, |\mathbf{r}|^2} \right] \left[ \frac{1}{\beta} \sum_{\omega_n} e^{-\sqrt{-1-i\omega_n/\epsilon_F}\, |\mathbf{r}|} g(i\omega_n) \right]^2 \, . \qquad (5.27)$$

Accordingly, the zero-frequency correlation function is given by

$$\tilde{\chi}_{\text{nl}}^{\text{sp}}(\mathbf{r}, \mathbf{r}, i\Omega = 0) = \frac{1}{2} \left[ \frac{\pi N_\sigma(\epsilon_F)\Gamma(0)}{k_F^2 \, |\mathbf{r}|^2} \right] \frac{1}{\beta^2} \sum_{\omega_n \omega_n'} e^{-2\sqrt{-1-i\omega_n/\epsilon_F}\, k_F|\mathbf{r}|} g(i\omega_n)g(i\omega_n) \times$$
$$\times \left[ \Gamma^{\uparrow\uparrow}(i\omega_n, i\omega_n, i\omega_n', i\omega_n') - \Gamma^{\uparrow\downarrow}(i\omega_n, i\omega_n, i\omega_n', i\omega_n') \right] g(i\omega_n')g(i\omega_n') \, . \qquad (5.28)$$

and

$$\chi_{\text{nl}}^{0\,\text{sp}}(\mathbf{r}, \mathbf{r}, i\Omega = 0) = -\frac{1}{2} \left[ \frac{\pi N_\sigma(\epsilon_F)\Gamma(0)}{k_F^2 \, |\mathbf{r}|^2} \right] \frac{1}{\beta} \sum_{\omega_n} \left[ e^{-\sqrt{-1-i\omega_n/\epsilon_F}\, |\mathbf{r}|} g(i\omega_n) \right]^2 \, . \qquad (5.29)$$

The charge correlation function is obtained by changing the sign in the angular brackets in (5.28) and multiplying the result by a factor of 4.

These equations are readily implemented using the scheme described in chapter 7. The single- and two-particle Green functions are obtained using the weak-coupling continuous-time quantum Monte Carlo solver as described in chapter 3.

## 5.3 Local Susceptibility

The local susceptibility of the impurity is defined as $\chi_{\text{dd}} := \int_0^\beta d\tau \langle S_z(\tau)S_z(0) \rangle$ (Eq. 3.79). The index 'd' is used to label correlation functions defined in terms of impurity operators[2] and 'c' for conduction electron operators. $\chi_{\text{dd}}$ is shown in Fig 5.3 for three

---

[2]This commonly used designation is historical; the AIM was introduced to study impurities with open d-shells embedded in a metallic host.

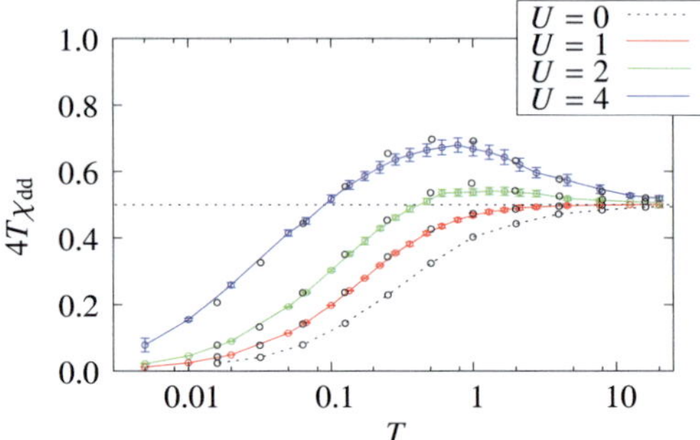


*Figure 5.3:* Local susceptibility $4T\chi_{dd}$ (in units of $(g\mu_B)^2$) as a function of temperature $T$ obtained using CTQMC. $T\chi_{dd}$ plays the role of the effective local moment of the impurity which is screened at low temperatures. At high temperatures all four impurity states $|0\rangle$, $|\uparrow\downarrow\rangle$, $|\uparrow\rangle$ and $|\downarrow\rangle$ contribute equally. The latter two contribute $1/4$ to the moment $\langle S_z^2\rangle$, which therefore approaches $1/8$, or $4T\chi_{dd} = 1/2$. Open black circles show the data taken from Ref. [99].

different values of the interaction $U$ as a function of temperature. These results were obtained according to Eq. 3.80. This figure reproduces results that have been previously obtained by Gubernatis *et al.* using the Hirsch-Fye quantum Monte Carlo method in Ref. [99] (open black circles). Their data was retrieved using an image digitizer tool [100]. The comparison provides a test for the calculation of (static) susceptibilities in the continuous-time algorithm. The overall agreement is good. From the errorbars one can see that this quantity converges more slowly than the equal-time correlation function $\mu^2 := \langle S_z^2\rangle$ (local moment) shown in Fig. 3.6. The largest deviations to the data from Ref. [99] occur for large $U$ and at intermediate temperatures, where the local moment is formed and starts being screened. The data are however consistent within the errorbar. The CTQMC data may be slightly underestimated. A very large frequency cutoff has to be used to get a converged result, although the asymptotic high-frequency behavior above the cutoff is taken into account. For example, the cutoff $N_\omega$ was consistently increased (proportional to $\beta$) from 1024 to 2048 when going from $T = 0.01$ ($\beta = 100$) to $T = 0.005$ ($\beta = 200$). For the consistent cutoff the susceptibility saturates at low temperatures (see chapter 9), instead of showing a slight downturn for a fixed cutoff. In general, the susceptibility can be substantially underestimated if the QMC algorithm looses ergodicity and the solution becomes polarized. This is however not the case here. The deviations at very low temperatures are likely due to the rather coarse imaginary time discretization down to $\Delta\tau = 0.25$ in the Hirsch-Fye calculations.

The local moment forms on a temperature scale set by the width of the resonance $T \sim \Gamma(0) = 1/2$. For relatively high temperatures, it behaves as a free spin and correspondingly the susceptibility behaves as $\langle S_z^2\rangle/T$. The increase in $T\chi_{\rm dd}$ at high temperatures therefore is due to the formation of the local moment.

Apart from being a consistency check, it is illustrative to point out the physical meaning of these results. First note that according to Sec. 3.3.10, the total susceptibility of the whole system can in general be written

$$\chi_{\rm tot} = \int_0^\beta d\tau \int d^3r \langle [S_z(\tau) + S_{z,{\rm c}}(\mathbf{r},\tau)]^2\rangle = \int_0^\beta d\tau \langle [S_z(\tau) + S_{\rm c}(\tau)]^2\rangle , \qquad (5.30)$$

which is the static, homogeneous component $\chi(\mathbf{q} = 0, \omega = 0)$ of the dynamical susceptibility. Here the total conduction electron spin operator has been defined such that $S_{z,{\rm c}} = \int d^3r S_{z,{\rm c}}(\mathbf{r})$. The $z$-component of the spin of the whole system $S_z^{\rm tot} = S_z + S_{z,{\rm c}}$ is a constant of the motion, i.e. $[S_z^{\rm tot}, H] = 0$. It follows from the imaginary time Heisenberg representation that $S_z^{\rm tot}(\tau) = e^{H\tau}S_z^{\rm tot}e^{-H\tau} = S_z^{\rm tot}(0)$ does not depend on time. Hence

$$\chi_{\rm tot} = \int_0^\beta d\tau \langle [S_z(\tau) + S_{z,{\rm c}}(\tau)]^2\rangle = \frac{\langle [S_z + S_{z,{\rm c}}]^2\rangle}{T} , \qquad (5.31)$$

where $T = \beta^{-1}$ is the temperature. The same reasoning holds for the system of conduction electrons in the absence of the impurity, which allows to write $\chi_{\rm c}^0(T) = \langle (S_{z,{\rm c}})^2\rangle_0/T$. The susceptibility of the impurity is usually defined as the excess susceptibility of the system due to the presence of the impurity $\chi_{\rm imp} := \chi_{\rm tot} - \chi_{\rm c}^0$ and is expressed in terms of equal-time expectation values as

$$\chi_{\rm imp}(T) := \frac{\langle (S_z + S_{z,{\rm c}})^2\rangle - \langle (S_{z,{\rm c}})^2\rangle_0}{T} = \frac{\langle S_z^2\rangle + 2\int\langle S_z S_{z,{\rm c}}(\mathbf{r})\rangle d^dr}{T} + \frac{\langle (S_{z,{\rm c}})^2\rangle - \langle (S_{z,{\rm c}})^2\rangle_0}{T}. \qquad (5.32)$$

The second term on the right is negligible meaning that the conduction electron response for small fields is unaffected by the presence of the impurity. This statement is the Clogston-Anderson compensation theorem which is valid for a wide flat band, where $\sum_{\mathbf{k}} |V|^2 /(i\omega_n - \epsilon_{\mathbf{k}})^2$ is negligible (see, e.g., Ref. [4]). It is important to note that the thus defined impurity susceptibility in general is not exactly equal to $\chi_{\rm dd}$ defined above (the latter is the response to a magnetic field acting at the site of the impurity only). This is obvious by expanding the total response Eq. 5.30:

$$\chi_{\rm tot} = \int_0^\beta d\tau \langle S_z^2 + S_z S_{z,{\rm c}} + S_{z,{\rm c}} S_z + S_{z,{\rm c}}^2\rangle =: \chi_{\rm dd} + \chi_{\rm dc} + \chi_{\rm cd} + \chi_{\rm cc} . \qquad (5.33)$$

The term $\chi_{\rm dc}$ arises from the coupling of the bath to the external field. It describes the feedback of a change in the conduction electron response onto the impurity and can be obtained from the correlation function (5.16). $\chi_{\rm dc}$ changes the response of the conduction electrons due to the field coupled to the impurity and is equal to $\chi_{\rm cd}$[3]. The last

---

[3]In general, the relation $\chi_{\rm cd}(\mathbf{q}, i\Omega_m) = \chi_{\rm dc}^*(-\mathbf{q}, -i\Omega_m)$ holds [130].

term is the response of the conduction electrons in presence of the impurity. Equating the impurity response with the local response $\chi_{dd}$ hence corresponds to neglecting $\chi_{dc} + \chi_{cd} + \chi_{cc} - \chi_c^0$ as implied by the compensation theorem ($\chi_{dc} = \chi_{cd} \approx 0$ and $\chi_{cc} \approx \chi_c^0$). While this is warranted in the wide band limit, this need not be a good approximation for a general density of states as has been clarified in [130] using a perturbative approach.

By virtue of (5.32), the interpretation of $T\chi_{dd} \approx T\chi_{imp}$ is hence that of an effective moment of the impurity $\mu_{eff}^2 = \langle S_z^2 \rangle + 2 \int \langle S_z S_{z,c}(\mathbf{r}) \rangle d^3 r$ which is screened by the conduction electrons at low temperatures, while the impurity moment $\langle S_z^2 \rangle$ remains finite according to Fig. 3.6. The screening of the local moment for $U = 0$ was attributed to Fermi-hole correlations [99].

One is tempted to identify e.g. $\langle S_z S_{z,c} \rangle / T$ with $\chi_{dc}$ or $\langle S_z^2 \rangle / T$ with $\chi_{dd}$ (the latter would imply a finite susceptibility in contrast to the results in Fig. 5.3). Note that $S_z$ does not commute with the hybridization term of the Hamiltonian (Eq. 2.25), i.e. in the presence of the bath and hence is not a constant of the motion, so that the analog of (5.31) does not hold.

## 5.4   Spatial Correlations

The parameters for which the following calculations were carried out are summarized in table 5.1. The Kondo temperatures have been given in Ref. [75]. They are obtained from the Bethe-Ansatz solution for the impurity susceptibility using the relation $\chi(T = 0) = 0.103/T_K$ from a renormalization group analysis (see e.g. [123] and references therin) and depend exponentially on $U$. The Kondo temperature introduces a length scale $\xi_K = v_F/T_K$, where $v_F$ is the Fermi velocity. The Kondo screening length rescaled by the Fermi wavevector can be written as $k_F\xi_K = k_F v_F/T_K = 2\epsilon_F/T_K$.

| $\epsilon_F$ | $U$ | $E_d$ | $T_K$ | $k_F\xi_K$ |
|---|---|---|---|---|
| 6.0 | 1.0 | -0.5 | 0.168 | $\sim 70$ |
| 6.0 | 2.0 | -1.0 | 0.0865 | $\sim 140$ |
| 6.0 | 4.0 | -2.0 | 0.0216 | $\sim 555$ |

*Table 5.1:* Parameters used for the numerical calculations. $\epsilon_F$, $U$ and $E_d$ denote the Fermi energy, the interaction and the position of the impurity level (located at $-U/2$ for the symmetric AIM), respectively. $T_K$ is the Kondo temperature and $\xi_K$ the Kondo screening length.

If not otherwise stated, the following results were obtained at inverse temperature $\beta = 200$, which corresponds to $T/T_K \sim 0.03, 0.06, 0.23$ for $U = 1, 2, 4$, respectively. We further take $\epsilon_F$ in accordance with Ref. [99]. Fig. 5.3 shows data for inverse temperatures up to $\beta = 200$, for which the effective moment for $U = 1$ is quenched down

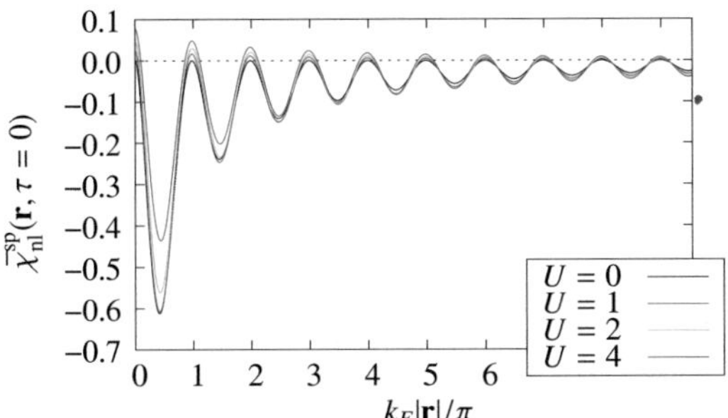

*Figure 5.4:* Nonlocal equal time impurity-bath correlation function $\chi_{\mathrm{nl}}^{\mathrm{sp}}(\mathbf{r}, \tau) = \chi_{\mathrm{nl}}^{0\,\mathrm{sp}}(\mathbf{r}, \tau) + \tilde{\chi}_{\mathrm{nl}}^{\mathrm{sp}}(\mathbf{r}, \tau)$ as given by Eqs. 5.26 and 5.27. In the graph $\bar{\chi}_{\mathrm{nl}}^{\mathrm{sp}}(\mathbf{r}, \tau) := [\pi N_\sigma(\epsilon_{\mathrm{F}})\Gamma(0)/(2k_{\mathrm{F}}^2\,|\mathbf{r}|^2)]^{-1}\chi_{\mathrm{nl}}^{\mathrm{sp}}(\mathbf{r}, \tau)$. The correlation of the conduction electron spins with the impurity spin is predominantly antiferromagnetic ($\chi < 0$).

to $\sim 1.17\%$ of its maximally possible value. The following calculations were thus performed well below the Kondo temperature and in the coherence regime.

Fig. 5.4 shows the nonlocal equal-time spin correlation function as a function of distance from the impurity. This function has been shown for inverse temperatures up to $\beta = 32$ ($T/T_{\mathrm{K}} \sim 0.19$) in Ref. [99]. It shows no qualitative differences as expected from the absence of another, yet smaller energy scale. For $U = 0$, it is strictly smaller than zero, while for finite $U$, it is predominantly so. This shows the antiferromagnetic correlation between the impurity spin and the conduction electron spins. For $U = 0$, this has been attributed to Fermi-hole correlations as a consequence of the Pauli exclusion principle, which leads to a reduced probability to find electrons with a given spin in the vicinity of the impurity if it is occupied by an electron with the same spin. The phase and period of the oscillations are not changed by the interaction. For small distances the AF correlations are reduced by increasing $U$, but increase monotonically with $U$ at larger distances. The ferromagnetic correlations can be understood from the fact that the occupation of the impurity with an electron of a given spin repels electrons with opposite spin (this is contrary to the effect arising from the exclusion principle). These are largely independent of temperature, while the AF correlations become more pronounced as the temperature is lowered from above to below the Kondo temperature. As was confirmed numerically, spatial integration of this function according to (5.32) approaches $-\langle S_z^2 \rangle$ for large distances, which shows that indeed the effective moment $\mu_{\mathrm{eff}}^2$ of the impurity is quenched by the conduction electrons. For large distances, of the order of the Kondo screening length, the amplitude of the AF correlations approaches that of the FM correlation since it decays much faster. Hence the screening occurs only

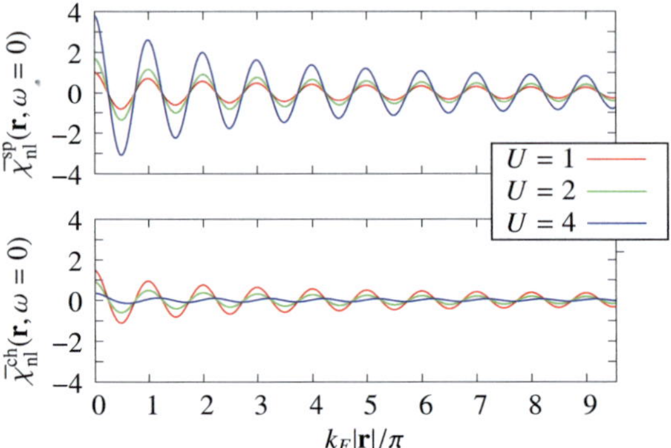

*Figure 5.5:* Static nonlocal spin (5.28) and charge correlations between the impurity and conduction electrons at distance $|\mathbf{r}|$ from the impurity. The correlation functions $\bar{\chi}_{\mathrm{nl}}^{\mathrm{sp/ch}}(\mathbf{r}, \omega = 0)$ are rescaled by the same prefactor as in Fig. 5.4.

within a finite range around the impurity. The AF correlations at small distances are smaller with increasing $U$, since the spin compensation has to occur over a wider range due to a larger screening length (see table 5.1). These observations are consistent with the findings reported in [99].

In Fig. 5.5 the nonlocal, static correlation function is shown as a function of distance from the impurity. It exhibits the same phase and period as the equal-time correlations. The period of the oscillations is given by $\pi k_{\mathrm{F}}^{-1}$. In contrast to Fig. 5.4, the positive and negative envelopes however are approximately symmetric around zero, showing ferromagnetic and antiferromagnetic correlations in equal parts. This illustrates that the impurity spin is screened dynamically. Higher frequency components eliminate the ferromagnetic correlations. This is expected on a scale set by the Kondo temperature $\omega_{\mathrm{K}} \sim T_{\mathrm{K}}^{-1}$. Note also that the integral of $\chi_{\mathrm{nl}}^{\mathrm{sp}}(\mathbf{r}, \omega = 0)$ over all space yields $\chi_{\mathrm{dc}}$, which is therefore seen to be small, as required by the compensation theorem. The nonlocal charge susceptibility has the same phase as the spin correlation. Unlike the spin correlations, which monotonically increase with $U$, the charge density oscillations are strongly damped as $U$ increases. This can be understood from the fact that the local charge fluctuations on the impurity are suppressed for large $U$, so that the charge degrees of freedom of the bath and the impurity become uncorrelated.

Fig. 5.6 shows the local bath correlation functions which enter the expression (5.2) for the conductance fluctuations. The local spin correlation function is predominantly positive, as expected. The frequency of these oscillations is doubled compared to the

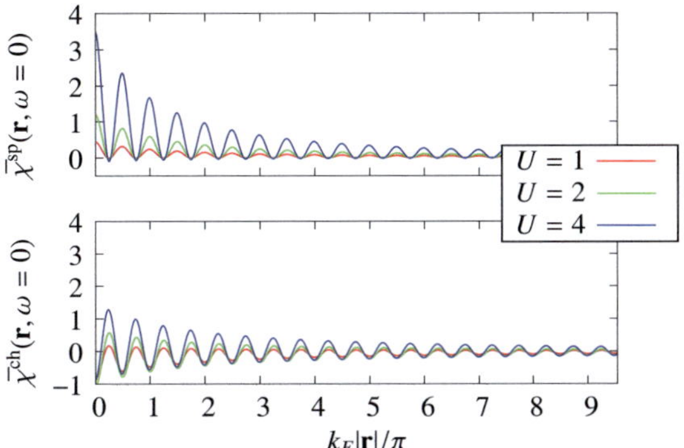

*Figure 5.6:* Local static conduction electron correlation function $\tilde{\chi}^{\text{sp/ch}}(\mathbf{r}, \omega = 0)$ at distance $|\mathbf{r}|$ from the impurity (5.24,5.25). The plotted correlation functions are rescaled according to $\bar{\chi}_{\text{nl}}^{\text{sp}}(\mathbf{r}, \omega) := [\pi N_\sigma(\epsilon_{\text{F}})\Gamma(0)/(2k_{\text{F}}^2 |\mathbf{r}|^2)]^{-2}\tilde{\chi}_{\text{nl}}^{\text{sp}}(\mathbf{r}, \omega)$. These quantities enter the expression for the conductance fluctuations, Eq. 5.2.

nonlocal correlations and the period is given by $\pi/(2k_{\text{F}})$, which is half the value as for Friedel oscillations. The charge correlation function is phase shifted by that from the spin correlation by $\pi/2$. Here the supression of the local charge correlations on the impurity site for increasing $U$ lead to an enhancement of the spin and charge density correlations in contrast to the case for the nonlocal charge correlation function.

In order to show that the decay of the correlations is set by the Kondo length scale $\xi_{\text{K}}$, the envelope of the local bath correlation function $\tilde{\chi}^{\text{sp}}(\mathbf{r}, \omega = 0)$ is plotted on a double-logarithmic scale in Fig. 5.7. The decay is nonalgebraic for small distances and changes to a power-law behavior (i.e. a straight line) at intermediate distances $|\mathbf{r}| \sim \xi_{\text{K}}$. At large distances correlations decay exponentially. This occurs on a length scale $\xi_T = v_{\text{F}}/T$ associated with the finite temperature $T$. The lengthscale increases as the temperature is lowered as seen in the figure. At the lowest temperature $T = 10^{-3}$, the thermal lengthscale is shifted far enough so that the envelope is well approximated by a fit proportional to $(|\mathbf{r}|/\xi_{\text{K}})^{-2}$ over a finite range. At zero temperature the nonalgebraic decay below the Kondo scale $\xi_{\text{K}}$ changes to a powerlaw behavior for all distances above this scale.

For small distances from the impurity, the curves deviate and fall to lower values as the temperature is increased. If the number of frequencies is increased proportional to the inverse temperature, the results at different temperatures coincide at small distances and deviate only for large distances as determined by the thermal length scale. Hence we do not attribute this to a physical effect, but to the finite frequency cutoff $N_\omega$ for

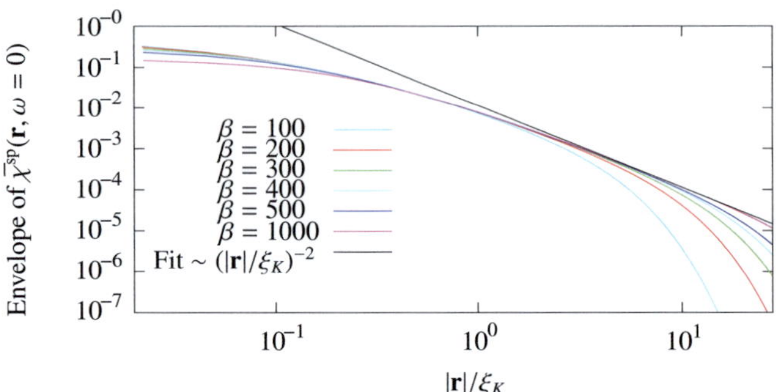


*Figure 5.7:* Envelope of the local conduction electron spin correlation function of Fig. 5.6 for $U = 1$, as a function of distance from the impurity. For distances smaller than the Kondo screening length $\xi_K$, the correlations decay nonalgebraically. This changes into a power-law behavior at large distances and low temperatures. At higher temperatures, and large distances, the correlations decay exponentially, as determined by the temperature scale $\xi_T = v_F/(k_B T_K)$.

the vertex, which here was taken to be $N_\omega = 110$. This requires the measurement of the order of $4(2N_\omega)^3$ observables and several hundred MB of memory for the full vertex (see Sec. 3.3.9). The number of frequencies is therefore limited by computational resources and memory requirements. Including asymptotic corrections for large frequencies to the nontrivial part of the correlation function, which is plotted here, would require the knowledge of the asymptotic behavior of the vertex.

Here the full vertex was measured in order to additionally compute the equal-time correlation function shown in Fig. 5.4. If only the static correlation function is desired, it is sufficient to measure the vertex for zero bosonic frequency, i.e. for two independent fermionic frequencies only, which would allow $N_\omega \sim 10^3$ with the same computational effort. This would suffice for inverse temperatures roughly up to $\beta = 10^3$.

For the lowest temperature ($\beta = 10^3$), the present cutoff is clearly too low. However, we have verified that the finite frequency cutoff does not invalidate our conclusion about the decay of the correlations within the Kondo screening length. At the highest temperatures ($\beta = 100, 200$) the result is essentially converged with respect to the cutoff frequency. A small error due to the cutoff cannot be excluded for very small distances $|\mathbf{r}|/\xi_K \lesssim 5 \cdot 10^{-2}$ (an extrapolation is unreliable since the correlations depend nonlinearly on the cutoff), but the nonalgebraic decay is visible for larger distances up to the Kondo length $\xi_K$. Remarkably, deviations from reducing the frequency cutoff only occur for distances up to the Kondo length scale. This is true even for $N_\omega$ as low as 10. For the nonlocal correlation function, this is not the case. Here the behavior is influenced by the cutoff at all distances from the impurity, including length scales of the order of the thermal length. In any case, the behavior of the local bath correlation function at large distances and very low temperatures, where it gradually approaches $|\mathbf{r}|^{-2}$ dependence, is reliable.

## 5.5 Conclusions and Outlook

The efficient separation of the impurity two-particle Green function in the expressions for the nonlocal correlation functions together with CTQMC have been employed to obtain results for spatial correlations around a Kondo impurity. The numerical results demonstrate that indeed a signature of the Kondo cloud is contained in the local spin-density correlation function of the conduction electrons. In principle, it can be extracted from the measurement of conductance fluctuations by a local probe, such as an STM, allowing an experimental verification of the Kondo screening cloud.

The framework in which these calculations have been carried out allows to treat multiple impurities, which may have internal structure and for example interact via an RKKY-exchange interaction (the general formulation is given in Ref. [129]). It is interesting to apply this to a system of at least two impurities which may reveal new effects, for example due to the interference of the two Kondo clouds. Alternatively one may investigate the spatial correlations in the vicinity of a multiorbital Kondo impurity.

The steps in $\partial I/\partial V$ spectra observed in inelastic tunneling spectroscopy measurements [131] have recently been explained through spin-assisted tunneling using a phenomenological approach [132]. Using the weak tunneling formalism similar to the case considered here, it is possible to relate the quantities of interest to response functions of an AIM in equilibrium, which can be computed using CTQMC. This yields a rigorous approach to model spin-assisted inelastic tunneling spectroscopy experiments.

# Chapter 6

# Dual Fermion Approach

The dual fermion approach has been introduced recently by Rubtsov *et al.* [41] following an earlier suggestion for bosonic fields [133, 134] and has been applied and further developed in subsequent publications [135, 136, 137, 138, 139, 140]. It allows to include spatial correlations beyond dynamical mean-field theory (DMFT). In contrast to related approaches [39, 40, 38], it relies on an expansion around an arbitrary impurity problem, which serves as a reference system. Moreover, it is based on the introduction of auxiliary fermionic degrees of freedom in the path integral representation of the partition function via a continuous Hubbard-Stratonovich transformation (HST). The special properties of this approach, which arise from these differences, will be investigated in the following. The transformation is essentially a mapping from lattice fermions with strong local correlations to weakly correlated, delocalized auxiliary fermions, which warrants the designation of "dual" fermions.

Mathematically, the HST is a Gaussian integral identity and as such is used as a standard tool in quantum field theory, either to introduce auxiliary degrees of freedom or to integrate out Gaussian ensembles. It is applied in several instances in this thesis (cf., e.g., Sec. 2.4 and chapter 5). It has, in fact, been applied previously in a similar context, namely to derive a strong-coupling expansion for the Hubbard model [29, 141, 142, 30, 31]: The unperturbed system, described by the Hamiltonian taken in the atomic limit, contains two-body operators so that Wick's theorem cannot be applied. This difficulty is circumvented by introducing auxiliary degrees of freedom through the HST. Integrating out the original fermions completes the change of variables. The result is an action that contains $n$-particle interactions up to all orders, which are given by the connected correlation functions of the unperturbed system. This work is related to the dual fermion approach, but the latter differs from it in two important respects: Instead of expanding around the atomic limit, the dual fermion approach introduces a local quantum impurity problem as a reference problem. Secondly, by imposing a condition to choose the impurity problem in an optimal way, a relation to the DMFT is established, namely, DMFT appears as the lowest (zero-) order approximation in this approach.

## 6.1 Formalism

As part of the work for this thesis, the approach was generalized in order to treat multi-orbital systems. The equations are therefore given here within the general formulation. As the derivation of the formalism is somewhat involved, only the essential steps are given here. The details of the derivation are deferred to appendix A.

The theory is introduced in the Matsubara formalism. The goal is to find an (approximate) solution to the general multiband lattice problem described by the imaginary time action

$$S[c^*, c] = - \sum_{\omega \mathbf{k} \sigma\, mm'} c^*_{\omega \mathbf{k} \sigma m} \left[ (i\omega + \mu)\mathbb{1} - h_{\mathbf{k}\sigma} \right]_{mm'} c_{\omega \mathbf{k} \sigma m'} + \sum_i S_{\text{loc}}[c_i^*, c_i] \,. \tag{6.1}$$

Here $h_{\mathbf{k}\sigma}$ is the one-electron part of the Hamiltonian, $\omega_n = (2n + 1)\pi/\beta, n = 0, \pm 1, \dots$ are the Matsubara frequencies, $\beta$ and $\mu$ are the inverse temperature and chemical potential, respectively, $\sigma = \uparrow, \downarrow$ labels the spin projection, $m, m'$ are orbital indices and $c^*, c$ are Grassmann variables. The index $i$ labels the lattice sites and the k-vectors are quasimomenta. In order to keep the notation simple, it is useful to introduce the combined index $\alpha \equiv \{m\sigma\}$. If indices on Green functions, self-energies, etc. are omitted, these are considered matrices in spin/orbital space and their product denotes the usual matrix product. All quantities are assumed to be diagonal in spin indices. For clarity, the dependence on frequency is written explicitly. Translational invariance is assumed for simplicity in the following. For applications it is important to note that $S_{\text{loc}}$ may contain *any* type of local interaction. The only requirement is that it is local within the multiorbital atom or cluster. For realistic calculations in the context of LDA+DMFT, the local Hamiltonian is given by (2.94), which contains the full Coulomb interaction matrix.

The original idea of the dual fermion approach is to formulate a perturbation expansion around DMFT. In this spirit, a local quantum impurity problem (Anderson impurity model) is introduced in the form

$$S_{\text{imp}}[c^*, c] = - \sum_{\omega\, \alpha\beta} c^*_{\omega\alpha} \left[ (i\omega + \mu)\mathbb{1} - \Delta_\omega \right]_{\alpha\beta} c_{\omega\beta} + S_{\text{loc}}[c^*, c] \,, \tag{6.2}$$

where $\Delta_{\omega\sigma}$ is an as yet unspecified hybridization matrix describing the interaction of the (cluster) impurity with an electronic bath. Apart from the connection to DMFT, another motivation for rewriting the lattice action in this form is to express it in terms of a reference problem that can be solved accurately for an arbitrary hybridization function using the methods described in chapter 3. Using the locality of the hybridization function, the lattice action (6.1) is rewritten *exactly* by adding and subtracting a hybridization function $\Delta_\omega$ at each lattice site:

$$S[c^*, c] = \sum_i S_{\text{imp}}[c_i^*, c_i] - \sum_{\omega \mathbf{k}\, \alpha\beta} c^*_{\omega \mathbf{k} \alpha} \left( \Delta_\omega - h_{\mathbf{k}} \right)_{\alpha\beta} c_{\omega \mathbf{k} \beta} \,. \tag{6.3}$$

This step is exact and leaves the hybridization function unspecified. It will be used later to optimize the approach. The lattice may be viewed as a collection of impurities, which are coupled through the bilinear term to the right of this equation. The effect of spatial correlations enters here and renders an exact solution (of large systems) impossible. A perturbative treatment is desirable, but not straightforward as the impurity action is non-Gaussian and hence there is no Wick theorem. Therefore, at this point, the dual fermions are introduced in the path integral representation of the partition function

$$Z = \int \exp\left(-S[c^*, c]\right) \mathcal{D}[c^*, c],\tag{6.4}$$

through the HST (cf. appendix A.1)

$$\exp\left(c^*_\alpha b_{\alpha\beta}(a^{-1})_{\beta\gamma} b_{\gamma\delta} c_\delta\right) = \frac{1}{\det a} \int \exp\left(-f^*_\alpha a_{\alpha\beta} f_\beta + f^*_\alpha b_{\alpha\beta} c_\alpha + c^*_\alpha b_{\alpha\beta} f_\beta\right) \prod_\gamma df^*_\gamma df_\gamma.\tag{6.5}$$

The goal is to transform the exponential of the bilinear term in (6.3). The choice of the matrices $a$, $b$ is not unique; in analogy to Refs. [41, 139], they are chosen as

$$a = g_\omega^{-1}\left(\Delta_\omega - h_{\mathbf{k}}\right)^{-1} g_\omega^{-1},\qquad b = -g_\omega^{-1},\tag{6.6}$$

where $g_\omega$ is the interacting Green function of the local impurity problem. With this choice, the lattice action transforms to

$$S[c^*, c, f^*, f] = \sum_i S_{\text{site},i} + \sum_{\omega\mathbf{k}\alpha\beta} f^*_{\omega\mathbf{k}\alpha}[g_\omega^{-1}(\Delta_\omega - h_{\mathbf{k}})^{-1} g_\omega^{-1}]_{\alpha\beta} f_{\omega\mathbf{k}\beta}.\tag{6.7}$$

Hence the coupling between sites is transferred to a *local* coupling to auxiliary fermions:

$$S_{\text{site},i} = S_{\text{imp}}[c^*_i, c_i] + \sum_{\omega\alpha\beta} f^*_{\omega i\alpha}\, g_{\omega\,\alpha\beta}^{-1} c_{\omega i\beta} + c^*_{\omega i\alpha}\, g_{\omega\,\alpha\beta}^{-1} f_{\omega i\beta}.\tag{6.8}$$

Since $g_\omega$ is local, the sum over all states labeled by $\mathbf{k}$ could be replaced by the equivalent summation over all sites by a change of basis in the second term. The crucial point is that the coupling to the auxiliary fermions is purely local and $S_{\text{site}}$ decomposes into a sum of purely local terms. The lattice fermions can therefore be integrated out from $S_{\text{site}}$ for each site $i$ separately. This completes the change of variables:

$$\int \exp\left(-S_{\text{site}}[c^*_i, c_i, f^*_i, f_i]\right) \mathcal{D}[c^*_i, c_i] = Z_{\text{imp}} \exp\left(-\sum_{\omega\alpha\beta} f^*_{\omega i\alpha}\,(g_\omega^{-1})_{\alpha\beta} f_{\omega i\beta} - V_i[f^*_i, f_i]\right).\tag{6.9}$$

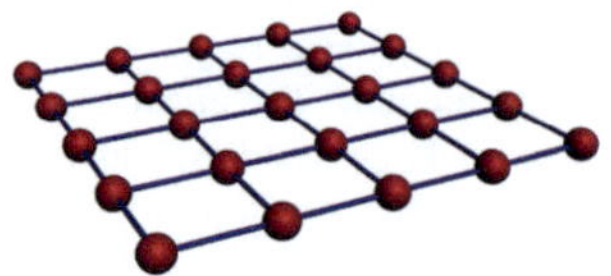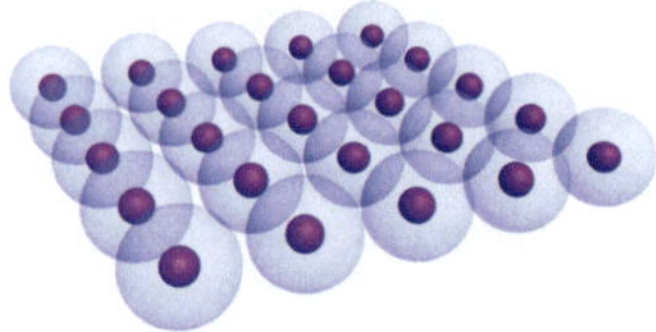

*Figure 6.1:* Construction of the dual fermion approximation: In a first step, the original lattice problem (left) with bonds (blue lines) is replaced by a collection of decoupled impurities exerted to an electronic bath, as indicated by the blue spheres (right).

The above equation may be viewed as the defining equation for the dual potential $V[f^*, f]$. The choice (6.6) ensures a particularly simple form for this potential. Expanding both sides of Eq. 6.9 and equating the resulting expressions by order, one finds that the $n$th-order term in the infinite series for the dual potential is $[(-1)^{n-1}/(n!)^2]\gamma^{(2n)}_{12\ldots2n} f_1^* f_2 \ldots f_{2n}$ for $n \geq 2$. The quantities $\gamma^{(2n)}$ are the exact fully antisymmetric, reducible vertices of the local quantum impurity problem. Formally this can be done up to all orders and in this sense the transformation to the dual fermions is exact. In practice the series has to be terminated. Up to the order which will be considered here, it is explicitly given by

$$V[f^*, f] = -\frac{1}{4}\gamma^{(4)}_{1234} f_1^* f_2 f_3^* f_4 + \frac{1}{36}\gamma^{(6)}_{123456} f_1^* f_2 f_3^* f_4 f_5^* f_6 \mp \ldots , \qquad (6.10)$$

where the combined index $1 \equiv \{\omega\alpha\}$ comprises frequency, spin and orbital degrees of freedom. The dual potential hence involves the connected impurity correlation functions at all orders. The two-particle vertex is given by

$$\gamma^{(4)}_{1234} = g^{-1}_{11'}g^{-1}_{33'} \left[ \chi^{\text{imp}}_{1'2'3'4'} - \chi^{\text{imp},0}_{1'2'3'4'} \right] g^{-1}_{2'2}g^{-1}_{4'4} , \qquad (6.11)$$

with the two-particle impurity Green function being defined as (see also Sec. 3.3.9)

$$\chi^{\text{imp}}_{1234} := \langle c_1 c_2^* c_3 c_4^* \rangle_{\text{imp}} = \frac{1}{Z_{\text{imp}}} \int c_1 c_2^* c_3 c_4^* \exp\left(-S_{\text{imp}}[c^*, c]\right) \mathcal{D}[c^*, c] . \qquad (6.12)$$

The disconnected part reads

$$\chi^{\text{imp},0}_{1234} = g_{12}g_{34} - g_{14}g_{32} . \qquad (6.13)$$

The definition of the three-particle Green function is given in appendix A.3. The Green functions can in general only be obtained numerically. For the results presented here, they have been calculated using the weak-coupling continuous-time quantum Monte

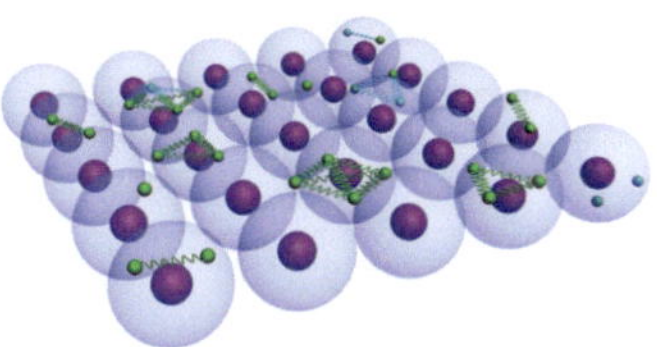

*Figure 6.2:* Illustration of the dual fermion approach. Spatial correlations in the original lattice problem are mediated through dual fermions between the impurities of Fig. 6.1. The dual fermions interact via *n*-particle interactions. Illustration by I. Labuhn.

Carlo solver as described in Secs. 3.3.8 and 3.3.9. Alternatively, they can be computed using the strong-coupling CTQMC algorithm (Sec. 3.4.2), the Lehmann representation in exact diagonalization (see Sec. 2.5 and 9.3), or by approximate methods (e.g. within iterated perturbation theory (IPT) [40]).

Note that here the vertex is written such that it describes scattering of particle-hole pairs. This is different from the usual field-theoretical definition, which is taken to be normal ordered and describes scattering of particles. The two representations are linked through the anticommutation properties of the Grassmann fields. In the latter notation the lowest order becomes

$$V[f^*, f] = +\frac{1}{4}\gamma^{(4)}_{1234} f_1^* f_2^* f_4 f_3 , \tag{6.14}$$

where now

$$
\begin{aligned}
\gamma^{(4)}_{1234} &= g_{11'}^{-1} g_{22'}^{-1} \left[ \chi^{\mathrm{imp}}_{1'2'3'4'} - \chi^{\mathrm{imp},0}_{1'2'3'4'} \right] g_{3'3}^{-1} g_{4'4}^{-1} , \\
\chi^{\mathrm{imp}}_{1234} &= \langle c_1 c_2 c_3^* c_4^* \rangle_{\mathrm{imp}} , \\
\chi^{\mathrm{imp},0}_{1234} &= g_{14} g_{23} - g_{13} g_{24} .
\end{aligned}
\tag{6.15}
$$

For the evaluation of the two-particle Green function in CTQMC it is more natural to use the definition (6.11). Furthermore, when calculating susceptibilities or corrections to the self-energy, the focus will be on particle-hole excitations. Therefore, it is more convenient to use Eqs. 6.11-6.13.

After integrating out the lattice fermions, the dual action depends on the new variables only and can be written as

$$S_{\mathrm{d}}[f^*, f] = -\sum_{\omega \mathbf{k} \alpha \beta} f^*_{\omega \mathbf{k} \alpha} [G^{\mathrm{d},0}_\omega(\mathbf{k})]^{-1}_{\alpha\beta} f_{\omega \mathbf{k} \beta} + \sum_i V[f_i^*, f_i] \tag{6.16}$$

and the bare dual Green function is found to be

$$G^{\mathrm{d},0}_\omega(\mathbf{k}) = -g_\omega \left[ g_\omega + (\Delta_\omega - h_\mathbf{k})^{-1} \right]^{-1} g_\omega , \tag{6.17}$$

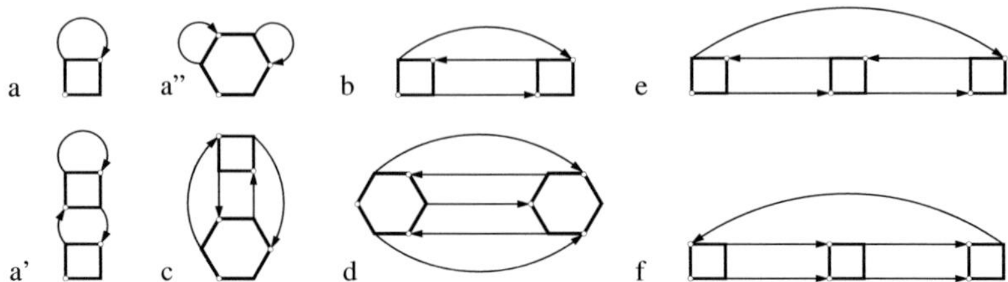

*Figure 6.3:* Diagrams contributing to the dual self-energy $\Sigma^d$. Diagrams a), a'), a") and c) give local, the other ones nonlocal contributions. The three diagrams labeled by a) do not contribute in case the condition (6.44) is fulfilled.

which involves the local single-particle Green function $g_\omega$ of the impurity problem.

Up to now (6.16) is merely a reformulation of the original problem. In practice, approximate solutions are constructed by treating the dual problem perturbatively. To this end, the perturbation series expansion and the series for the dual potential need to be terminated at some point.

Several diagrams that contribute to the dual self-energy are shown in Fig. 6.3. The diagrams are constructed from the impurity vertices and dual Green functions as lines. The first diagram (a) is purely local, while higher orders contain nonlocal contributions, e.g. diagram b). Using self-energy diagrams instead of those for the Green function allows to sum an infinite partial series for Green's function by application of the Dyson equation, in the usual way. Inserting the renormalized Green function into diagram a) includes contributions such as the one in a'). Approximations to the self-energy are constructed in terms of skeleton diagrams. The lines shown in Fig. 6.3 are therefore understood to be interacting (renormalized) dual Green functions. The use of skeleton diagrams is necessary to ensure the resulting theory to be conserving in the Baym-Kadanoff sense [61], i. e. it fulfills the basic conservation laws for energy, momentum, spin and particle number. In general, an approximation for the single-particle Green function obtained from Dyson's equation is conserving if the self-energy can be written as a functional derivative of a generating functional [143]. The Hartree-Fock and the fluctuation-exchange approximation (FLEX, see chapter 11) are such conserving approximations. For dual fermions

$$\Sigma^d = \frac{\delta \Phi^d}{\delta G^d} \, .$$
(6.18)

The first two lowest-order contributions to the dual Luttinger-Ward functional $\Phi^d[G^d; V]$ are shown in Fig. 6.4, together with the corresponding self-energy diagrams. Diagrammatically, the functional derivative with respect to Green's function corresponds to

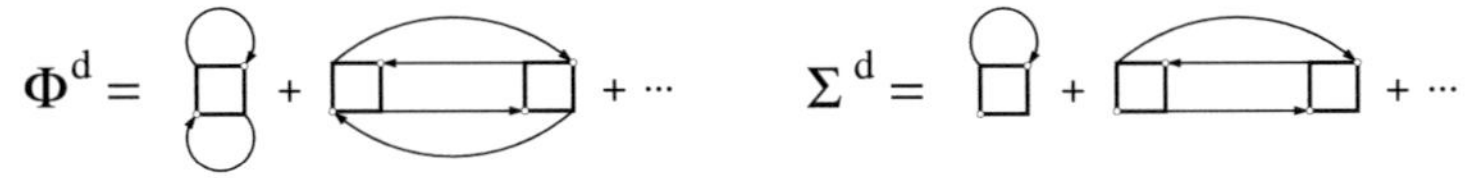

$$\Phi^{\mathrm{d}} = \quad + \quad + \cdots \qquad \Sigma^{\mathrm{d}} = \quad + \quad + \cdots$$

*Figure 6.4:* Diagrammatic representation of the dual Luttinger Ward functional $\Phi^{\mathrm{d}}[G^{\mathrm{d}}; V]$ (left) and corresponding dual self-energy obtained by functional derivative with respect to $G^{\mathrm{d}}$.

removing a line from the diagram. It is an important consequence of the exact transformation (6.5) that for a theory which is conserving in terms of dual fermions, the result is also conserving in terms of lattice fermions [41, 139]. This allows to construct general conserving approximations within the dual fermion approach. Numerically, the self-energy is obtained in terms of skeleton diagrams by performing a self-consistent renormalization as described in Sec. 6.5.

Once an approximate dual self-energy is found, the result may be transformed back to a physical result in terms of lattice fermions using exact relations (see Sec. 6.3). Due to the abstract nature of the dual variables, the truncation of the series for the dual potential and the perturbation series may appear as arbitrary and uncontrolled approximations. As will be shown in the following, these can be well justified and in fact, diagrammatic approximations to the dual self-energy may be constructed based on physical considerations much like in the usual way.

For a perturbation theory, it is important to analyze the convergence properties. In the weak-coupling limit ($U \to 0$), the vertices appear as a small parameter in the expansion, because they vanish at least proportional to $U$: $\gamma^{(4)} \sim U$, $\gamma^{(6)} \sim U^2$,... (see, e.g. the explicit expression for $\gamma^{(4)}$ in the atomic limit in chapter 9). For an expansion around the atomic limit (i.e. $\Delta \equiv 0$), the dual Green function is small near this limit: For $h_{\mathbf{k}}$ small, the bare dual Green function can be approximated as

$$G_{\omega}^{\mathrm{d}0}(\mathbf{k}) \approx g_{\omega} h_{\mathbf{k}} g_{\omega} . \tag{6.19}$$

This enforces the convergence of the series in the opposite strong coupling limit. The fact that the local frequency dependent vertex appears as the interaction in this approach allows to account for strong correlations. In contrast, IPT or FLEX, which operate with the bare interaction $U$, have to break down at intermediate to large $U$. In the general case, a fast convergence cannot be proven rigorously. In addition, the convergence in opposite regimes does not necessarily imply such a behavior for intermediate coupling. However, in chapter 11 it is demonstrated that this approach has superior convergence properties compared to standard diagrammatic techniques. The diagrammatic expansion yields reasonable results even at intermediate coupling and indeed converges in the strong coupling limit.

## 6.2  Dual Perturbation Theory

The action (6.16) allows for a standard Feynman-type diagrammatic expansion in powers of the dual potential $V$. The first few lowest-order diagrams for the dual self-energy are depicted in Fig. 6.3. The perturbation expansion is explicitly carried out in appendix A.4 and general expressions are given there. The rules are similar to those of the antisymmetrized diagrammatic technique [93]. Extension of the rules to include generic $n$-particle interaction vertices is straightforward. In summary, the rules for the perturbation theory of the dual self-energy in momentum space are the following:

- Draw all topologically distinct, connected diagrams involving any $n$-body interaction $\gamma^{(2n)}$ depicted by regular polygons with $2n$ edges or endpoints, whereof $n$ are outgoing (incoming) endpoints, where a directed line originates (terminates)

- connect the vertex endpoints with directed lines, compliant with the designation of ingoing and outgoing endpoints

- with each line associate a dual Green function $G^{\mathrm{d}}$

- Assign a frequency, momentum, orbital and spin label to each endpoint, taking into account energy- momentum- and spin-conservation at each vertex

- Sum/integrate over all internal variables

- For each tuple of $n$ equivalent lines (such as equally directed lines connecting the same two vertices), associate a factor $1/n!$

- Multiply the resulting expression by a factor $(T/N)^m S^{-1}$, where $m$ is the number of independent frequency/momentum summations and $S$ is the symmetry factor (see Sec. A.4.3)

- For diagrams containing two-particle interactions only, determine the sign according to Sec. A.4.4.

Some notes are in place. In momentum space, each endpoint of a vertex is associated with a frequency, wave vector, spin and orbital index. According to the definition of the Fourier transform (C.3), frequencies at an incoming (outgoing) endpoint of a vertex enter with a positive (negative) sign. Hence, energy conservation is associated with a $\delta$-function $\delta(\omega_1 + \omega_3 + \ldots + \omega_{2n-1} - \omega_2 - \omega_4 - \ldots - \omega_{2n})$ at each vertex. This holds for momentum and spin conservation accordingly. The external labels are kept fixed and all internal indices are summed over.

The final expression is multiplied by $(T/N)^m$, where $m$ is the number of *independent* frequency and momentum summations, i.e. the ones that remain after eliminating the $\delta$-functions. $T = \beta^{-1}$ is the temperature and $N$ the number of k-points in the first Brillouin zone. The general expressions in appendix A.4.2 are translated to final expressions in momentum space by writing momentum and frequency sums explicitly and including this factor. In real space, each vertex is associated with a coordinate $\mathbf{x}$ and the internal coordinates are summed over (without factors $1/N$). Assuming translational invariance, the diagram depends on the difference $\mathbf{x} - \mathbf{x}'$ of the external coordinates only.

Exchange of two equivalent endpoints of the vertex or exchanging vertices of the same sort in a diagram does not change its topology. This has to be taken into account via combinatorial prefactors. For a tuple of $n$ equivalent lines, the expression has to be multiplied by a factor $1/n!$. For certain diagrams, e.g. the generic ring diagrams contributing to the Luttinger Ward functional (see chapter 11), additional symmetry factors have to be taken into account. It is instructive to exemplify the role of the combinatorial prefactors using the lowest order contributions to the dual Luttinger-Ward functional and self-energy as an example. The first diagram in $\Phi^d$ in Fig. 6.4 carries a factor $1/2$ because the two lines are equivalent. Consequently, this factor cancels by the product rule when taking the functional derivative. The prefactor of the second diagram is $1/8$. It has two pairs of equivalent lines, which give a factor $1/4$. The symmetry factor of this diagram is $S = 2$, since exchanging the vertices yields the same diagram. The corresponding self-energy diagram therefore carries a factor $1/2$. The sign of a diagram which involves two-particle interaction vertices only is determined by graphically replacing the squares by a wiggly interaction line as shown in Sec. A.4.4 and counting the number $n_L$ of closed fermion loops. The sign is then given by $(-1)^{n_L}$.

Two examples follow. Writing spin indices explicitly for clarity, the self-energy correction of the Hartree-Fock-like diagram a) in Fig. 6.3 which is constant in reciprocal space and reads

$$[\Sigma^{d\,(a)}_{\omega\sigma}]_{\alpha\beta,} = -T \sum_{\omega'} \sum_{\sigma'} \gamma^{\sigma\sigma\sigma'\sigma'}_{\alpha\beta\gamma\delta}(\omega,\omega,\omega',\omega')\,[G^{d\,\mathrm{loc}}_{\omega'\sigma'}]_{\delta\gamma}\,, \tag{6.20}$$

where $G^{d\,\mathrm{loc}} = (1/N)\sum_{\mathbf{k}} G^d(\mathbf{k})$ denotes the local part of the dual Green function and summation over repeated indices is implied. This diagram is local in real space, i.e. proportional to $\delta_{\mathbf{x},\mathbf{x}'}$. The second-order contribution represented by diagram b) evaluates to

$$\begin{aligned}
[\Sigma^{d\,(b)}_{\omega\sigma}(\mathbf{k})]_{\alpha\nu} = &-\frac{1}{2}\left(\frac{T}{N}\right)^2 \sum_{\mathbf{k}_1\mathbf{k}_2}\sum_{\omega_1\omega_2}\sum_{\sigma'} \gamma^{\sigma\sigma\sigma'\sigma'}_{\alpha\beta\gamma\delta}(\omega,\omega_1,\omega_2,\omega_3)[G^d_{\omega_1\sigma}(\mathbf{k}_1)]_{\beta\mu}[G^d_{\omega_2\sigma'}(\mathbf{k}_2)]_{\lambda\gamma}\times \\
&\times [G^d_{\omega_3\sigma'}(\mathbf{k}_3)]_{\delta\kappa}\gamma^{\sigma'\sigma'\sigma\sigma}_{\kappa\lambda\mu\nu}(\omega_3,\omega_2,\omega_1,\omega) \\
&-\frac{1}{2}\left(\frac{T}{N}\right)^2 \sum_{\mathbf{k}_1\mathbf{k}_2}\sum_{\omega_1\omega_2} \gamma^{\sigma\bar\sigma\bar\sigma\sigma}_{\alpha\beta\gamma\delta}(\omega,\omega_1,\omega_2,\omega_3)[G^d_{\omega_1\bar\sigma}(\mathbf{k}_1)]_{\beta\mu}[G^d_{\omega_2\bar\sigma}(\mathbf{k}_2)]_{\lambda\gamma}\times \\
&\times [G^d_{\omega_3\sigma}(\mathbf{k}_3)]_{\delta\kappa}\gamma^{\sigma\bar\sigma\bar\sigma\sigma}_{\kappa\lambda\mu\nu}(\omega_3,\omega_2,\omega_1,\omega)\,, \tag{6.21}
\end{aligned}$$

where $\omega_3 = \omega + \omega_2 - \omega_1$ and $\mathbf{k}_3 = \mathbf{k} + \mathbf{k}_2 - \mathbf{k}_1$ by energy/momentum conservation. Here $\bar\sigma = -\sigma$ and the sum over spins runs over $\sigma' =\uparrow,\downarrow$. This expression covers all spin configurations that are compliant with spin conservation. As described in Sec. 6.6 below, it is more efficient to evaluate the lowest order diagrams in real space. As can be straightforwardly verified by Fourier transform, the same diagram in real space is given by

$$
[\Sigma_{\omega\sigma}^{\mathrm{d}\,(b)}(\mathbf{x})]_{\alpha\nu} = -\frac{1}{2}T^2 \sum_{\omega_1\omega_2} \sum_{\sigma'} \gamma_{\alpha\beta\gamma\delta}^{\sigma\sigma\sigma'\sigma'}(\omega,\omega_1,\omega_2,\omega_3)[G_{\omega_1\sigma}^{\mathrm{d}}(\mathbf{x})]_{\beta\mu}[G_{\omega_2\sigma'}^{\mathrm{d}}(-\mathbf{x})]_{\lambda\gamma}\times
$$
$$
\times [G_{\omega_3\sigma'}^{\mathrm{d}}(\mathbf{x})]_{\delta\kappa}\gamma_{\kappa\lambda\mu\nu}^{\sigma'\sigma'\sigma\sigma}(\omega_3,\omega_2,\omega_1,\omega)
$$
$$
-\frac{1}{2}T^2 \sum_{\omega_1\omega_2} \gamma_{\alpha\beta\gamma\delta}^{\sigma\bar\sigma\bar\sigma\sigma}(\omega,\omega_1,\omega_2,\omega_3)[G_{\omega_1\bar\sigma}^{\mathrm{d}}(\mathbf{x})]_{\beta\mu}[G_{\omega_2\bar\sigma}^{\mathrm{d}}(-\mathbf{x})]_{\lambda\gamma}\times
$$
$$
\times [G_{\omega_3\sigma}^{\mathrm{d}}(\mathbf{x})]_{\delta\kappa}\gamma_{\kappa\lambda\mu\nu}^{\sigma\bar\sigma\bar\sigma\sigma}(\omega_3,\omega_2,\omega_1,\omega)\,. \tag{6.22}
$$

## 6.3   Exact Relations

After an approximate result for the dual self-energy or the dual Green function is obtained, it has to be transformed back to the corresponding quantities for the original fermions, in terms of which a solution is being sought. For the latter the designation lattice fermions is used in the following. The fact that the dual fermions are introduced through the exact HST (6.5) allows to establish exact identities between dual and lattice quantities. This is important, because the transformation in this case does not involve any additional approximations.

### 6.3.1   Single-Particle Quantities

The dual and lattice Green functions are defined via the imaginary time path integral as

$$
[G_\omega(\mathbf{k})]_{\alpha\beta} = -\frac{1}{\mathcal{Z}} \int c_{\omega\mathbf{k}\alpha} c_{\omega\mathbf{k}\beta}^* \exp\left(-S[c^*,c]\right) \mathcal{D}[c^*,c]\,,
$$
$$
[G_\omega^{\mathrm{d}}(\mathbf{k})]_{\alpha\beta} = -\frac{1}{\mathcal{Z}_{\mathrm{d}}} \int f_{\omega\mathbf{k}\alpha} f_{\omega\mathbf{k}\beta}^* \exp\left(-S_{\mathrm{d}}[f^*,f]\right) \mathcal{D}[f^*,f]\,. \tag{6.23}
$$

The lattice Green function is obtained by a suitable functional derivative of the partition function (6.4) with respect to a source, where the bare dispersion $h_{\mathbf{k}}$ in (6.1) is employed as the source field. As detailed in appendix A.6, the identity between dual and lattice Green functions is established by applying the same functional derivative to the partition function in the equivalent form

$$
\mathcal{Z} = \mathcal{Z}_f \int\int \exp\left(-S[c^*,c;f^*,f]\right) \mathcal{D}[c^*,c;f^*,f]\,, \tag{6.24}
$$

where

$$\mathcal{Z}_f = \prod_{\omega \mathbf{k}} \det\left[g_\omega \left(\Delta_\omega - h_\mathbf{k}\right) g_\omega\right] . \tag{6.25}$$

Note that in contrast to the calculation of expectation values, the determinant *is* relevant in this context. The resulting identity reads

$$G_\omega(\mathbf{k}) = (\Delta_\omega - h_\mathbf{k})^{-1} + \left[g_\omega (\Delta_\omega - h_\mathbf{k})\right]^{-1} G_\omega^\mathrm{d}(\mathbf{k}) \left[(\Delta_\omega - h_\mathbf{k}) g_\omega\right]^{-1}$$
$$= (\Delta_\omega - h_\mathbf{k})^{-1} + L_\omega(\mathbf{k}) \, G_\omega^\mathrm{d}(\mathbf{k}) \, R_\omega(\mathbf{k}) , \tag{6.26}$$

where for later purposes the matrix valued functions

$$L_\omega(\mathbf{k}) := (\Delta_\omega - h_\mathbf{k})^{-1} g_\omega^{-1} , \qquad R_\omega(\mathbf{k}) := g_\omega^{-1} (\Delta_\omega - h_\mathbf{k})^{-1} . \tag{6.27}$$

have been defined. These are identical for a diagonal basis. Further exact relations can be established between the self-energies of dual and lattice fermions. To this end, it is useful to define the (in general k-dependent) correction to the impurity self-energy as

$$\Sigma_\omega'(\mathbf{k}) := \Sigma_\omega(\mathbf{k}) - \Sigma_\omega^\mathrm{imp} . \tag{6.28}$$

As shown in appendix A.2, a simple $T$-matrix-like relation holds between this quantity and the dual self-energy:

$$\Sigma_\omega'(\mathbf{k}) = \left[\mathbb{1} + \Sigma_\omega^\mathrm{d}(\mathbf{k}) g_\omega\right]^{-1} \Sigma_\omega^\mathrm{d}(\mathbf{k}) . \tag{6.29}$$

In terms of this quantity, the lattice Green function can generally be expressed as

$$G_\omega(\mathbf{k}) = \left[g_\omega^{-1} - \Sigma_\omega'(\mathbf{k}) + \Delta_\omega - h_\mathbf{k}\right]^{-1} . \tag{6.30}$$

The dual Green function is related to the self-energy correction through

$$G_\omega^\mathrm{d}(\mathbf{k}) = -g_\omega \left\{[\Delta_\omega - h_\mathbf{k}]^{-1} + \left[g_\omega^{-1} - \Sigma_\omega'(\mathbf{k})\right]^{-1}\right\}^{-1} g_\omega . \tag{6.31}$$

It reduces to the bare dual Green function (6.17) for $\Sigma' \equiv 0$ as expected, because this is equivalent to $\Sigma^\mathrm{d} \equiv 0$ according to (6.29).

## 6.3.2 Higher-Order Cumulants

In analogy to the procedure outlined in the previous section, an exact relation can be obtained between two-particle Green functions, which are defined as

$$\chi_{1234} := \langle c_1 c_2^* c_3 c_4^* \rangle = \frac{1}{\mathcal{Z}} \int c_1 c_2^* c_3 c_4^* \exp\left(-S[c^*, c]\right) \mathcal{D}[c^*, c] ,$$
$$\chi_{1234}^\mathrm{d} := \langle f_1 f_2^* f_3 f_4^* \rangle_\mathrm{d} = \frac{1}{\mathcal{Z}_\mathrm{d}} \int f_1 f_2^* f_3 f_4^* \exp\left(-S_\mathrm{d}[f^*, f]\right) \mathcal{D}[f^*, f] . \tag{6.32}$$

This relation is derived in detail in appendix A.7. When written in terms of the nontrivial (connected) part of the two-particle Green function, instead of the two-particle Green function itself, it takes a particularly simple form:

$$\chi_{1234} - [G \otimes G]_{1234} = L_{11'} L_{33'} \left[ \chi^{\mathrm{d}} - G^{\mathrm{d}} \otimes G^{\mathrm{d}} \right]_{1'2'3'4'} R_{2'2} R_{4'4} , \tag{6.33}$$

where the antisymmetrized matrix product is defined as

$$[G \otimes G]_{1234} := G_{12} G_{34} - G_{14} G_{32} . \tag{6.34}$$

Since the functions $L$ and $R$ only contain single-particle quantities, one may conclude that two-particle excitations of dual and lattice fermions are the same. This fact is of paramount importance for the construction of practical approximations, as will become apparent in subsequent chapters.

The above relation only involves the connected part of the two-particle Green functions. It therefore retains a simple character when translated into a relation between vertex functions. Using the relation between the (dual) two-particle vertex and the (dual) two-particle Green function,

$$\left[ \chi^{(\mathrm{d})} - G^{(\mathrm{d})} \otimes G^{(\mathrm{d})} \right]_{1234} = G^{(\mathrm{d})}_{11'} G^{(\mathrm{d})}_{33'} \Gamma^{(\mathrm{d})}_{1'2'3'4'} G^{(\mathrm{d})}_{2'2} G^{(\mathrm{d})}_{4'4} , \tag{6.35}$$

one is lead to define the functions

$$L'_{12} := \left[ G^{-1} L \, G^{\mathrm{d}} \right]_{12} = -[\mathbb{1} + \Sigma^{\mathrm{d}} g]^{-1}_{12} , \qquad R'_{12} := \left[ G^{\mathrm{d}} R \, G^{-1} \right]_{12} = -[\mathbb{1} + g \Sigma^{\mathrm{d}}]^{-1}_{12} , \tag{6.36}$$

which allow to express the relation between vertices in the form

$$\Gamma_{1234} = L'_{11'} L'_{33'} \Gamma^{\mathrm{d}}_{1'2'3'4'} R'_{2'2} R'_{4'4} . \tag{6.37}$$

Note that this transformation is the identity for $\Sigma^{\mathrm{d}} \equiv 0$. In view of this simple form, one may anticipate similar relations for higher-order vertices. Indeed, such relations have been established in Ref. [139]. Here a different derivation is given. The fact that the above relation only involves the connected parts of the two-particle Green functions suggests to employ the cumulant (linked cluster) expansion for this purpose. This is an elegant way to recover the identities (6.26) and (6.33). The general relation between the $n$-particle vertices of dual and lattice fermions is established by means of different, but equivalent representations of the following functional:

$$F[J^*, J; K^*, K] := \ln \mathcal{Z}_f \int \exp\left( -S[c^*, c; f^*, f] + J_1^* c_1 + c_2^* J_2 + K_1^* f_1 + f_2^* K_2 \right) \mathcal{D}[c^*, c; f^*, f]. \tag{6.38}$$

Integrating out the lattice fermions from this functional in analogy to the procedure in Sec. A.3 (this can be done with the sources $J$ and $J^*$ set to zero) yields

$$F[K^*, K] = \ln \tilde{\mathcal{Z}}_f \int \exp\left( -S_{\mathrm{d}}[f^*, f] + K_1^* f_1 + f_2^* K_2 \right) \mathcal{D}[f^*, f]. \tag{6.39}$$

with $\tilde{\mathcal{Z}}_f = \mathcal{Z}/\mathcal{Z}_d$ (see appendix A.2). The dual Green function and higher-order cumulants are obtained from (6.39) by suitable functional derivatives, for example

$$
G^d_{12} = -\langle f_1 f_2^* \rangle = -\frac{\delta^2 F}{\delta K_2 \delta K_1^*}\bigg|_{K^*=K=0} \quad , \quad \left[ \chi^d - G^d \otimes G^d \right]_{1234} = \frac{\delta^4 F}{\delta K_4 \delta K_3^* \delta K_2 \delta K_1^*}\bigg|_{K^*=K=0} .
$$

$$(6.40)$$

Conversely, integrating out the dual fermions from (6.38) using the HST, one obtains an alternative representation, which more clearly reveals a connection of the functional derivatives with respect to the sources $J, J^*$ and $K, K^*$. The result is

$$
F[J^*, J; K^*, K] = K_1^*[g(\Delta - h)g]_{12}K_2 + \ln \int \exp\Big( - S[c^*, c] + J_1^* c_1 + c_2^* J_2 +
$$
$$
+ K_1^*[g(\Delta - h)]_{12}c_2 + c_1^*[(\Delta - h)g]_{12}K_2 \Big) \mathcal{D}[c^*, c].
$$

$$(6.41)$$

The cumulants in terms of lattice fermions are obviously obtained by functional derivative with respect to the sources $J$ and $J^*$, in analogy to (6.40). On the other hand, applying the derivatives with respect to $K, K^*$ to (6.41) and (6.39) yields the following identity linking the single-particle Green functions:

$$
G^d_{12} = -[g(\Delta - h)g]_{12} + [g(\Delta - h)]_{11'}G_{1'2'}[(\Delta - h)g]_{2'2}. \tag{6.42}
$$

Solving for $G$ immediately recovers (6.26). For higher-order cumulants the additive term in (6.41) does not contribute and the relation between the fourth-order cumulants evaluates to

$$
\left[ \chi^d - G^d \otimes G^d \right]_{1234} =
$$
$$
[g(\Delta - h)]_{11'} [g(\Delta - h)]_{33'} [\chi - G \otimes G]_{1'2'3'4'} [(\Delta - h)g]_{2'2} [(\Delta - h)g]_{4'4} , \tag{6.43}
$$

which is obviously equivalent to (6.33). From the structure of the functional it is apparent that similar relations hold for higher-order cumulants. The cumulant of order $2n$ in terms of lattice fermions is obtained from the dual cumulant by multiplying with $n$ factors $L$ $(R)$ (6.27) from left (right) and factors $L'$ $(R')$ (6.36) for the corresponding vertex functions, respectively. Hence one may conclude that $n$-particle collective excitations (and corresponding instabilities) are the same for dual and lattice fermions.

## 6.4   Self-Consistency Condition and Relation to DMFT

The hybridization function $\Delta$, which has not been specified so far, allows to optimize the starting point of the perturbation theory and should be chosen in an optimal way. The condition of the first diagram (Fig. 6.3 a) in the expansion of the dual self-energy to be equal to zero at all frequencies fixes the hybridization. The reasoning behind this, on the one hand, is to eliminate the leading order diagrammatic correction to the self-energy, which one may expect to be dominant. This is possible because this correction is purely local and $\Delta$ is a local quantity.

Since $\gamma^{(4)}$ is local, this condition amounts to demanding that the local part of the dual Green function be zero: $\sum_{\mathbf{k}} G_\omega^{\mathrm{d}}(\mathbf{k}) = 0$. The simplest nontrivial approximation is obtained by taking the leading-order correction, diagram a), evaluated with the bare dual propagator (6.17). Using that also $g$ is local, the condition in this case reduces to

$$\frac{1}{N} \sum_{\mathbf{k}} G_\omega^{\mathrm{d}0} = 0 \qquad \text{or} \qquad \frac{1}{N} \sum_{\mathbf{k}} \left[ g_\omega + (\Delta_\omega - h_{\mathbf{k}})^{-1} \right]^{-1} = 0 \,. \tag{6.44}$$

As shown in appendix (A.2), a simple relation holds between the bare dual Green function and the DMFT Green function, i.e. $G_0^{\mathrm{d}} = G^{\mathrm{DMFT}} - g$. It immediately follows that (6.44) is exactly equivalent to the self-consistency condition of DMFT (cf. Sec. 2.6.2):

$$g_\omega = \frac{1}{N} \sum_{\mathbf{k}} G_\omega^{\mathrm{DMFT}}(\mathbf{k}). \tag{6.45}$$

Hence DMFT appears as the lowest-order approximation in the dual fermion approach. The fact that a relation to DMFT can be established is not surprising, since the approach is based on an arbitrary impurity problem. Hence it must be possible to impose a condition on the hybridization function which recovers the DMFT result. It should be emphasized that the statement of DMFT appearing as the lowest-order approximation holds irrespective of the spatial dimension (it is in fact a local approximation). A formal relation to DMFT can be established using the Feynman variational functional approach. In this context, DMFT appears as the optimal approximation to a Gaussian ensemble of dual fermions [139].

The hybridization function is not known a priori and needs to be found self-consistently. Therefore, a formula for an iterative update of the hybridization function is needed for practical calculations. In appendix A.2, it is shown that the rule

$$\Delta_\omega^{\mathrm{new}} = \Delta_\omega^{\mathrm{old}} + \xi \left[ g_\omega^{-1} G_\omega^{\mathrm{d,\,loc}} (G_\omega^{\mathrm{loc}})^{-1} \right] \tag{6.46}$$

is exactly equivalent to the self-consistency condition (2.65) for DMFT calculations. The lattice Green function is obtained from the dual Green function via the identity (6.26). The same update of the hybridization function may conveniently be used in dual fermion calculations, as self-consistency is reached when $G^{\mathrm{d}}_{\mathrm{loc}} \equiv 0$. Note that a relation analogous to (6.45) does not hold as soon as diagrams are taken into account. This implies that in general the local part of the lattice Green function is no longer determined from the local Green function of the impurity problem. Note further that the condition (6.44) eliminates diagrams such as a), a') and a") of Fig. 6.3, i.e. all diagrams with a local, closed loop. Eliminating such diagrams hence corresponds to the summation of an infinite partial series.

The fact that DMFT is the lowest-order approximation allows to perform an expansion around the dynamical mean-field theory and to systematically construct diagrammatic corrections to DMFT. Spatial correlations beyond DMFT are included through nonlocal dual self-energy diagrams. This is only the case as long as the hybridization function of the underlying impurity problem is equal to its DMFT value, $\Delta = \Delta^{\mathrm{DMFT}}$. The hybridization function is not restricted to its DMFT value through the self-consistency condition (6.46). As soon as nonlocal diagrams are taken into account beyond DMFT with $\Delta = \Delta^{\mathrm{DMFT}}$, the condition (6.44) will in general be violated. Note, for example, that the self-energy correction of diagram b) in general gives a local correction to the dual Green function even when the local part of $G^{\mathrm{d}}$ is exactly zero. The condition is reinforced by changing the hybridization function.

Once the DMFT hybridization is modified, the approach should be viewed as a perturbation theory around an optimal reference problem rather than around DMFT. The hybridization then plays the role of a free optimization parameter. It turns out that the self-consistency condition remains a sensible choice. It allows a feedback of spatial correlations onto the local degrees of freedom. Physically this means that spatial correlations lead to a modification of the local environment. Adjusting the hybridization function causes a renormalization of the local interaction and improves the starting point of the perturbation theory. In general, this property is crucial in situations where DMFT delivers a poor description of the local environment and therefore is particularly important for low-dimensional systems. As will be demonstrated in the following, this feature represents a considerable advantage compared to alternative diagrammatic expansions around DMFT, such as the dynamical vertex approximation [39] and related approaches [40].

If the hybridization is viewed as an auxiliary field, which is used to optimize the approach, there is no need to optimize it over the entire function space. One may restrict the optimization by projection onto the subspace spanned by the functions $\sum_{k\gamma} V_{k\alpha\gamma} V^{*}_{k\beta\gamma}/(i\omega - \epsilon_{k\gamma})$, where $k$ runs over the bath states. The impurity problem is again viewed as a reference problem which may be solved efficiently using exact diagonalization with a small number of bath parameters. This is the basis for an efficient approximation for quantum lattice models proposed in Sec. 11.6. Choosing the reference system to be the atomic limit (i.e. $\Delta_{\omega} \equiv 0$) recovers the strong coupling expansion of Refs. [29, 30].

The condition that the local part of the dual Green function be zero implies that the imaginary part of the dual Green function on the real frequency axis cannot always be positive. This, however, does not imply that causality is violated (a theory is causal if the retarded function is zero for negative time-differences). Note that this condition is also fulfilled in DMFT, which is a manifestly causal theory. Until now, no rigorous proof of the causality of the dual fermion approach has been given [139], but no apparent causality violations have appeared in the calculations so far. The above mentioned property of the dual Green function shows that it is a purely auxiliary quantity which has no physical interpretation.

## 6.5　Calculation Procedure

The dual fermion calculation procedure is depicted in Fig. 6.5. In a DMFT calculation, only the bare dual Green function is evaluated. The difference between a DMFT and a dual fermion (DF) calculation is that in the latter diagrammatic corrections are included into the dual Green function. Therefore an additional loop for the self-consistent renormalization of the dual Green function enters (inner loop). The self-consistency condition for DMFT is identical to the one for dual fermion calculations, so that both calculations share the outer loop for the adjustment of the hybridization function.

Independent dual fermion calculations are usually started with a DMFT calculation, where the DMFT solution serves as an initial guess. This allows to systematically study the modification of the DMFT solution through nonlocal corrections. For the DMFT part of the calculation, the self-consistency loop is initialized with an initial guess for the hybridization function $\Delta$. A generic choice is to take the hybridization function which corresponds to the noninteracting case. One may straightforwardly verify that the self-consistency condition (6.46) for $g_\omega = (i\omega + \mu)\mathbb{1} - t_{\text{loc}}$, $\Delta_\omega^{\text{old}} \equiv 0$ and $\xi = 1$ generates the initial guess (2.66). The local Green function $g$ is subsequently calculated in the impurity solver (IS) step. From this, the bare dual Green function (6.17) is evaluated and used to construct a new guess for the hybridization function according to (6.46). Note that this self-consistency loop is exactly equivalent to the one depicted in Fig. 2.3.

After a self-consistent solution of the DMFT equations has been obtained, the vertex $\gamma^{(4)}$ (and possibly $\gamma^{(6)}$) is additionally computed in the impurity solver step. After constructing the bare dual Green function, the inner loop is entered and an approximation to the self-energy is obtained by summing up a certain set of diagrams. Note that the local diagram a) in Fig. 6.3 is always evaluated, because its contribution is nonzero as long as the hybridization function has not been determined self-consistently. Using Dyson's equation, the renormalized dual Green function is calculated, which is subsequently used in the diagrams. This inner loop is exited when self-consistency is reached within some predetermined accuracy $\epsilon$. The loop is very stable and can be executed automatically. As an abort criterion one can use the distance function

$$d = \| G_{\text{new}}^{\text{d}} - G_{\text{old}}^{\text{d}} \| := [1/(N\,N_\omega^{\text{max}})] \sum_{n=0}^{N_\omega^{\text{max}}-1} \sum_{\mathbf{k}} \sum_{\alpha\beta} |G_{\alpha\beta}^{\text{d}}(\mathbf{k}, i\omega_n)_{\text{new}} - G_{\alpha\beta}^{\text{d}}(\mathbf{k}, i\omega_n)_{\text{old}}| < \epsilon.$$

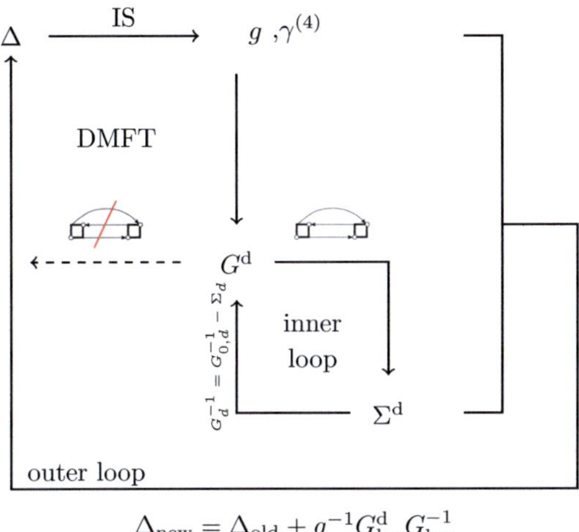



$$\Delta_{\mathrm{new}} = \Delta_{\mathrm{old}} + g^{-1} G_{\mathrm{loc}}^{\mathrm{d}} G_{\mathrm{loc}}^{-1}$$

*Figure 6.5:* Illustration of the dual fermion calculation procedure. For DMFT calculations no diagrammatic corrections are evaluated and the new guess for the hybridization function is constructed from the bare dual Green function (indicated by the dashed line). In dual fermion calculations, the interacting dual Green function enters the self-consistency condition of the outer loop. The only difference is therefore the self-consistent renormalization of the dual Green function in the inner loop.

After a few iterations, $d$ decreases exponentially. The loop converges within typically 20–30 iterations down to an accuracy of $10^{-10}$. Near phase transitions, the distance function passes over more slowly into the exponential decay and the number of iterations to achieve a given accuracy increases. The self-consistent renormalization is computationally a rather fast operation. For standard dual fermion calculations with diagrams a) and b) where the self-energy is evaluated using fast Fourier transforms (see below), the time spent for the inner loop is only a few percent of the total computation time. The evaluation of the vertex function within CTQMC strongly dominates the overall computational effort.

The renormalized dual Green function is used in the self-consistency condition to enter the next iteration in the outer self-consistency loop. Such calculations are referred to as fully self-consistent in the following. Once a self-consistent hybridization function is found, additional quantities, such as the lattice self-energy, are evaluated. The lattice Green function is always calculated, as it enters the self-consistency condition. This is not strictly necessary, but also does not constitute a serious computational burden. The convergence of the outer loop has similar practical convergence as is found for DMFT calculations. In cases where DMFT does not provide an adequate initial guess, it is advantageous to use the hybridization function from a converged dual fermion calculation with similar values of the model parameters instead, to enhance convergence. This is particularly important near phase transitions, where the convergence is strongly decelerated (see Sec. 6.8).

## 6.6 Implementation Notes

The implementation of the dual fermion approach was developed as part of this work. This section covers some technical details regarding the efficient implementation of the actual equations. The general program layout is described in chapter 7.

The implementation is independent of the impurity solver, so that the latter can be replaced by a different method. The self-consistent renormalization is performed within the dual fermion program, which is used to evaluate the self-consistency condition for both DMFT and DF calculations. The outer loop is controlled using Bash scripts. From an approximation to the dual self-energy $\Sigma_\omega^d(\mathbf{k})$, the renormalized dual Green function can be obtained directly from the Dyson equation, or through Eqs. 6.29 and 6.31 via the self-energy correction $\Sigma'$. Some care must be taken to implement all formulas in a form which avoids numerical divergence or instabilities. For example, instead of using the hybridization update formula $\Delta_\omega^{\text{new}} = \Delta_\omega^{\text{old}} + \xi g_\omega^{-1}[(G_\omega^{d,\,\text{loc}})^{-1} + g_\omega^{-1}]^{-1} g_\omega^{-1}$, one should use the equivalent form (6.46).

The self-energy and Green functions are evaluated on a lattice with periodic boundary conditions, i.e. for a discrete set of k-points. In reciprocal space, the evaluation of the second-order contribution to the dual self-energy (6.21) requires the evaluation of two internal momentum sums for each k-point, so that the computational effort scales as $N^3$, where $N = L^d$ with $d$ being the spatial dimension and $L$ the linear dimension of the lattice. This becomes computationally prohibitive already for relatively small lattice sizes. It is much more efficient to evaluate the diagram in real space according to (6.22). The dual Green function is transformed to real space using a discrete multidimensional fast Fourier transform (FFT), and the inverse transform is used for the self-energy. The FFT provided by Intel's Math Kernel Library is employed here [144]. The use of the FFT reduces the computational effort from $O(N^3)$ down to $O(N \log N)$ operations. Lattices up to $256 \times 256$ can thus be readily handled, even though internal frequency, spin and orbital summations have to be performed. The maximum lattice size is basically restricted only by memory requirements to store the dual Green function and self-energy.

Diagram e) requires $O(N^4)$ operations in reciprocal space and $O(N^2)$ operations in real space due to an internal summation over coordinates. The computational effort can still be reduced down to $O(N \log N)$ operations by subsequent Fourier transforms. Omitting frequency, spin and orbital sums for simplicity, the k-dependent part of this diagram has the structure

$$\Sigma(\mathbf{k}) \sim \frac{1}{N^3} \sum_{\mathbf{k}'\mathbf{k}''\mathbf{q}} [G^{\mathrm{d}}(\mathbf{k}')G^{\mathrm{d}}(\mathbf{k}' + \mathbf{q})][G^{\mathrm{d}}(\mathbf{k}'')G^{\mathrm{d}}(\mathbf{k}'' + \mathbf{q})]G^{\mathrm{d}}(\mathbf{k} + \mathbf{q}) \,. \qquad (6.47)$$

The two convolutions $F[\mathbf{q}] := (1/N) \sum_{\mathbf{k}} G^{\mathrm{d}}(\mathbf{k})G^{\mathrm{d}}(\mathbf{k}+\mathbf{q})$ can be performed independently using so-called fast convolution, i.e by evaluating $F(\mathbf{q})$ as the FFT of $G^{\mathrm{d}}(\mathbf{r})G^{\mathrm{d}}(-\mathbf{r})$. The remaining convolution $(1/N) \sum_{\mathbf{q}} F^2(\mathbf{q})G^{\mathrm{d}}(\mathbf{k} + \mathbf{q})$ is evaluated by fast convolution of $F^2(\mathbf{r})G^{\mathrm{d}}(-\mathbf{r})$. The convolutions have to be stored and transformed for all frequencies and orbital indices. For higher-order diagrams this becomes impractical and a different scheme is used as described in chapter 11.

For single-particle quantities, a frequency cutoff $N_\omega$ is introduced. Since this cutoff does not dominate the computational effort, it can be chosen large, for most cases at least such that $\omega_{N_\omega} \sim 5W$ ($W$ is the bandwidth). The decisive factor is the calculation of the vertex, for which a separate, smaller cutoff is introduced. The diagrammatic corrections are evaluated up to this cutoff. For a too small cutoff, a kink appears in the single-particle quantities at the cutoff frequency, best visible in the self-energy. Increasing the cutoff beyond the minimal value at which the kink disappears, the results are found to be essentially independent of the cutoff. It has to be increased with increasing $U$ and decreasing temperature. For most calculations shown here, a cutoff of $\sim 2W$ has proven sufficient. For low temperatures, an interpolation scheme for functions of Matsubara frequencies may be used [145].

## 6.7 Numerical Results

In order to illustrate the method, several numerical results are shown in this section for the Hubbard model. The model is governed by the Hamiltonian

$$H = \sum_{ij\sigma} t_{ij}c_{i\sigma}^\dagger c_{j\sigma} + U \sum_i n_{i\uparrow}n_{i\downarrow} \,. \qquad (6.48)$$

Unless otherwise stated, the two lowest-order diagrams a) and b) of Fig. 6.3 have been used in all calculations. To begin with, consider the one-dimensional (1D) model. It may be considered as a benchmark system for the approach, because the importance of nonlocal correlations is expected to increase by reducing the dimensionality. This is clearly an unfavorable situation for DMFT, which completely neglects spatial correlations. Analytical results which are available via the Bethe Ansatz and the possibility to obtain numerically accurate results in Density Matrix Renormalization Group (DMRG) calculations [146] allows to assess the quality of approximate solutions.

Fig. 6.6 shows results for the one-dimensional Hubbard model at half filling and nearest-neighbor hopping, $t_{ij} = t$ for $j = i \pm 1$ and $t_{ij} = 0$ else. The corresponding bare

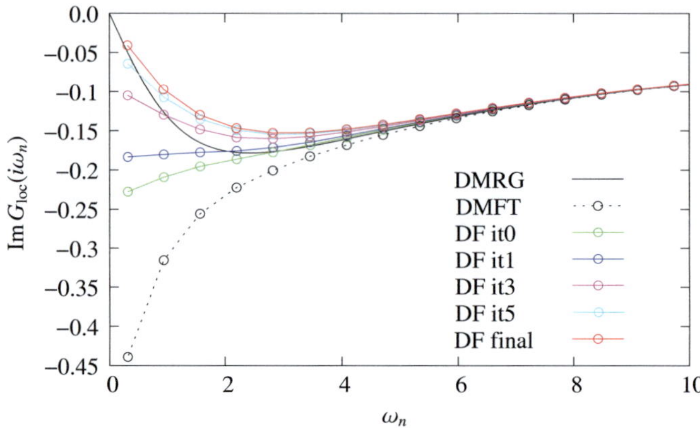

*Figure 6.6:* Imaginary part of the local Green function obtained within a dual fermion calculation from successive outer loop iterations in comparison with DMFT and a result from a zero-temperature DMRG calculation. The data labeled 'it0' corresponds to a diagrammatic expansion around DMFT. Parameters are $U/t = 6$, $T/t = 0.1$.

dispersion is $h_\mathbf{k} = -2t \cos k$ (the lattice constant is taken to be unity, $a = 1$). DMFT fails to describe the physics of the model even qualitatively, as it predicts a metallic phase. An insulating solution within DMFT is obtained only for very large values of the Coulomb repulsion, although the model is known to be an insulator for any finite value of $U$. The first correction arising from a self-consistent renormalization of the dual Green function with diagrams a) and b) is significant, but cannot recover insulating behavior. However, as the hybridization function is adjusted successively to reinforce the condition (6.44), the Green function becomes closer to the DMRG result and the system becomes insulating. This illustrates the importance of the self-consistency condition and the modification of the dynamical field. It is particularly important for low-dimensional systems, where DMFT gives a rather poor description of the local environment (e.g. whether it is metallic or insulating) and an unfavorable starting point of the perturbation theory. Although the final Green function qualitatively accounts for the insulating state, it still deviates significantly from the DMRG result. In particular, the width of the gap is underestimated (see also chapter 8). Higher-order corrections are important in this case. Although DMRG is a zero-temperature method, the role of temperature is negligible here and not the origin of the discrepancy. A calculation for $T/t = 0.05$ gave essentially the same results. However, additionally including the diagrams c–f of Fig. 6.3, does not appreciably improve the result. While the bare dual Green function extends over

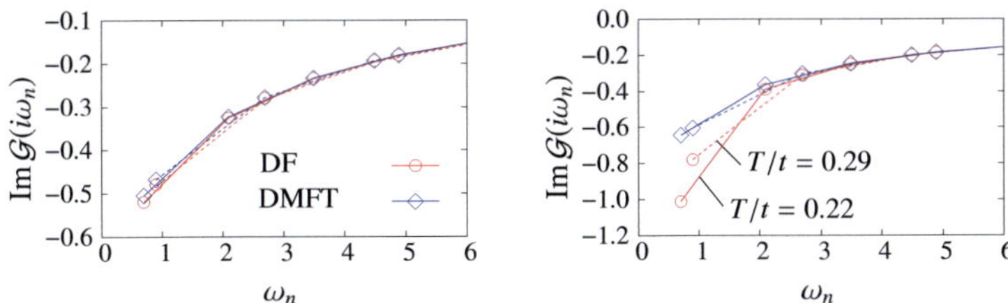

*Figure 6.7:* Imaginary part of the DMFT Weiss field $\mathcal{G}(i\omega) = [i\omega + \mu - \Delta(i\omega)]^{-1}$ and its modification after a fully self-consistent dual fermion calculation, for $U/t = 4$ (left) and $U/t = 8$. Dashed lines correspond to inverse temperature $\beta t = 3.5$ and solid lines to $\beta t = 4.5$.

several lattice sites, the fully renormalized Green function of the converged calculation is found to decay rapidly with distance. In such a case, the self-energy diagram b) is seen to include only short-range corrections to the self-energy.

The ladder approximation introduced in chapter 11 is expected to include long-range correlations, but it is not applicable in this particular case. An eigenvalue analysis (see Sec. 11.3) reveals that the ladder is divergent in various channels. These instabilities are present in the DF and DMFT calculations and reflect the tendency of the one-dimensional system towards spin- or charge-density wave order: excitations with a given momentum have a well defined energy. As a result, the approximation cannot be significantly improved diagrammatically. Better agreement for this system can be achieved by instead improving the reference system. This leads to the cluster dual fermion approach (see chapter 8).

The relevance of the modification of the DMFT hybridization function is further illustrated by results for the 2D Hubbard model. In 2D, the bare dispersion is given by $h_{\mathbf{k}} = -2t(\cos k_x + \cos k_y)$, which has bandwidth $W = 8t$. Fig. 6.7 compares the corresponding Weiss fields $\mathcal{G}(i\omega) = [i\omega + \mu - \Delta(i\omega)]^{-1}$ after self-consistent DMFT and DF calculations for the half-filled model. For $U$ well below the bandwidth, $U = W/2$, there is only a small change of $\mathcal{G}(i\omega)$ in the DF calculation. The Weiss field of the DMFT calculation hardly depends on temperature. It remains independent of temperature for $U = W$. However, there is a strong, temperature dependent modification of the DMFT dynamical field in the DF calculation at these parameters. The origin of this temperature dependence is due to short-range antiferromagnetic correlations, which are absent in DMFT and which become stronger as the temperature is lowered.

From the modification of the Weiss field, one may expect that also spectral properties are significantly affected. Fig 6.8 shows the k-resolved spectral function $A(\mathbf{k}, \omega)$ for $U = W$ and $T/t = 0.235$, as obtained from maximum-entropy analytical continuation of the k-resolved imaginary time Green function. The spectral function is accessible to experimental observation via angular resolved photoemission spectroscopy. The results of the two theories are indeed qualitatively different. The DMFT spectral function displays a quasiparticle band, while in the DF calculation, spectral weight is transferred away from the Fermi level. Recalling the nesting condition $\epsilon_{\mathbf{k}+\mathbf{Q}} = -\epsilon_{\mathbf{k}}$ for the antiferromagnetic wave vector $\mathbf{Q} = (\pi, \pi)$, the locus of these features allows to interpret them as shadow bands due to dynamical short-range antiferromagnetic correlations. The strength of these correlations increases as the temperature is lowered.

As a final illustration, consider the case of finite doping. Fig. 6.9 shows results for the spectral function $A(\mathbf{k}, \omega = 0)$ at the Fermi level of the hole-doped system for $\delta = 7\%$. The chemical potential was adjusted iteratively during the outer-loop iterations until the desired occupancy was reached, similarly to the procedure in DMFT. In general, the chemical potential within dual fermion calculations $\mu_{\mathrm{DF}}$ for a given occupancy can differ substantially from its DMFT value $\mu_{\mathrm{DMFT}}$. Here next-nearest-neighbor hopping $t'/t = -0.3$ relevant to the real experimental situation in the cuprates is included (the asymmetry of the hole- and electron-doped materials can be modeled through this parameter). The spectral function was obtained by polynomial extrapolation of the lattice Green function to zero frequency using a polynomial of degree 5. The error of the extrapolation is large ($\sim 10\%$), but the results were found to be qualitatively unchanged using polynomials of varying degree. One can see that in DMFT, the spectral function is flat everywhere along the Fermi surface, as required for a constant (k-independent) self-energy. By contrast, including nonlocal corrections to the self-energy, the spectral function shows the $k$-selective destruction of quasiparticles in the antinodal region (in the proximity of the $X$-point) and the formation of

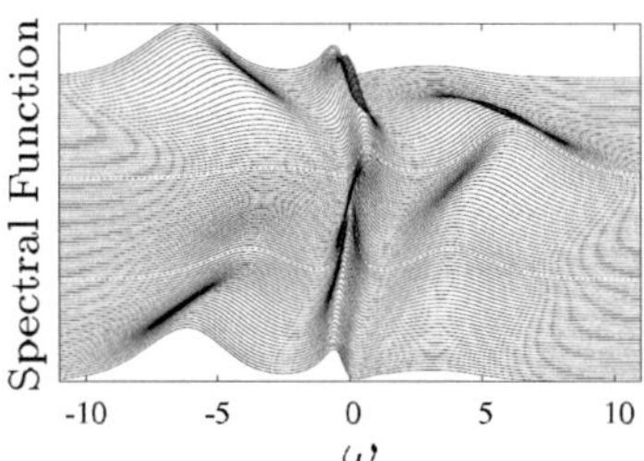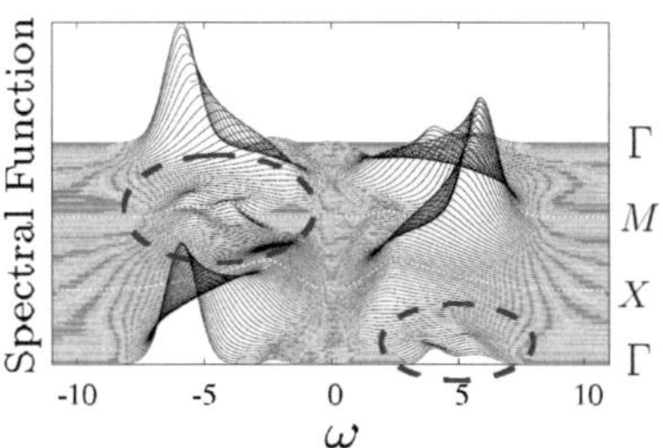

*Figure 6.8:* Spectral function $A(\mathbf{k}, \omega)$ for the 2D Hubbard model at half-filling obtained within DMFT (left) and dual fermion calculations (right) for $U = 8t$ and $T/t = 0.235$. From bottom to top, the curves are plotted along the high-symmetry lines $\Gamma \rightarrow X \rightarrow M \rightarrow \Gamma$. The high-symmetry points $X = (0, \pi)$ and $M = (\pi, \pi)$ are marked by dashed lines. The structures encircled in blue can be attributed to dynamical short-range antiferromagnetic correlations.

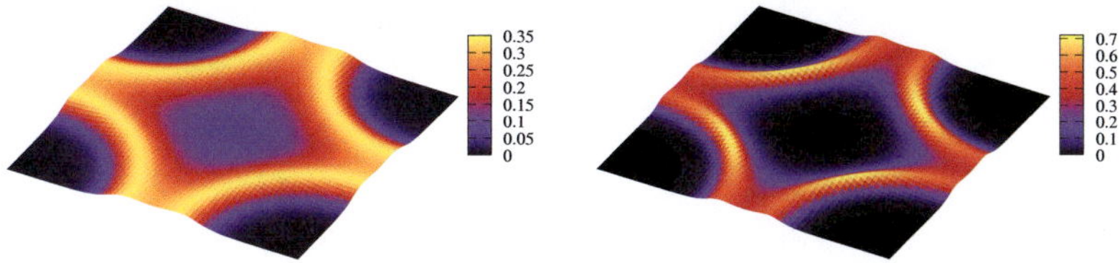

*Figure 6.9:* Spectral function $A(\mathbf{k}, \omega = 0)$ ($\Gamma$-centered) obtained by polynomial extrapolation of the lattice Green function $G(\mathbf{k}, \omega_n)$ for finite doping $\delta = 7\%$, $U/t = 10$, $T/t = 0.05$ and next-nearest-neighbor hopping $t'/t = -0.3$. While the DMFT spectral function (left) is constant along the Fermi surface (lines of highest intensity) as expected, the dual fermion result (right) shows the destruction of quasiparticles in the antinodal region.

Fermi arcs. A concomitant feature appears in the effective quasiparticle dispersion law $h_{\mathbf{k}}^{\mathrm{eff}} = \mathrm{Re}\{[h_{\mathbf{k}} - \mu + \Sigma(\mathbf{k}, \omega = 0)]/[1 + i\partial/\partial_{\omega}\Sigma(\mathbf{k}, \omega = 0)]\}$, which is flattened in the antinodal region. This behavior was predicted earlier as being due to non-Fermi-liquid behavior as the Van-Hove singularity crosses the Fermi level, see Ref. [139] and references therein. As noted in Ref. [50], this can be attributed to short-range correlations, which is confirmed in the present calculations. Results from recent DCA calculations suggest that this phenomenon can be understood in terms of an orbital selective Mott transition in momentum space [147]. The present translationally invariant solution does not indicate a full gap-opening in the antinodal region. The extrapolated self-energies at the node and the antinode differ by approximately a factor of two.

## 6.8   Mott Transition Revisited

The occurrence of a strong modification of the DMFT solution for $U$ equal to the bandwidth suggests that it may be related to Mott physics. An early application of DMFT established the Mott metal-insulator transition (MIT) in the 2D Hubbard model [15]. The following calculations were inspired by the results of Ref. [148]. A qualitative change of the mechanism driving the MIT was found when including momentum dependence into the self-energy through a cluster extension of DMFT. More recently, this issue was addressed in Ref. [149], where a detailed study of the $U$-$T$ phase diagram of the paramagnetic model was carried out within CDMFT. This allows to compare results from two complementary approaches. Compared to (single-site) DMFT, the MIT remains first order, but shows a qualitative modification of the transition lines. In single-site DMFT, the insulating state is favored over the metallic state as the temperature is increased from low temperatures, due to a large residual entropy of the insulator. The converse is true for the CDMFT, where the entropy of the Mott insulator is lowered due

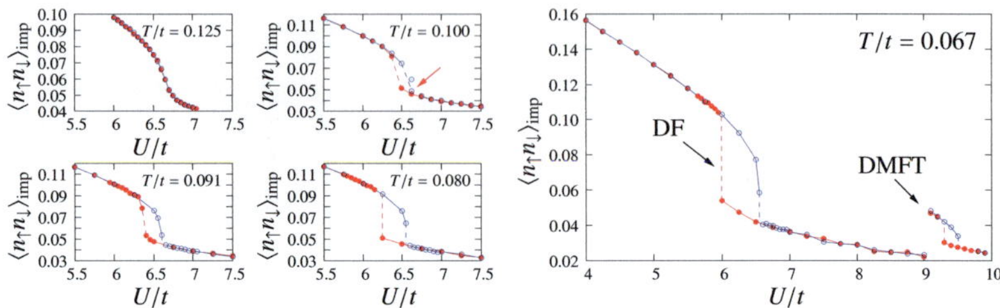

*Figure 6.10:* Left: Double occupancy of the impurity as a function of $U$ for various temperatures. The hysteresis at the first-order Mott transition is clearly visible. Right: Direct comparison of the hysteresis obtained from dual fermion (DF) and DMFT calculations. The critical value of the interaction is significantly reduced from $U > W$ in DMFT to $U < W$ in DF calculations.

to short-range singlet formation. The transition remains first order at $T = 0$ in VCA [150].

The double occupancy $\langle n_\uparrow n_\downarrow \rangle$ can be used to visualize the hysteresis at the Mott transition. Within the DF (and DMFT) calculations, the double occupancy of the impurity problem was calculated using the methods of Sec. 3.3.10. For DF calculations, this is not exactly equal to the double occupancy of the lattice problem, but is nevertheless well suited to map out the coexistence region at the first-order MIT. The left panel of Fig. 6.10 shows the double occupancy as a function of $U$ for different temperatures. To obtain these results, two independent calculations were performed, where the interaction $U$ was varied in small steps from above and below. In each case, the hybridization function of a converged calculation for the previous value of $U$ was used as an initial guess for the next. It should be emphasized that the DF calculations are therefore not related to DMFT (diagrams were summed in all iterations of the self-consistency loop). At high temperatures a crossover behavior is obtained, which changes to hysteretic behavior below a critical temperature $T_c/t \sim 0.11$. As the temperature is lowered, the width of the coexistence region increases.

The right panel of Fig. 6.10 compares the double occupancy of the impurity obtained within DF and DMFT calculations for a fixed temperature $T/t = 0.067$, below the critical point. Obviously, the critical $U$ is drastically reduced as spatial correlations are taken into account. This is in accordance with the findings of Ref. [149], where the critical $U$ was found to be reduced from $U_c/t = 9.35$ in DMFT down to $U_c = 6.05$ in CDMFT calculations.

For the interpretation of the results it must be noted that in order to keep the calculations computationally manageable, the transition was roughly located by performing calculations with the number of iterations fixed to 8. The results were then refined by

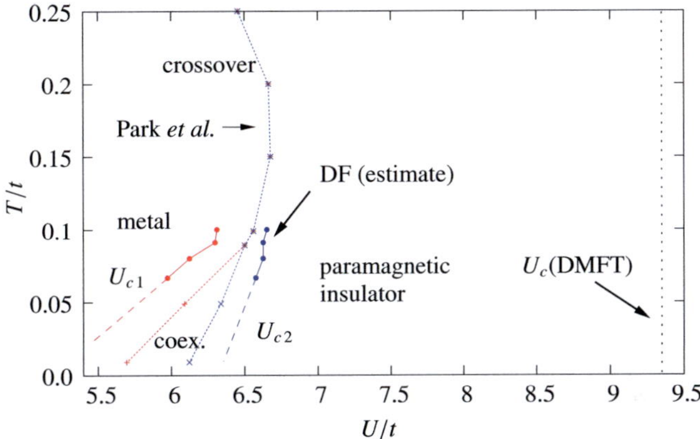


*Figure 6.11:* *U-T* phasediagram of the paramagnetic two-dimensional Hubbard model at half-filling. The dual fermion result for the transition lines was obtained from the hysteresis loops of Fig. 6.10 and should be viewed as upper and lower bounds (see text). The data from Park *et al.* were taken from Ref. [149]. The dual fermion (DF) approach qualitatively captures the reduction of the critical $U$ and the modification of the transition lines.

reducing the distance between points (and increasing the number of iterations up to 10). Close to the transition, this is still insufficient. For example, the data point at the red arrow in Fig. 6.10 for the hysteresis loop at $T/t = 0.1$ was obtained after performing ten additional iterations. This shows that the calculations in the immediate vicinity of the transition are not fully converged. Since the number of iterations was kept fixed, the transition lines determined from the border of the hysteresis loops, where the lines merge, can nevertheless be regarded as upper and lower bounds.

The phasediagram obtained from the hysteresis loops is shown in Fig. 6.11. One can see that indeed the CDMFT transition lines are bracketed by the DF estimate. Agreement is found for the approximate position and the shape of the transition lines, which clearly favor the metal at higher temperatures below the critical point. A closer analysis reveals the the eigenvalue in the $\mathbf{q} = (\pi, \pi)$ spin scattering channel strongly dominates[1]. Thus the diagrammatic correction of Fig. 6.3 b) includes short-range antiferromagnetic correlations responsible for the rather drastic reduction of the critical $U$ compared to DMFT.

---

[1]It is larger than one at the temperatures considered here, showing that an infinite ladder diagram summation diverges; the lowest orders nevertheless yield sensible results.

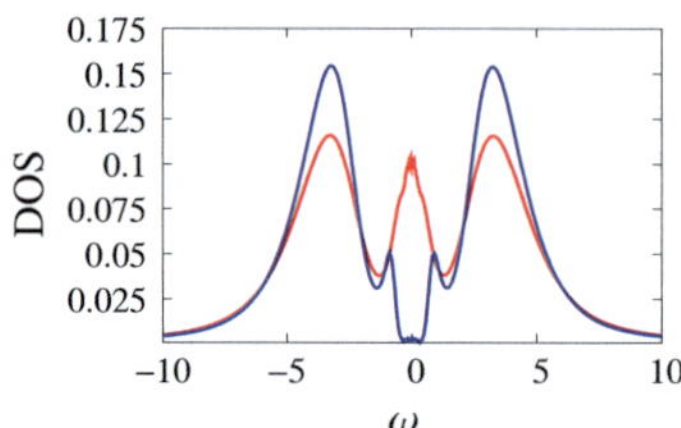
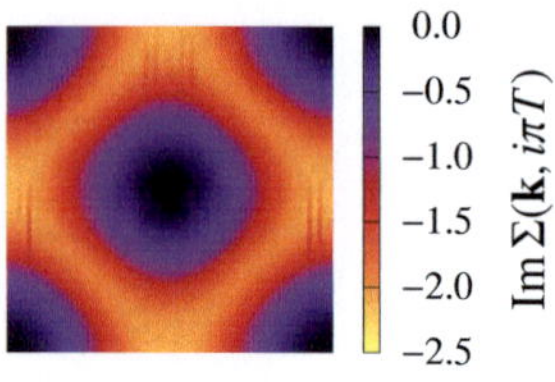

*Figure 6.12:* Left: Metallic and insulating local density of states obtained in the coexistence region of the Mott transition for $U/t = 6.5$ and $T/t = 0.08$. The insulating solution exhibits characteristic peaks at the gap edge. Right: $k$-dependence of the imaginary part of the self-energy of the metallic solution from within the coexistence region at $U/t = 6.55$ and $T/t = 0.08$ on the first Matsubara frequency. As the transition is approached, the self-energy diverges along the noninteracting Fermi surface. Note that within DMFT (for fixed frequency) this is a constant.

The significance of short-range AF correlations is underlined by examining the local density of states of the insulating and metallic solutions within the coexistence region. These are shown in Fig. 6.12. The insulating solution exhibits characteristic peaks at the gap edge. The antiferromagnetic correlations lead to an antiferromagnetic-gap-like behavior [151]. The metallic solution exhibits shoulders on the peak at the Fermi level. These results are in qualitative agreement with the ones presented in Ref. [149].

It should be emphasized that in order to obtain these results, an adjustment of the hybridization function is essential. The behavior is not correctly captured by the diagrammatic corrections when the hybridization is fixed to its DMFT value [40]. Here this would result in a pseudogap for example at $U/t = 8$. This is not unexpected since the Mott transition cannot be described perturbatively [14]. Note that the elimination of the local dual fermion propagator through the self-consistency condition corresponds to the summation of an infinite partial series (see Sec. 6.4). Therefore, straightforward perturbative expansions around DMFT cannot provide a consistent description in this parameter regime. While providing qualitatively the same picture, the DF approach differs from CDMFT in that it does not break translational invariance of the lattice. As the short-range correlations are treated perturbatively, true singlet correlations are not included. The latter can be achieved by extending the dual fermion approach to clusters (see chapter 8), however at the cost of breaking translational invariance.

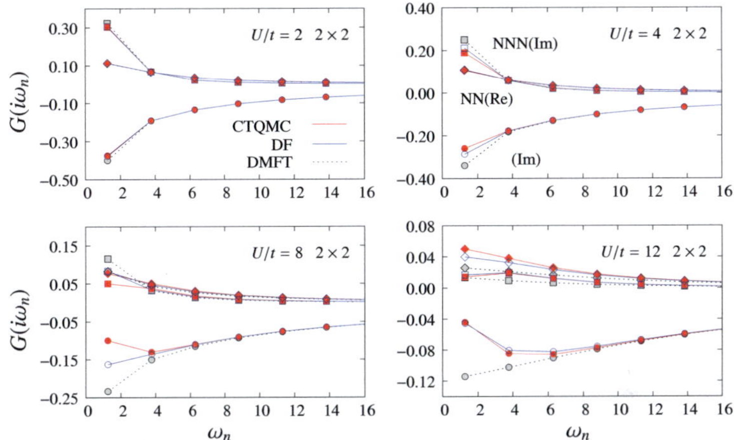

*Figure 6.13:* Comparison of DMFT and DF results for the local, nearest-neighbor (NN) and next-nearest-neighbor (NNN) Green functions (depicted by circles, diamonds and squares, respectively) in comparison to CTQMC for a $2 \times 2$ Hubbard cluster with periodic boundary conditions. Results are shown for different values of $U$ at fixed temperature $T/t = 0.4$. The dual perturbation theory works particularly well for weak and strong coupling.

## 6.9 Comparison to CTQMC

In order to gain insight into the behavior of the DF approximation, it is elucidative to compare it to essentially exact results. The Hubbard model on the $2 \times 2$ plaquette can be solved exactly within ED. Here instead the weak-coupling CTQMC is employed as a lattice QMC solver, in order to obtain the two-particle vertex in addition to the single-particle Green functions. The computational effort to obtain the vertex in ED is comparable for this system. The two-particle quantities are compared in chapter 10. To ensure comparability, periodic boundary conditions have been imposed for the CTQMC calculation (this merely amounts to changing $t \to 2t$ for this cluster).

Representative results from numerical tests are displayed in Fig. 6.13. One can see that the deviation of the DMFT Green functions to the CTQMC result systematically increases with interaction $U$. On the contrary, the DF results are close not only for small, but also large $U$. This is in favor of the arguments regarding convergence of the approach given at the end of Sec. 6.1. However, the nonlocal Green functions still show significant deviations even in the large $U$ limit. The convergence properties of the approach are reexamined in chapter 11.

# 6.10    Generalizations of the Dual Fermion Approach

The extension of the formalism to clusters, the calculation of susceptibilities and the construction of more elaborate conserving approximations, in particular the infinite-order ladder approximation, are described in detail in the subsequent chapters. Here it is outlined how the dual fermion approach can be generalized to treat symmetry-broken phases and combined with bandstructure calculations for the application to real materials.

## 6.10.1    Antiferromagnetism

In the antiferromagnetic state on a bipartite lattice, the spatial modulation of the magnetization is described by

$$\mathbf{M}(\mathbf{x}) = \mathbf{M}e^{i\mathbf{Q}\mathbf{x}} \, . \tag{6.49}$$

Translational invariance is broken and momentum is conserved only up to the ($d$-dimensional) wave vector of the modulation $\mathbf{Q} = (\pi, \ldots, \pi)$. The unit cell of the lattice is no longer identical to the magnetic unit cell, which is doubled, and the volume of the magnetic Brillouin zone of the lattice is correspondingly reduced by half. The notation is the same as in Sec 2.7. A lattice vector of the original lattice is decomposed as $\mathbf{x} = \tilde{\mathbf{x}} + \mathbf{X}$, where the vectors $\tilde{\mathbf{x}}$ connect the origin of the magnetic unit cells and span the superlattice. The vectors $\mathbf{X}$ label sites within the cell. The two sites correspond to integer components $\mathbf{X} = (n, 0)$, $n = 0, 1$ (in units of the lattice constant). The wave vectors are decomposed accordingly, $\mathbf{k} = \mathbf{K} + \tilde{\mathbf{k}}$, with $\mathbf{K} = (2\pi n)/L_c$. Since translational invariance *within* the cell is broken, the Green function must depend on two cell momenta separately, which now can take on two values: $\mathbf{K}_1, \mathbf{K}_2 = \mathbf{0}, \mathbf{Q}$. The Fourier transform with respect to intracluster sites for this case reads (cf. Eq. 2.67):

$$G(\mathbf{K}_1, \mathbf{K}_2, \tilde{\mathbf{k}}) = \frac{1}{N_c} \sum_{\mathbf{X}_1 \mathbf{X}_2} G(\mathbf{X}_1, \mathbf{X}_2, \tilde{\mathbf{k}}) e^{-i[(\mathbf{K}_1 + \tilde{\mathbf{k}})\mathbf{X}_1 - (\mathbf{K}_2 + \tilde{\mathbf{k}})\mathbf{X}_2]} \, . \tag{6.50}$$

The normalization enters because of two independent sums[2]. In matrix notation $G(\mathbf{X}_1, \mathbf{X}_2, \tilde{\mathbf{k}})$ for given $\tilde{\mathbf{k}}$ is diagonal with $G^\sigma := G_{00}^\sigma = G_{11}^{-\sigma}$. Inserting this into (6.50) gives

$$\bar{G}_{\mathbf{K}_1 \mathbf{K}_2}(\tilde{\mathbf{k}}) = \begin{pmatrix} \frac{1}{2}[G^\uparrow(\tilde{\mathbf{k}}) + G^\downarrow(\tilde{\mathbf{k}})] & \frac{1}{2}[G^\uparrow(\tilde{\mathbf{k}}) - G^\downarrow(\tilde{\mathbf{k}})] \\ \frac{1}{2}[G^\uparrow(\tilde{\mathbf{k}}) - G^\downarrow(\tilde{\mathbf{k}})] & \frac{1}{2}[G^\uparrow(\tilde{\mathbf{k}}) + G^\downarrow(\tilde{\mathbf{k}})] \end{pmatrix} \, . \tag{6.51}$$

The case of antiferromagnetic long-range order is readily implemented using the multiorbital implementation. The above matrix is treated in the same way as a matrix in orbital space and the computations are performed with $\bar{G}$ in this form. $\bar{G}(\tilde{\mathbf{k}})$ is diagonal

---

[2]If translational invariance is regained, the Green function depends on the difference $\mathbf{X}_1 - \mathbf{X}_2$ only and one of the sums can be performed to give $N_c \delta(\mathbf{K}_1 - \mathbf{K}_2)$.

with respect to $\tilde{\mathbf{k}}$ and the superlattice Fourier transform is performed in the usual way. If this matrix were to be used for evaluating the diagrams however, the interaction would have to be transformed accordingly. The diagrams are instead evaluated in real space, which is in fact done for efficiency reasons (cf. Sec. 6.6). The equations then take a simple form, since $\gamma^{(4)}$ and $\bar{G}$ are site-diagonal,

$$\bar{G}_{X_1 X_2}(\tilde{\mathbf{x}}) = \left( \begin{array}{cc} G^{\uparrow}(\tilde{\mathbf{x}}) & 0 \\ 0 & G^{\downarrow}(\tilde{\mathbf{x}}) \end{array} \right). \tag{6.52}$$

Two independent equations for the two spin-components have to be evaluated as in the usual case. When transforming back to k-space however, one has to use the proper cluster Fourier transform that relates (6.52) with (6.51).

A simple test is provided by repeating the calculations of Fig. 11.10 for lower temperature, e.g. $T/t = 0.2$. For sufficiently large $U$, the eigenvalue exceeds unity in the paramagnetic phase. Using the implementation that allows for a symmetry broken solution, the eigenvalue closely follows this result for small $U$. Instead of exceeding unity for larger interaction it drops as the system enters the symmetry broken state. Numerical results for DF calculations in the symmetry broken state have been presented in Ref. [139].

### 6.10.2  Superconductivity

The generalization to superconducting phases bears some similarity to the case of DMFT. As described in Sec. 2.8, one introduces anomalous Green functions $F(\mathbf{k}, i\omega) = -\langle c_{\mathbf{k}\omega\sigma} c_{-\mathbf{k}\omega\bar{\sigma}} \rangle$ and the matrix Green function (2.90). The action is rewritten in terms of Nambu spinors. The dual self-energy should contain the relevant contributions, such as the spin fluctuation diagrams of chapter 11. In general, the anomalous averages of the impurity problem may become nonzero. Hence these should be retained. This is accomplished by introducing a local matrix Green function in the form (2.90) and expanding Eq. A.51 in the derivation of the dual potential in the same way as (2.88) (these equations have the same structure).

In DMFT, the superconducting state cannot break the symmetry of the original lattice. Hence d-wave pairing cannot be accounted for in single-site DMFT and requires the extension to clusters [32]. This is not the case for dual fermion calculations. Phases with finite values of the d-wave order parameter are possible already for calculations with a single-site reference problem. For d-wave symmetry, the anomalous impurity averages will however be zero. A different approach to superconductivity is based on the calculation of the corresponding pairing susceptibilities. This is described in Sec. 10.4.

## 6.11   Application to Real Materials

The dual fermion approach is a promising tool for the study of real materials beyond dynamical mean-field theory. Merging the LDA with the dual fermion approach is straightforward within the multiorbital formulation given in Sec. 6.1. It is similar to the LDA+DMFT scheme described in Sec. 2.9 and basic considerations, e.g. regarding double-counting corrections, remain valid. The DMFT self-consistency loop of Sec. 2.6.2 is replaced with the equivalent one in terms of dual quantities as described in Sec. 6.5. The main difference for dual fermion calculations is therefore the summation of diagrams for the dual Green function. The approach has the convenient property that the complex local interaction enters only via the impurity problem, which can be treated by the efficient continuous-time quantum Monte Carlo solvers (see chapter 3). A technical challenge is the calculation of the impurity vertex for multiorbital models, which requires to measure a huge number of observables. Corresponding work is currently in progress.

# Chapter 7

# Implementation

A significant part of the work for this thesis was devoted to the development of the implementation of the dual fermion approach. In this chapter, the basic principles which have guided the object-oriented C++ implementation of the dual fermion formalism are presented. Notion of basic concepts of C++ and its syntax are presumed. An introduction to the C++ programming language can e.g. be found in Ref. [152].

The implementation is general and allows to accommodate paramagnetic and spin-polarized multiorbital calculations and also the case of spin-orbit coupling. This is controlled by mainly three different variables. The variable $n\_zone$ determines the number of spins in the CTQMC calculation, which is $n\_zone = 2$ if all quantities are spindiagonal. For paramagnetic calculations, the results for the two spin projections are averaged to reduce the statistical noise of the QMC calculation. After averaging it is therefore sufficient to evaluate all quantities for one of the spin projections, which is accommodated by setting $n\_spin = 1$. Correspondingly, $n\_spin = 2$ is used for spinpolarized calculations. If spin-orbit coupling is present, the quantities are no longer diagonal in spin space. This can be accommodated by introducing a superindex $\alpha \equiv \{m\sigma\}$, where $m$ labels orbitals and $\sigma$ the spin projection. The dimension of the orbital matrices is simply enlarged by a factor of two with $n\_zone$ set to one. The different modes of calculation are summarized in the following table:

| $n\_spin$ | $n\_zone$ | $n\_orb$ | calculation |
|:---:|:---:|:---:|:---:|
| 1 | 2 | n | paramagnetic |
| 2 | 2 | n | spinpolarized |
| 1 | 1 | 2n | spin-orbit coupling |

An additional parameter $n\_corb$ is introduced to distinguish correlated from uncorrelated orbitals for realistic LDA+DF calculations.

For an approach based on perturbation theory, the necessity to construct various different approximations arises naturally. Therefore, a highly flexible implementation is desirable. The underlying idea of the implementation is to exploit object-oriented features and the concepts of generic programming to build a high-level programming environment for the implementation of approximations based on diagrammatic methods.

The implementation is based on the observation that all objects that appear in the perturbation theory (Green functions, vertices, susceptibilities, etc.) have a well-defined structure. They may therefore be implemented as classes with a well-defined interface. This ensures the reusability of large parts of the code. The building blocks of the perturbation theory are implemented as functors, more precisely as container classes with overloaded function call operators. Operations on these objects, such as Fourier transform and file I/O, are implemented as member functions. The interaction between the objects is governed by the equations of the theory, which are implemented as functors. The driver routine, which controls the actual program flow is implemented in the main program within relatively few lines of code and can therefore be easily overviewed.

Representative for other objects, the following (simplified) code listing shows an excerpt of the container class definition for Green functions. Other container classes are implemented in a similar fashion, with the same key features:

```cpp
template<>
class CONTAINER<GREENFUNCTION>{
public:
    //constructor
    CONTAINER<GREENFUNCTION>(const int n_spin_, const int size_w_, const KMESH &mesh_);
    void allocate_memory(void);
    //file import/export
    void import_f(const char *fname, ios::openmode mode = ios::in);
    void export_f(const char *fname, ios::openmode mode = ios::out) const;

    //element access
    smat::Matrix operator()(const int s, const int w, const KPOINT k) const;

    template <class G>
    void operator=(const G &g); //assignment; hidden loop over
    //all degrees of freedom (spin, frequency, orbitals...)

    void init_fft(const int direction, CONTAINER<GREENFUNCTION> &target_);
    void fft(void);

    ...

private:
    std::complex<double> *array; //storage

    ...
};
```

Here GREENFUNCTION is a dummy class to label the containers. The constructor of the class takes the number of spin degrees of freedom, the frequency cutoff and the set of k-points as input and performs certain initializations. Passing the KMESH class instead of the number of k-points is used to specify the subset of points on which the Green function is stored. This allows to exploit symmetries, i.e. by storing the Green function only on the irreducible wedge of the Brillouin zone. Another example where this is useful is the calculation of susceptibilities, for which it is often sufficient to calculate them along the high-symmetry lines of the Brillouin zone.

Memory allocation for the data storage is not performed by the constructor but through a separate method. This allows to save memory in case not all containers in the program are used for a particular calculation depending on the program control parameters (e.g. several arrays need not be allocated for DMFT calculations). The next two methods allow file import and export in regular mode and can simply be adapted to binary storage. Elements are accessed through functor calls (line 12). The functor provides access to all elements whether or not they are stored, by generating unstored elements automatically using symmetry operations. The functor explicitly depends on the degrees of freedom for which the Green function is diagonal (if the Green function is evaluated in real space it is not diagonal, but still depends on a single variable, i.e. the coordinate difference). For a given combination of the diagonal degrees of freedom, the functor returns a matrix in orbital space. This allows to write equations in an intuitive manner, as described below. The matrix class is optimized for efficient handling of small matrices. The implementation of the building blocks of the perturbation theory as functors instead of regular functions is also important for performance reasons, because often functor calls can be inlined. Replacing a regular function for the vertex by a container class resulted into a reduction of computation time by up to 20%.

Element access via the functor is restricted to read access. Values always have to be assigned to an object for all degrees of freedom simultaneously (spin, frequency, etc.). This is facilitated through an overloaded assignment operator. It takes any kind of object as input that provides a functor with the same interface as the container class itself. It is implemented as a template to allow assignment from independent objects which need not be inherited from each other or the same base class. The functor internally performs a nested for-loop over all values of the arguments for which the Green function has to be stored. The method in line 18 allows to initialize the routine for the fast Fourier transform. The direction specifies whether to transform forward from real space to reciprocal space or backwards. The FFT is implemented as a member function because it is adapted to the particular storage scheme of that class.

The private part of the class contains several variables such as the frequency cutoff and a pointer to the array for element storage. This is a simplified schematic example; the structure of the element storage is adapted to the specific requirements. Apart from the building blocks of the perturbation theory, it is further advantageous to introduce some specialized containers, for example for the particle-hole bubble, susceptibilities and the k-dependent vertex. Where useful, inheritance is only used through templates, for performance reasons.

In order to illustrate how equations are implemented, take the Dyson equation for $G$, $G = [G_0^{-1} - \Sigma]^{-1}$ as a simple example:

```
1   template <class G0, class S>
2   class DYSON{
3   public:
4       DYSON(G0 &g0_,S &sigma_):g0(g0_),sigma(sigma_){}; //constructor
5       smat::Matrix operator()(int s, int wn, KPOINT k) const; //element access
6
7   private:
8       G0 &g0;
9       S &sigma;
10  };
11
12  template <class G0, class S>
13  smat::Matrix DYSON<G0,S>::operator()(int s, int wn, KPOINT k) const{
14      return Inverse( Inverse(g0(s,wn,k))-sigma(s,wn,k) );//matrix inverse
15  }
```

The class definition is given in the lines 1–10. The constructor takes two objects, the bare Green function and self-energy, as input. Due to the dependence on frequency and momentum, the memory requirements of these objects are typically large. It is inefficient to create local copies. The constructor therefore takes and stores references to the input objects. Copy constructors are never used in the implementation.

The actual equation is implemented within the overloaded function call operator in line 14. The use of such operators obviously allows to implement equations in a very intuitive manner. This is significant for complex diagrammatic approximations as the ones described in Sec. 10.4 and chapter 11. The explicit inversion of the matrices in orbital space is performed through the **Inverse** member function of the matrix class.

The implementation of the Dyson equation as a functor involves some overhead for the actual class definition. Note, however, that the equation is conveniently evaluated for all degrees of freedom through the assignment operator of the container classes. This assignment is performed internally. Otherwise several nested for-loops would have to be written explicitly and repeatedly throughout the code. The implementation of the Dyson equation as a template further allows the objects G0 and S to be of different type. They can be derived from container classes or be functors themselves. The use of the functor in a driver program is illustrated in the following code excerpt:

```
1   typedef dual::CONTAINER<GREENFUNCTION> CONTAINER_GREEN;
2   CONTAINER_GREEN g(n_spin, size_w, kmesh);
3   g.allocate_memory();
4   ... //the same for sigma; gd0 is a functor
5   ... //calculate self-energy
6
7   //instantiate functor for Dyson equation
8   DYSON<GDUAL0,CONTAINER_GREEN> > dyson(gd0,sigma);
9
10  g=dyson; //assignment
11  g.export_f("g.dat"); //save to file
12  g.init_fft(FFT_BACKWARD, g); //in-place transform
13  g.fft();
14  g.export_f("g_realspace.dat");
```

The typedef statement in the first line allows a fast replacement of the container with a modified version (with the same interface). In the following two lines, a container for the storage of the Green function is instantiated and memory is allocated.

The same is done for the self-energy which is then calculated within some approximation (not shown). Here, values of the dual Green function are provided by the functor GDUAL0 (classes are denoted by uppercase and their instances by lowercase letters). This functor in turn depends on a container for the local impurity Green function $g_\omega$, the hybridization function $\Delta_\omega$ and a functor for the bare Hamiltonian $h_\mathbf{k}$ (also not shown). This example shows how the hierarchy of equations is organized in a way that does not affect performance and remains manageable. Line 8 instantiates the functor for Dyson's equation (this does not depend on whether the objects have been initialized). The Dyson equation is evaluated for all degrees of freedom through the single statement in line 10. The Green function can then be exported to file for analysis. If the Green function is desired in real space, it can be Fourier transformed and written to file in merely three additional lines of code.

From the above example it should be clear that this implementation is highly flexible and allows to quickly implement different approximations. The code remains highly transparent because the program flow is visible from a few lines of code and the data handling is fully encapsulated. For example, it avoids a multiple, error-prone implementation of nested for-loops and the computation of memory offsets.

Another advantage is that code improvement, such as exploiting certain symmetries, can be done locally within the particular object for which this symmetry holds, and not in multiple instances where its elements are accessed. This is important because for realistic calculations the use of the particular symmetries of the problem is imperative and the code has to be adapted to a given problem. Complicated equations can first be implemented and tested for the simpler single-oribital case using the same infrastructure and driver program. The extension to the multiorbital case only affects the particular functor for this equation.

Once the container classes are implemented and tested, debugging of newly implemented approximations is simple, since it is restricted to the actual equations and program flow. The equations can be implemented in a very intuitive manner, i.e. as matrix equations. Simple equations are therefore automatically valid for the multiorbital case. A disadvantage in this respect is that one has to keep track of whether memory has been allocated or what is currently stored in a container (for example the Green function in real or reciprocal space). This is aided by status flags. Debugging is further supported by the possibility to quickly include file export statements to view intermediate results.

An advantage is that entire, previously used and validated code blocks which perform a certain task, can be reused in a different driver program. For example, the code-block which extracts the local irreducible vertex needed for the calculation of the DMFT susceptibilities could be reused completely unchanged for the implementation of the DΓA equations. The implementation of the DΓA, FLEX and other perturbative approximations (see chapter 11) becomes very simple due to the separation of the container classes from the equations in the driver program. This separation also reduces the compilation time, since container classes can be compiled separately.

# Chapter 8

# Cluster Dual Fermion Approach

H. Hafermann, S. Brener, A. N. Rubtsov
M. I. Katsnelson and A. I. Lichtenstein
JETP Lett. **86**, 677 (2007)

The dual fermion approach in its translationally invariant formulation is based on a single-site impurity problem and treats correlations on all length scales diagrammatically. As found in Sec. 6.7, the approximation appears to be insufficient for the one-dimensional case. The solution cannot be improved through a ladder approximation to the self-energy, because instabilities are present in various channels. For systems with strong short-range, e.g. singlet, correlations, it is further desirable to extend the dual fermion approach to clusters, so that such correlations can be treated explicitly.

The construction of the cluster dual fermion approach (CDFA) is based on a tiling of the lattice into clusters of equal size. The derivation proceeds in the same way as for the single-site approach. A cluster impurity problem is introduced at the position of each cluster and the lattice degrees of freedom are integrated out locally on each cluster. The perturbation is performed around the cluster as the reference problem. One may readily verify that the resulting self-consistency condition (6.44) is exactly equivalent to the respective condition in cellular DMFT (CDMFT, cf. Sec. 2.7).

This construction breaks translational invariance, as CDMFT does. As outlined in Sec. 2.7, the CDMFT solution may be viewed as a superlattice of clusters, each embedded in a selfconsistently determined host. The intercluster self-energy is zero, but is formally set equal to the (averaged) intracluster self-energy to regain translational invariance. Self-energy contributions at distances exceeding the maximum distance of sites inside the cluster are neglected. In the cluster dual fermion approach, the intercluster self-energy is reintroduced perturbatively.

A related approach is the multiscale method proposed in Ref. [38], which is rather different in spirit. In the multiscale approach, a large cluster is embedded in a self-consistent bath. The size of this cluster determines the range of correlations included.

Since it is chosen much larger that what can be handled by nonperturbative solvers, the self-energy on this cluster is constructed diagrammatically. This construction is similar to that of the ladder dual fermion approach described in chapter 11. The irreducible vertex and the self-energy are approximated by the corresponding quantities of a subcluster small enough to be solved nonperturbatively within the dynamical cluster approximation (DCA). To further reduce the computational effort, The vertex is only calculated for zero transferred frequency and momentum corresponding to a long wavelength approximation. Short-wavelength correlations are reintroduced via an analytic ansatz for the large cluster self-energy. For the calculations presented here, the full vertex of the cluster impurity problem is determined from CTQMC without approximations.

In the multiscale approach, the self-consistency scheme is considerably more complicated because it involves self-consistency on three different levels. The large cluster is self-consistently embedded into a bath, the small cluster is embedded self-consistently into the large one and the self-energy on the large cluster is determined by a self-consistent renormalization. In contrast, the dual fermion approach employs a simpler two-level self-consistency (see Sec. 6.5). In addition, since the CDMFT solution captures the usually dominant short-range correlations, the adjustment of the hybridization function in the outer self-consistency loop is expected to be much less important compared to the single site case.

This chapter contains additional details to those published previously.


## 8.1   Construction of Lattice Self-Energy

The CDFA breaks translational invariance with respect to the cluster sites. Using the notation of Sec. 2.7, a translationally invariant self-energy can be constructed as follows. Unlike CDMFT, the CDF self-energy additionally depends on the distance $\tilde{\mathbf{x}} - \tilde{\mathbf{x}}'$ between clusters in the superlattice. Due to translational invariance of the superlattice, all quantities are diagonal with respect to superlattice momenta $\tilde{\mathbf{k}}$. The momentum representation of the CDF self-energy is related to the self-energy in real space representation through the Fourier transform

$$\Sigma(\mathbf{K}, \mathbf{K}', \tilde{\mathbf{k}}) = \sum_{\tilde{\mathbf{x}}} \sum_{\mathbf{X}\mathbf{X}'} e^{-i\tilde{\mathbf{k}}(\tilde{\mathbf{x}}-\tilde{\mathbf{x}}')} e^{-i(\mathbf{K}+\tilde{\mathbf{k}})\mathbf{X}} \Sigma(\mathbf{X}, \mathbf{X}', \tilde{\mathbf{x}} - \tilde{\mathbf{x}}') e^{i(\mathbf{K}'+\tilde{\mathbf{k}})\mathbf{X}'} , \qquad (8.1)$$

where the frequency dependence is suppressed for brevity. Since CDF breaks translational invariance with respect to the cluster sites, the lattice self-energy in general depends on two independent momenta $\mathbf{k}$ and $\mathbf{k}'$. These differ only through the difference between $\mathbf{K}$ and $\mathbf{K}'$, which in turn can differ by a reciprocal lattice vector $\mathbf{Q}$. The lattice self-energy is hence given by

$$\Sigma(\mathbf{k}, \mathbf{k}') = \frac{1}{N_c} \sum_{\mathbf{Q}} \sum_{\mathbf{X}\mathbf{X}'} e^{-i\mathbf{k}\mathbf{X}} \Sigma(\mathbf{X}, \mathbf{X}', \mathbf{k}) e^{i\mathbf{k}'\mathbf{X}'} \delta(\mathbf{k} - \mathbf{k}' - \mathbf{Q}) . \qquad (8.2)$$

Here the superlattice Fourier transform has been carried out and $\Sigma(\mathbf{X}, \mathbf{X}', \tilde{\mathbf{k}})$ has been replaced by $\Sigma(\mathbf{X}, \mathbf{X}', \mathbf{k})$. This can be done because $\mathbf{K}$ is a reciprocal lattice vector so that the phase $\tilde{\mathbf{k}}(\tilde{\mathbf{x}} - \tilde{\mathbf{x}}')$ can be replaced by $\mathbf{k}(\tilde{\mathbf{x}} - \tilde{\mathbf{x}}')$ in (8.1). As in CDMFT, it is natural to approximate the lattice self-energy by the homogeneous $\mathbf{Q} = 0$ component. This yields the translationally invariant self-energy

$$\Sigma(\mathbf{k}) = \frac{1}{N_c} \sum_{\mathbf{X}\mathbf{X}'} e^{-i\mathbf{k}(\mathbf{X}-\mathbf{X}')} \Sigma(\mathbf{X}, \mathbf{X}', \mathbf{k})$$

$$= \frac{1}{N_c} \sum_{\mathbf{X}\mathbf{X}'} \sum_{\tilde{\mathbf{x}}} e^{-i\mathbf{k}(\mathbf{X}-\mathbf{X}')} e^{-i\mathbf{k}\tilde{\mathbf{x}}} \Sigma(\mathbf{X}, \mathbf{X}', \tilde{\mathbf{x}}) . \tag{8.3}$$

The meaning of this equation is more obvious after transforming back to real space,

$$\Sigma(\mathbf{x} - \mathbf{x}') = \frac{1}{N} \sum_{\mathbf{k}} \Sigma(\mathbf{k}) e^{i\mathbf{k}(\mathbf{x}-\mathbf{x}')}$$

$$= \frac{1}{N N_c} \sum_{\tilde{\mathbf{x}}} \sum_{\mathbf{X}\mathbf{X}'} \sum_{\mathbf{k}} e^{-i[\mathbf{k}(\mathbf{X}-\mathbf{X}')+\mathbf{k}\tilde{\mathbf{x}}-\mathbf{k}(\mathbf{x}-\mathbf{x}')]} \Sigma(\mathbf{X}, \mathbf{X}', \tilde{\mathbf{x}})$$

$$= \frac{1}{N_c} \sum_{\tilde{\mathbf{x}}} \sum_{\mathbf{X}\mathbf{X}'} \Sigma(\mathbf{X}, \mathbf{X}', \tilde{\mathbf{x}})\, \delta_{(\mathbf{X}-\mathbf{X}'+\tilde{\mathbf{x}}),\mathbf{x}-\mathbf{x}'} . \tag{8.4}$$

The construction of the CDF self-energy for the $N_c = 2$ cluster in a 1D system is exemplified by the following table and illustrated in Fig. 8.1.

| Coordinates | | | | | $\Sigma(\mathbf{x} - \mathbf{x}')$ |
|:---:|:---:|:---:|:---:|:---:|:---:|
| $\mathbf{x} - \mathbf{x}'$ | $\tilde{\mathbf{x}}$ | $\mathbf{X}$ | $\mathbf{X}'$ | $\mathbf{X} - \mathbf{X}'$ | |
| 0 | 0 | 0 | 0 | 0 | $[\Sigma_{00}(0) + \Sigma_{11}(0)]/2$ |
| 0 | 0 | 1 | 1 | 0 | |
| 1 | 0 | 1 | 0 | 1 | $[\Sigma_{10}(0) + \Sigma_{01}(2)]/2$ |
| 1 | 2 | 0 | 1 | −1 | |
| 2 | 2 | 0 | 0 | 0 | $[\Sigma_{00}(2) + \Sigma_{11}(2)]/2$ |
| 2 | 2 | 1 | 1 | 0 | |
| 3 | 2 | 1 | 0 | 1 | $[\Sigma_{10}(2) + \Sigma_{01}(4)]/2$ |
| 3 | 4 | 0 | 1 | −1 | |

Here the matrix notation $\Sigma_{\mathbf{X},\mathbf{X}'}(\tilde{\mathbf{x}} - \mathbf{x}') := \Sigma(\mathbf{X}, \mathbf{X}', \tilde{\mathbf{x}})$ is used. The translationally invariant self-energy in real space at a given distance is obtained by averaging all contributions $\Sigma(\mathbf{X}, \mathbf{X}', \tilde{\mathbf{x}})$ corresponding to that distance. Note that the $N_c = 2$ cluster constitutes a special case. The local cluster quantities are averaged over all phases and hence

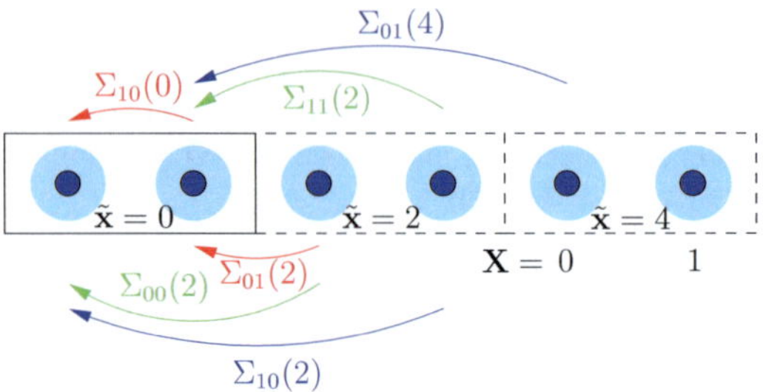


*Figure 8.1:* Construction of the translationally invariant CDF self-energy for an $N_c = 2$ cluster. Only nonlocal contributions are shown. Self-energy contributions that are averaged according to 8.4 are marked by the same color. Arrows are plotted from $\mathbf{X}$ to $\mathbf{X}'$. The nearest-neighbor contribution involves averaging of intra- and intercluster components. All other contributions are obtained perturbatively.

are symmetric, i.e. $\Sigma_{c\,00} = \Sigma_{c\,11}$. Consequently, the $\mathbf{Q} = (\pi)$ contribution $(\Sigma_{00} - \Sigma_{11})/2$ is exactly zero. For an $N_c = 3$ cluster, contributions to the local and nonlocal self-energy from bulk and surface sites are averaged by choosing the $\mathbf{Q} = (0)$ component. Apart from averaging the intracluster and intercluster components, in practice, the nonlocal symmetry-equivalent contributions on the impurity should be averaged to reduce the Monte Carlo error, e.g. $\Sigma_{10}(0)$ and $\Sigma_{01}(0)$ for the $N_c = 2$ cluster.

The number of contributions to the self-energy at a given distance is of course exactly equal to $N_c$. Hence there is no underestimation of the nonlocal contributions and no need for reweighing as in CDMFT (see Sec. 2.7).

## 8.2   Application to the 1D Hubbard Model

In this section, the CDF approach is applied to the 1D Hubbard model using an $N_c = 2$ cluster. The Hamiltonian of the one-dimensional Hubbard model for a chain of $N$ sites is given by

$$H = H_0 + H_{\text{int}} = -t \sum_{i=1}^{N} \left( c_{i+1}^{\dagger} c_i + c_i^{\dagger} c_{i+1} \right) + U \sum_{i=1}^{N} n_{i\uparrow} n_{i\downarrow} \, . \tag{8.5}$$

Periodic boundary conditions are assumed with respect to the chain. In real space, the bare dispersion $h(\tilde{\mathbf{x}})$ reads:

$$h(\tilde{\mathbf{x}} = 2) = \begin{pmatrix} 0 & 0 \\ -t' & 0 \end{pmatrix}, \qquad h(\tilde{\mathbf{x}} = -2) = \begin{pmatrix} 0 & -t' \\ 0 & 0 \end{pmatrix}, \qquad h(\tilde{\mathbf{x}} = 0) = \begin{pmatrix} 0 & -t \\ -t & 0 \end{pmatrix}$$

$$\tag{8.6}$$

and $h(\tilde{\mathbf{x}}) = 0$ for $|\tilde{\mathbf{x}}| > 2$. The matrix $h_{ij}$ is a matrix in the cluster sites $\mathbf{X}$, $\mathbf{X}'$. The $\mathbf{X} = 0$ site hybridizes with a cluster site to the right and with the neighboring cluster to the left (and vice versa for $\mathbf{X} = 1$). Hence the sites are inequivalent. Applying the Fourier transform (2.67) with respect to the superlattice vectors yields the dispersion

$$h_{\tilde{k}} = - \begin{pmatrix} 0 & t + t'e^{i2\tilde{k}} \\ t + t'e^{-i2\tilde{k}} & 0 \end{pmatrix}. \tag{8.7}$$

Alternatively, the same result is obtained by applying the cluster transform (2.80) to the dispersion $h_k = -2t\cos(k)$. The intracluster hopping $t$ and intercluster hopping $t'$ are equal, $t' = t$, and all energies are measured in units of $t$ in the following. The inequivalence of the sites is reflected in the different phase factors. Note that the sign of the phases $2\tilde{k}$ in (8.7) is immaterial for CDMFT calculations, since local quantities are averaged over all phases. In CDF, the sign of the phase must be compatible with the definition of the superlattice Fourier transform. Otherwise, for example the two self-energies $\Sigma_{10}(\tilde{\mathbf{x}} = 2)$ and $\Sigma_{01}(\tilde{\mathbf{x}} = 2)$ shown in Fig. 8.1 will be interchanged.

## 8.3  Results

The CDF results shown in the following have been obtained from a fully selfconsistent calculation. The DMFT hybridization function remains nearly unchanged in the outer loop iterations and the results from different iterations are essentially the same as expected. This is important because the calculation of the full vertex for the cluster impurity problem requires a sizeable numerical effort. Fig. 8.2 shows the local Green function obtained within the CDFA and other approximations in comparison to results from a density matrix renormalization group (DMRG) calculation. The DMRG results may be viewed as essentially exact (for a review, see [146]). Already the cluster approximations for $N_c = 2$ cluster sites, i.e. the cellular DMFT and the variational cluster approach (VCA) considerably improve the result compared to the single-site DF calculation. This emphasizes the importance of nearest-neighbor correlations in the model. The $N_c = 2$ CDF Green function nevertheless considerably improves upon the CDMFT result and appears to be the best approximation that can be achieved for $N_c = 2$. On the other hand, as shown in the inset, it is also apparent that the VCA solution for $N_c = 4$ is superior compared to CDF for $N_c = 2$. This is an indication that correlations beyond the $N_c = 2$ cluster are not negligible and are insufficiently captured by the second-order CDF perturbation theory, which predominantly takes short-range correlations into account. Note that although the VCA and DMRG data correspond to $T = 0$, finite temperature effects are found to be negligible here. Indeed, the VCA and the finite temperature CDMFT calculation give very similar results for $N_c = 2$ and essentially the same results are obtained for CDMFT and the CDFA by reducing the temperature down to $T/t = 0.05$.

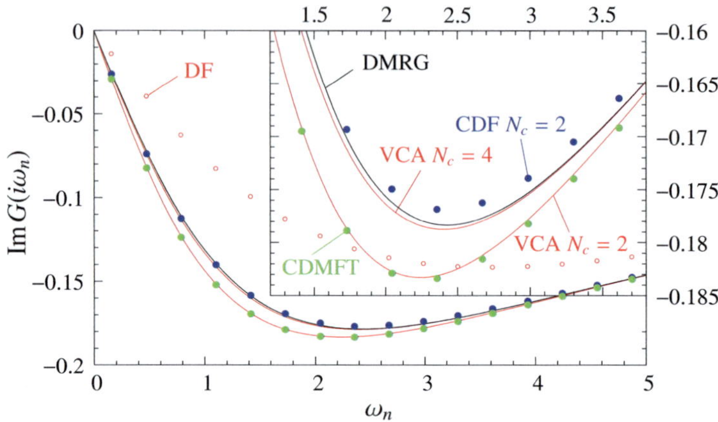


*Figure 8.2:* Imaginary part of the local $N_c = 2$ cluster dual fermion (CDF) Green function $G(i\omega_n)$ for the one-dimensional Hubbard model at $U/t = 6$ and $T/t = 0.1$ in comparison with density matrix renormalization group (DMRG) results and various other approximations: cellular DMFT (CDMFT) for $N_c = 2$, second-order dual perturbation theory (DF), and variational cluster approach (VCA) data from Ref. [153] for the indicated number of cluster sites in the reference system. VCA and DMRG results are for $T = 0$. The latter may be regarded as essentially exact.

The local density of states obtained by analytical continuation of the Matsubara Green functions from Fig. 8.2 are compared to the DMRG result in Fig. 8.3. The DF solution considerably underestimates the width of the gap. It is still somewhat underestimated in the cluster approaches. CDMFT agrees qualitatively, but the position of the Hubbard bands is not correctly captured and the width of the peaks at the gap edge is overestimated. The CDF result already resembles the DMRG solution rather well and provides a satisfactory description of the local properties. The height of the peaks however appears to be too small. This is an artefact of the analytical continuation procedure, which has problems to resolve such sharp features. Here the method of Ref. [107] was used, which for the present case is superior to the maximum entropy method (see Sec. 3.5). The peaks at the gap edge are overlooked completely in the maximum entropy density of states.

Results for the nearest-neighbor Green function are shown in Fig. 8.4. The CDMFT solution strongly overestimates this quantity. In the CDFA, the intracluster component of the nearest-neighbor Green function is improved compared to CDMFT. The intercluster component, which is not present in CDMFT, is obtained perturbatively in the

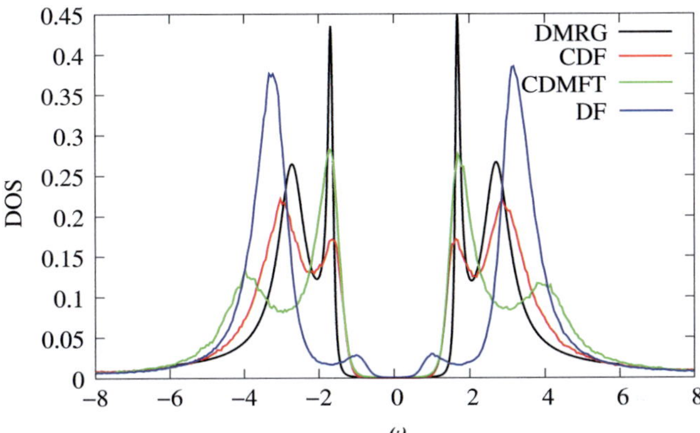

*Figure 8.3:* Comparison of the local density of states. For CDF, DF and CDMFT, the spectral function has been obtained by analytical continuation of the Matsubara Green functions from the previous figure using the method of Ref. [107]. Parameters are the same as in Fig. 8.2.

CDFA. It appears to be underestimated and decays quickly at high energy. This is not a consequence of the finite frequency cutoff for the vertex and the dual self-energy, which is considerably larger ($\sim 40t$). The nearest-neighbor component of the translationally invariant Green function is also shown. It is obtained by averaging the inter- and intracluster components according to (8.4): $G(1) = [G_{10}(0) + G_{01}(2)]/2$. For small frequencies, it approximates the DMRG solution rather well. However, due to the fast decay of the intercluster component, it does not have the correct high-frequency behavior. It is interesting to note that while the single-site dual fermion approach is clearly insufficient for the description of the local properties of the 1D system, it seems to capture the nearest-neighbor correlations rather well. This may be attributed to the fact that the single-site DFA does not break translational invariance. The CDFA shows a clear tendency to restore the translational invariance. This is however insufficient at this level of approximation to the dual self-energy, which is reflected in the large difference between the intra and intercluster components $G_{10}(0)$ and $G_{01}(2)$.

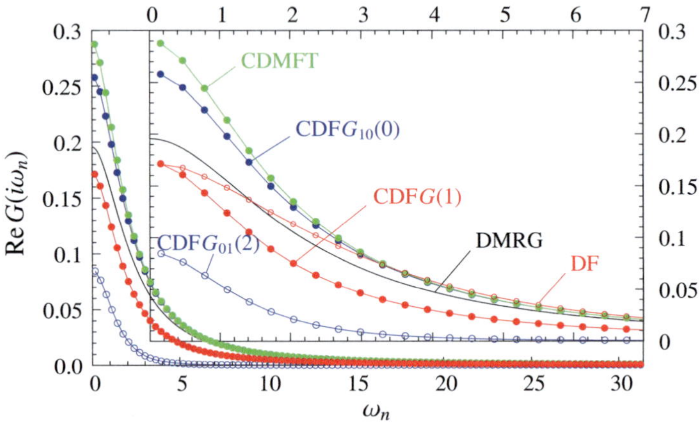



*Figure 8.4:* Real part of the nearest-neighbor Green function as a function of Matsubara frequencies in comparison to DMRG. The intra and intercluster Green functions $G_{10}(0)$ and $G_{01}(2)$ are averaged to give the translationally invariant solution $G(1)$.

## 8.4   Conclusions and Outlook

The results illustrate that for a given cluster size, the CDFA is superior to other cluster approaches. This is expected, since CDMFT (and VCA in the limit of of bath sites $n_{\mathrm{b}} \rightarrow \infty$ [34]) appear as the zero-order approximation in this approach. On the other hand, it is also found that already the VCA for $N_c = 4$ cluster sites seems to give somewhat better results. The same may be expected for the CDMFT. A rough estimate of how these approaches compare computationally can be obtained by considering the computational cost of the impurity solver step, which strongly dominates the calculation. Assuming QMC is used as the solver for the cluster approaches, the computational effort in CDF is dominated by the measurement of the vertex. Regardless of symmetries, the number of observables grows with the cluster size as $N_c^4$, while it increases as $N_c^2$ if only the Green function is measured. (In addition, the total number of frequencies measured for the vertex is considerably larger than for the Green function.) The computational effort of CDMFT for $N_c = 4$ and CDFA for $N_c = 2$ is comparable, since $4^2 = 2^4$. VCA (CDMFT) is preferable as it seems to give more accurate results. Nevertheless, one should also bear in mind that the one-dimensional case is the most unfavorable one. The case of higher dimensions is in favor of the CDFA. One may anticipate that in two and three dimensions the short-range, in particular nearest-neighbor singlet correlations, strongly dominate and correlations beyond the extension of the cluster can accurately be treated perturbatively.

The requirement to measure the vertex limits the CDF to small cluster reference problems. It is therefore desirable to better use the information contained in the cluster vertex by improving the approximation to the dual self-energy. Such an approach, the ladder dual fermion approximation (LDFA) is developed in chapter 11. It has not yet been applied to the cluster problem, but one may expect that the approximation improves considerably. In particular, the problem caused by breaking translational invariance can be expected to be alleviated by improving the approximation to the self-energy. A further possibility to improve the scheme would be to use the translationally invariant quantities in the selfconsistency loop to obtain a periodized scheme [32, 154]. The CDF approach has the advantage that it is readily generalized to model chains on a substrate [155], or to quasi-1D multiorbital systems, which is difficult within DMRG [146].

# Chapter 9

# Superperturbation Solver for Quantum Impurity Models

H. Hafermann, C. Jung, S. Brener, M. I. Katsnelson,
A. N. Rubtsov and A. I. Lichtenstein

EPL **85**, 27007 (2009)

## 9.1  Introduction

An important area of application of quantum impurity models are realistic LDA+DMFT calculations (for further examples see Sec. 2.4). Correlated materials with open d- or f-shells require the solution of the multiorbital DMFT impurity problem with up to seven correlated orbitals. For a realistic description, the complicated local interactions should be taken into account through the full Coulomb vertex (see Sec. 2.9). The dimension of the Hilbert space and hence the computational complexity grow exponentially with the number of orbitals ($2^{10}$ and $2^{14}$ states for d- and f-systems, respectively). Accurate and efficient methods are needed for a reliable solution of the impurity problem.

Monte Carlo methods are often used for this purpose. The continuous-time solvers described in chapter 3 provide a numerically exact solution of the problem, but require a sizeable numerical effort. At present, up to five orbitals (as required for d-systems) can be reliably handled with a modern implementation and up to seven using additional optimization, such as a truncation of the basis for the strong-coupling solver (see Sec. 3.4). Monte Carlo solvers work in the Matsubara representation. In order to obtain spectral functions that can be compared to experiment, these methods require analytical continuation of the imaginary time data to the real axis. This is an ill-posed problem even in absence of statistical errors as described in Sec. 3.5 and obtaining reliable spectra from quantum Monte Carlo data is an open problem. In addition, some restrictions on the local interaction Hamiltonian arise because for example spin-flip and pair-hopping terms are known to cause a severe sign problem for the Hirsch-Fye Monte Carlo solver. For the strong-coupling solver the computational complexity increases considerably by taking these terms into account, as simulations can no longer be performed in the efficient segment representation.

An alternative to Monte Carlo solvers is the exact diagonalization (ED) method. The ED for the impurity problem was introduced by Caffarel and Krauth [58]. In ED, the continuous density of states of the bath is represented by a discrete number of bath levels. A full diagonalization of the problem including impurity and bath states is required. The number of bath levels is severely limited by the exponential growth of the Hilbert space. At present, the limit on the total number of sites for a full diagonalization is $n_s = 6$–$8$ (depending on memory available). The ground state of the problem can be found using the Lanczos algorithm, which allows to treat larger systems up to $n_s \sim 16$. The spectral function can be obtained on the real axis and the approach allows to access low temperatures. General two-fermion interactions are readily incorporated in this approach, with an increase in computational cost that is not critical (certain interaction terms increase the symmetry sectors, see Sec. 2.5).

The discretization of the bath in ED introduces a finite temperature cutoff: Below a temperature scale set by the level spacing, the Kondo physics cannot be accessed in ED. The numerical renormalization group (NRG) initially developed by Wilson is a variant of the exact diagonalization, which overcomes this shortcoming. It is based on a logarithmic discretization of the bath with discretization points distributed according to $\Lambda^{-n}$ and a mapping of the system onto a semi-infinite chain. The chain is diagonalized iteratively. The iterative diagonalization employs a truncation scheme that discards high-energy states, which is possible because the hopping between successive sites in the chain falls off exponentially $t_n \sim \Lambda^{-n/2}$. With increasing chain length, successively lower energy scales are accessed. This nonperturbative method was successfully applied to the Kondo problem. It was later used for the Anderson impurity model (AIM) [123]. Due to the logarithmic discretization, the bath density of states is only well represented in the vicinity of the Fermi level and the truncation can lead to artefacts at high energies.

Several perturbative techniques have been applied to the AIM, such as weak-coupling perturbation theory (iterated perturbation theory in the context of DMFT), the fluctuation-exchange approximation (FLEX), non-crossing approximation (NCA) and slave-boson methods at low energy (see, e.g. Ref. [15] and references therein). The FLEX is essentially limited to weak coupling. NCA suffers from causality problems at low temperature. Extensions of the NCA, such as the one-crossing approximation, are able to alleviate such problems. A strong-coupling solver based on an expansion in the impurity-bath hybridization around the atomic limit has been proposed in order to address systems with open d- or f-shells in the Mott insulating phase [156].

In this chapter, an efficient approximate solver for the AIM is presented. The underlying idea is essentially to employ the dual fermion formalism to solve the local quantum impurity problem. Here the reference problem is chosen such that it can be solved (numerically) exactly and efficiently using ED for a small number of bath sites. The perturbation theory is formulated in terms of auxiliary (dual) degrees of freedom via the continuous Hubbard-Stratonovich transformation.

The perturbation theory is therefore not based on a simple limit, such as the non-interacting or atomic limit, but on a nontrivial, albeit numerically solvable reference problem. We refer to this kind of perturbation theory as a superperturbation. With the AIM in the atomic limit as the reference problem, the perturbation theory is shown to become equivalent to the hybridization expansion of Ref. [156] for weak hybridization. In the weak-coupling limit, it approaches conventional perturbation theory, if the hybridization is sufficiently strong. Similarly to the convergence properties of the dual fermion approach in the context of lattice models (see Sec 6.1), the perturbation theory is found to interpolate between these opposite limits, giving a sensible approximation in a wide parameter range.

Compared to Monte Carlo methods, the present solver is considerably faster and analytical continuation is more stable due to the absence of noise in the imaginary time data. It is demonstrated that it can also be applied in the context of DMFT, for example to study the Mott metal-insulator transition. It is further shown that dynamical quantities can be evaluated on (or close to) the real axis. Because of these properties, the solver is predestined for the study of multiplet effects in solids.

The results presented in this chapter have been obtained in close collaboration with C. Jung, who implemented the exact diagonalization part and carried out an appreciable part of the calculations performed to get an overview over the behavior of the solver. Some representative results are presented in the following. This chapter contains significant additional material to that contained in the above listed publication.


## 9.2 Formalism

The underlying formalism is very similar to that of the dual fermion approach. In particular, the derivation of the dual fermion formalism remains valid. The expressions can be transformed to the ones used here by writing them for a single k-point, setting $h_{\mathbf{k}} = 0$ and changing $\Delta_\omega \to (\Delta_\omega^{(N)} - \Delta_\omega)$, hence replacing every instance of $(\Delta_\omega - h_{\mathbf{k}})$ by $(\Delta_\omega^{(N)} - \Delta_\omega)$. Therefore only the particularities for the application to the local problem are given here. For details, regarding explicit expressions for the diagrams or the derivation, the reader is referred to chapter 6 and appendix A.

The general Hamiltonian of the AIM reads

$$H_{\mathrm{AIM}} = \sum_{k\alpha} \epsilon_{k\alpha} f_{k\alpha}^\dagger f_{k\alpha} + \sum_{k\alpha\beta} V_{k\alpha\beta} c_\alpha^\dagger f_{k\beta} + V_{k\beta\alpha}^* f_{k\alpha}^\dagger c_\beta + \sum_\alpha E_\alpha c_\alpha^\dagger c_\alpha + H_{\mathrm{loc}}[c^\dagger, c] , \qquad (9.1)$$

where Greek indices $\alpha$ label orbitals and spin, respectively: $\alpha \equiv \{m\sigma\}$. Whenever indices are omitted, the hybridization, Green functions, etc. are to be understood as matrices in spin and orbital space. Summation over repeated indices is always implied. The local impurity degrees of freedom represented by $c^\dagger, c$ couple to a bath of free conduction electrons with dispersion $\epsilon_k$ via a matrix hybridization $V_k$. $H_{\mathrm{loc}}$ stands for any local interaction. Because $H_{\mathrm{AIM}}$ is bilinear in the bath operators $b^\dagger, b$, they can be integrated out exactly. As shown in Sec. 2.4, this results in the following action in Matsubara representation (henceforth the path integral representation will be used):

$$S[c^*, c] = -\sum_{\omega\alpha\beta} c^*_{\omega\alpha} \left[(i\omega + \mu)\mathbb{1} - \Delta_\omega\right]_{\alpha\beta} c_{\omega\beta} + S_{\mathrm{loc}}[c^*, c] . \tag{9.2}$$

Here $\mu$ denotes the chemical potential and the sum is over Matsubara frequencies $\omega_n = (2n + 1)\pi/\beta$, where $\beta$ is the inverse temperature. In ED the continuous dispersion $\epsilon_k$ of the bath is approximated by a finite number $N$ of bath levels, which corresponds to replacing the hybridization function $\Delta(i\omega_n)$ by its discrete counterpart

$$\Delta^{(N)}_{\omega\,\alpha\beta}(i\omega) = \sum_{k=1}^{N} \sum_{\gamma} \frac{V_{k\alpha\gamma} V^*_{k\beta\gamma}}{i\omega - \epsilon_{k\gamma}} . \tag{9.3}$$

In this step, the problem arises to determine the parameters $V_{k\alpha\beta}$, $\epsilon_{k\alpha}$ such that $\Delta^{(N)}$ in a certain sense is the best approximation to the original hybridization $\Delta$. Mathematically, this corresponds to projecting the hybridization function onto a restricted subspace spanned by the functions (9.3). Various methods have been proposed for this purpose (see, e.g. [15, 157] and references therein). One way is to perform a conjugate gradient minimization of, e.g., the distance function

$$d = \frac{1}{N_\omega} \sum_{\omega}^{N_\omega} \sum_{\alpha\beta} |\omega|^{-s} \left| \Delta^{(N)}_{\omega\,\alpha\beta} - \Delta_{\omega\,\alpha\beta} \right|^2 , \tag{9.4}$$

where the sum is over Matsubara frequencies up to a cutoff $N_\omega$. In ED, the minimization is carried out on Matsubara frequencies even for calculations at zero temperature. In that case, a finite fictitious temperature is introduced, which plays the role of a low-energy cutoff. The parameter $s$, if chosen large (e.g. $s = 3$), enhances the importance of the lowest Matsubara frequencies in the minimization procedure. As the number of bath sites included should be kept as small as possible due to the exponential growth of the Hilbert space, this parameter is particularly important. It reflects the arbitrariness of approximating a given function with a small, finite number of parameters. For a given distance function, $d$ provides a measure of the quality of the approximation.

Instead of approximating $\Delta_\omega$ by a large number of bath sites as in conventional ED, a different route is followed here by rewriting the action, Eq. 9.2, in terms of an exactly solvable reference problem and an additional bilinear term:

$$S[c^*, c] = -\sum_{\omega\alpha\beta} c^*_{\omega\alpha}[(i\omega + \mu)\mathbb{1} - \Delta^{(N)}_\omega]_{\alpha\beta} c_{\omega\beta} + S_{\text{loc}}[c^*, c]$$

$$-\sum_{\alpha\beta} c^*_{\omega\alpha}(\Delta^{(N)}_{\omega\,\alpha\beta} - \Delta_{\omega\,\alpha\beta})c_{\omega\beta}\,. \tag{9.5}$$

The exactly solvable part (the first line of Eq. 9.5) will henceforth be referred to as $S^{(N)}$. It is useful to introduce the abbreviation

$$D_\omega := \Delta^{(N)}_\omega - \Delta_\omega\,. \tag{9.6}$$

Clearly, the difference $D_\omega$ can be made arbitrarily small by including more and more bath sites. The main point is that the number of bath sites which will be employed to solve (9.5) and the corresponding Hilbert space are much smaller than for the case of conventional ED. In analogy to the dual fermion approach, the continuous Hubbard-Stratonovich transformation (A.1) is applied to the exponential of the bilinear term in the path integral for the partition function. The problem is thereby reformulated in terms of auxiliary (dual) fermions, which is then solved perturbatively. The building blocks of the perturbation expansion are given by the Green function and vertices of the underlying reference problem. Choosing $a = (g_\omega D_\omega g_\omega)^{1-}$ and $b = -g_\omega^{-1}$ in (A.9), the transformation for the present case reads

$$\int \exp\left(-f^*_{\omega\alpha}[g_\omega D_\omega g_\omega]^{-1}_{\alpha\beta} f_{\omega\beta} - f^*_{\omega\alpha}(g_\omega)^{-1}_{\alpha\beta} c_{\omega\beta} - c^*_{\omega\alpha}(g_\omega)^{-1}_{\alpha\beta} f_{\omega\beta}\right) \mathcal{D}[f^*, f]$$

$$= \prod_\omega \det(g_\omega D_\omega g_\omega)^{-1} \exp\left(c^*_{\omega\alpha} D_{\omega\,\alpha\beta} c_{\omega\beta}\right)\,. \tag{9.7}$$

In the above equation $g_\omega$ denotes the impurity Green function of the reference problem with discrete hybridization $\Delta^{(N)}_\omega$ defined as $(g^{(N)}_\omega)_{\alpha\beta} := -\langle c_{\omega\alpha} c^*_{\omega\beta}\rangle^{(N)}$, which can be calculated exactly within ED. With this substitution, the action becomes

$$S[c^*, c; f^*, f] = S^{(N)}[c^*, c] + \sum_{\omega\alpha\beta} f^*_{\omega\alpha}(g_\omega)^{-1}_{\alpha\beta} c_{\omega\beta} + c^*_{\omega\alpha}(g_\omega)^{-1}_{\alpha\beta} f_{\omega\beta} + \sum_{\omega\alpha\beta} f^*_{\omega\alpha}[g_\omega D_\omega g_\omega]^{-1}_{\alpha\beta} f_{\omega\beta}\,. \tag{9.8}$$

From the resulting expression for the partition function

$$\mathcal{Z} = \mathcal{Z}^{(N)} \int\int \exp(-S[c^*, c; f^*, f])\mathcal{D}[c^*, c; f^*, f]\,, \tag{9.9}$$

the original fermionic degrees of freedom represented by $c^*$ and $c$ can formally be integrated out. The procedure is the same as in the dual fermion approach and is described

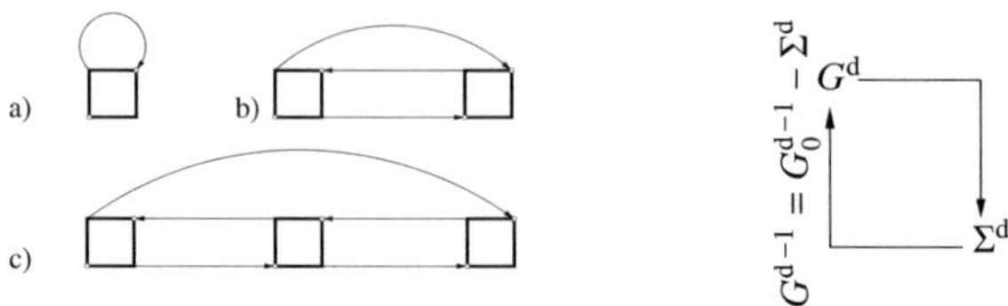

*Figure 9.1:* Lowest order diagrams for the self-energy $\Sigma^{\mathrm{d}}$ with the action given by Eq. 9.11 and illustration of the Dyson iterations: The self-energy is obtained from $G^{\mathrm{d}}$ by summing up the diagrams a)-c). $G^{\mathrm{d}}$ is in turn obtained from the Dyson equation, which is subsequently used in the diagrams.

in appendix A.3. The integral over $\exp(-S^{(N)})$ produces the correlation functions of the reference problem, such as the two-particle Green function

$$\chi^{(N)}_{1234} = \frac{1}{Z^{(N)}} \int c_1 c_2^* c_3 c_4^* \exp(-S^{(N)}[c^*,c])\mathcal{D}[c^*,c] \,. \tag{9.10}$$

Here a combined index $1 \equiv \{\alpha, \omega\}$ is used in order to simplify the notation. The index '$(N)$' emphasizes that this function is obtained exactly for the discrete hybridization (9.3) using ED. The Lehmann representation is given below. The interaction of the auxiliary fermions includes $n$-particle interactions up to all orders. Retaining the leading-order two-particle interaction, the resulting action in terms of the auxiliary fermions reads

$$S^{\mathrm{d}}[f^*,f] = -\sum_{\omega\alpha\beta} f_\alpha^* (G^{\mathrm{d}}_{0\,\omega})^{-1}_{\alpha\beta} f_\beta - \frac{1}{4}\gamma^{(N)}_{1234} f_1^* f_2 f_3^* f_4 \,. \tag{9.11}$$

Here $\gamma^{(N)}$ is the two-particle vertex constructed from the two-particle Green function as (henceforth the superscript '$(N)$' on $\chi$, $g$ and $\gamma$ is omitted)

$$\gamma_{1234} = g^{-1}_{11'} g^{-1}_{33'}(\chi_{1'2'3'4'} - \chi^{0}_{1'2'3'4'})g^{-1}_{2'2}g^{-1}_{4'4} \,, \tag{9.12}$$

with $\chi^{0}_{1234} = \beta(g_{12}g_{34} - g_{14}g_{32})$. The dual Green function $G^{\mathrm{d}}$ in matrix notation is given by

$$G^{\mathrm{d}}_{0\,\omega} = -g_\omega \left[ g_\omega + (\Delta^{(N)}_\omega - \Delta_\omega)^{-1} \right]^{-1} g_\omega \,. \tag{9.13}$$

This function and the corresponding perturbation theory have some remarkable properties. In order to relate these to known results, consider the case $\Delta^{(N)}_\omega = 0$ (expansion around atomic limit as the reference) and Hubbard interaction $H_{\mathrm{loc}} = U n_\uparrow n_\downarrow$. In this case, the two-particle vertex for the one-band model is explicitly given by

$$
\gamma^{\uparrow\downarrow}(\omega_1,\omega_2,\omega_3,\omega_4) = -U + \frac{U^3}{8}\frac{\omega_1^2 + \omega_2^2 + \omega_3^2 + \omega_4^2}{\omega_1\omega_2\omega_3\omega_4} + \frac{3U^5}{16\omega_1\omega_2\omega_3\omega_4}
$$
$$
+ \beta\frac{U^2}{4}\frac{1}{1+e^{\beta U/2}}\frac{2\delta_{\omega_2,-\omega_3}+\delta_{\omega_1,\omega_2}}{\omega_2^2\omega_3^2}\left(\omega_2^2+\frac{U^2}{4}\right)\left(\omega_3^2+\frac{U^2}{4}\right)
$$
$$
- \beta\frac{U^2}{4}\frac{1}{1+e^{-\beta U/2}}\frac{2\delta_{\omega_2,\omega_3}+\delta_{\omega_1,\omega_2}}{\omega_1^2\omega_3^2}\left(\omega_1^2+\frac{U^2}{4}\right)\left(\omega_3^2+\frac{U^2}{4}\right),
$$
$$
\gamma^{\uparrow\uparrow}(\omega_1,\omega_2,\omega_3,\omega_4) = \beta\frac{U^2}{4}\frac{\delta_{\omega_1,\omega_2}-\delta_{\omega_2,\omega_3}}{\omega_1^2\omega_3^2}\left(\omega_1^2+\frac{U^2}{4}\right)\left(\omega_3^2+\frac{U^2}{4}\right). \tag{9.14}
$$

With the atomic limit as the reference one has $D = -\Delta$ and $G^{\mathrm{d}}$ according to (9.13) can be approximated for sufficiently strong hybridization (in comparison to $g^{-1}$) by $G^{\mathrm{d}} \approx g$. The Green function $g$ approaches the bare Green function in the weak-coupling limit $U \to 0$. Since in this limit $\gamma^{\uparrow\downarrow} \sim U$ to leading order[1], the dual perturbation theory becomes equivalent to conventional perturbation theory.

On the other hand, in the case of strong coupling and weak hybridization $\Delta$, $G^{\mathrm{d}}$ itself is small and can obviously be approximated as $G^{\mathrm{d}}_\omega \approx g_\omega\Delta_\omega g_\omega$. This essentially generates a strong-coupling perturbation expansion around the atomic limit in powers of the hybridization. The limit of small hybridization is closely related to previous work, where an approximate solver for the Anderson impurity model based on a hybridization expansion of the Green function was developed [156]. By straightforwardly expanding the imaginary time Green function,

$$
G_{\alpha\beta}(\tau) = -\frac{\int c_\alpha(\tau)c_\beta^*(0)e^{-S[c^*,c]}\mathcal{D}[c^*,c]}{\int e^{-S[c^*,c]}\mathcal{D}[c^*,c]}, \tag{9.15}
$$

with $S$ given by (9.2), up to first order in the hybridization, one obtains the following expression for Green's function:

$$
G_{12} = g_{12} + g_{12}\beta\,\mathrm{Tr}[g\,\Delta] + \chi_{1234}\Delta_{43}. \tag{9.16}
$$

The term involving the trace stems from the expansion of the denominator. Here $g$ and $\chi$ are the single- and two-particle Green functions in the atomic limit ($\Delta^{(N)}_\omega \equiv 0$). This result is contained in the lowest order term of the perturbation expansion for the auxiliary propagator $G^{\mathrm{d}}$ using the action (9.11). To see this, approximate the dual Green function as

$$
G^{\mathrm{d}} \approx G^{\mathrm{d}}_0 + G^{\mathrm{d}}_0\Sigma^{\mathrm{d}}_{(a)}G^{\mathrm{d}}_0, \tag{9.17}
$$

---

[1] Note that to leading order $\gamma^{(N)} = -U$, because the dual interaction in (9.11) formally appears as an attractive interaction. $\gamma^{\uparrow\uparrow}$ vanishes faster than $U$, since the Hubbard interaction $Un_\uparrow n_\downarrow$ only couples densities for opposite spins.

where the self-energy correction of diagram a) in Fig. 9.1 is given by

$$(\Sigma^{\mathrm{d}}_{(a)})_{12} = -\gamma_{1234}(G^{\mathrm{d}}_0)_{43} \ . \tag{9.18}$$

In order to compare this to the above result, the auxiliary Green function has to be related to the Green function of the impurity, $G_{12} := -\langle c_1 c_2^* \rangle$, through the following identity (see A.6):

$$G = D^{-1} + (gD)^{-1}G^{\mathrm{d}}(Dg)^{-1} \ . \tag{9.19}$$

Inserting the approximation to $G^{\mathrm{d}}$ (9.17) into this equation using (9.18) yields the following expression for $G$ after some straightforward algebra:

$$G_{12} = [g(Dg+1)^{-1}]_{12} - [(1+gD)^{-1}]_{11'}\left[(\chi-\chi^0)_{1'2'3'4'}[(g+D^{-1})^{-1}]_{4'3'}\right][(Dg+1)^{-1}]_{2'2} \ , \tag{9.20}$$

where the vertex $\gamma$ is expressed in terms of the two-particle Green function through (9.12). The term containing $\chi^0_{1234} = \beta(g_{12}g_{34} - g_{14}g_{32})$ can be recast into the form

$$\beta(g_{12}g_{34}-g_{14}g_{32})[(g+D^{-1})^{-1}]_{43} = \beta g_{12}\,\mathrm{Tr}[g(g+D^{-1})^{-1}] - \beta g_{14}[(g+D^{-1})^{-1}]_{43}g_{32} \ . \tag{9.21}$$

Using $D = -\Delta$ as above, in the limit of small $\Delta$ one has $(1-\Delta g)^{-1} \to 1$ and $(g-\Delta^{-1})^{-1} \to -\Delta$. Note further that by expanding the first term in (9.20) to first order in $\Delta$ cancels the second term in (9.21). Gathering the results, $G$ is approximated in the limit of small $\Delta$ as

$$G_{12} \approx g_{12} + g_{12}\beta\,\mathrm{Tr}[g\,\Delta] + \chi_{1234}\Delta_{43} \ , \tag{9.22}$$

which exactly recovers expression (9.16) from Ref. [156].

The perturbation theory thus has the correct limiting behavior in the two opposite limits: In the strongly hybridized weak-coupling and the weakly hybridized strong-coupling limit. One can expect that this ensures a reasonable interpolation between these limits so that sensible results can be anticipated even in the intermediate coupling regime. For intermediate coupling, it is however crucial to exploit the possibility to improve the starting point of the perturbation theory beyond the atomic limit by expanding around the ED solution for a finite number of bath sites.


## 9.3 Exact Diagonalization

For superperturbation calculations, the dual fermion source code may conveniently be used without modifications: The input hybridization is taken to be $\Delta^{(N)} - \Delta$ and calculations are carried out for a single k-point with the bare Hamiltonian set to zero, $h_{\mathbf{k}} \equiv 0$.

The difference to dual fermion calculations is in the solution of the reference problem for the hybridization $\Delta^{(N)}$, which is solved much more efficiently using ED instead of CTQMC for a small number of bath levels. The formula for the evaluation of the single-particle Green function within ED is given by Eq. 2.38. The Lehmann representation for the two-particle Green function (2PGF) was given in Ref. [39]. Here we give a modified expression.

By definition, the 2PGF in Matsubara space is given by (see also Sec. 2.5)

$$\chi_{\alpha\beta\gamma\delta}(\omega_1, \omega_2, \omega_3) = \int_0^\beta d\tau_1 \int_0^\beta d\tau_2 \int_0^\beta d\tau_3 \, e^{i(\omega_1\tau_1 + \omega_2\tau_2 + \omega_3\tau_3)} \times$$

$$\times \langle T_\tau c_\alpha(\tau_1) c_\beta^\dagger(\tau_2) c_\gamma(\tau_3) c_\delta^\dagger(0) \rangle \,. \quad (9.23)$$

Energy conservation requires $\omega_1 + \omega_2 + \omega_3 + \omega_4 = 0$. Note that here the frequencies in the exponential corresponding to annihilation and creation operators have the same sign in contrast to the usual definition for the Fourier transform (cf. appendix C). Correspondingly, the condition for energy conservation is different from the one given in Sec. 3.3.9. By restricting the range of integration such that time ordering is explicit, one obtains 3! different terms. These can be brought into the same form by permuting the operators *and* corresponding frequencies. By the anticommutation relations, each term picks up the sign of the permutation. Note that an analogous argument was used in the derivation of the strong-coupling CTQMC in Sec. 3.4. After introducing the sum over eigenstates, the 2PGF can be written in the form

$$\chi_{\alpha\beta\gamma\delta}(\omega_1, \omega_2, \omega_3) = \frac{1}{Z} \sum_{klmn} \sum_{p \in \mathcal{P}} \phi(E_k, E_l, E_m, E_n, \omega_{p_1}, \omega_{p_2}, \omega_{p_3})$$

$$\text{sgn}(p)\langle k|O_{p_1}|l\rangle \, \langle l|O_{p_2}|m\rangle \, \langle m|O_{p_3}|n\rangle \, \langle n|c_\delta^\dagger|k\rangle \,, \quad (9.24)$$

where the first sum is over all eigenstates of the Hamiltonian and the second over all permutations $\mathcal{P}$ of the permutation group of the indices $i = \{123\}$. $p_i$ denotes the permutation of index $i$ and $O_1 := c_\alpha$, $O_2 := c_\beta^\dagger$ and $O_3 := c_\gamma$. The different choice of convention for the Fourier transform simplifies the notation, since otherwise the sign of the frequency associated with the creation operator would have to be permuted. The function $\phi$ is given by the integral

$$\phi(E_k, E_l, E_m, E_n, \omega_1, \omega_2, \omega_3) = \int_0^\beta d\tau_1 \int_0^{\tau_1} d\tau_2 \int_0^{\tau_2} d\tau_3 e^{-\beta E_k} e^{(E_k - E_l)\tau_1} e^{(E_l - E_m)\tau_2} \times$$

$$\times e^{(E_m - E_n)\tau_3} e^{i(\omega_1\tau_1 + \omega_2\tau_2 + \omega_3\tau_3)} \,.$$

$$(9.25)$$

The latter expression should be evaluated by distinguishing the case where two energies are degenerate. This results in

$$
\phi(E_k, E_l, E_m, E_n, \omega_1, \omega_2, \omega_3) = \frac{1}{i\omega_3 + E_m - E_n} \times
$$
$$
\left[ \frac{1 - \delta_{\omega_2,-\omega_3}\delta_{E_l,E_n}}{i(\omega_2 + \omega_3) + E_l - E_n} \left( \frac{e^{-\beta E_k} + e^{-\beta E_l}}{i\omega_1 + E_k - E_l} - \frac{e^{-\beta E_k} + e^{-\beta E_n}}{i(\omega_1 + \omega_2 + \omega_3) + E_k - E_n} \right) \right.
$$
$$
+ \delta_{\omega_2,-\omega_3}\delta_{E_l,E_n} \left( \frac{e^{-\beta E_k} + e^{-\beta E_l}}{(i\omega_1 + E_k - E_l)^2} - \beta \frac{e^{-\beta E_l}}{i\omega_1 + E_k - E_l} \right) - \frac{1}{i\omega_2 + E_l - E_m} \times
$$
$$
\left. \left( \frac{e^{-\beta E_k} + e^{-\beta E_l}}{i\omega_1 + E_k - E_l} - (1 - \delta_{\omega_1,-\omega_2}\delta_{E_k,E_m}) \frac{e^{-\beta E_k} - e^{-\beta E_m}}{i(\omega_1 + \omega_2) + E_k - E_m} + \beta e^{-\beta E_k}\delta_{\omega_1,-\omega_2}\delta_{E_k,E_m} \right) \right],
$$
$$
(9.26)
$$

where the $\delta$-functions arise from degenerate levels. The terms are collected in such a way that when evaluated on Matsubara frequencies $\omega_1, \omega_2, \omega_3$, the above expression has no singularities. If this is not done so, a small value has to be added to some eigenvalues in order to avoid exactly degenerate energies and the associated singularities. The main difference of this result to the Lehmann representation given in Ref. [39] is that the above formulae avoid singular terms and are written in unified notation using the sum over permutations.

The evaluation of diagrams b) and c) in Fig. 9.1 requires the knowledge of the full two-particle Green function, which has to be evaluated on $(2N_\omega)^3$ independent Matsubara frequencies $\omega_n$. $N_\omega$ is the frequency cutoff and $n$ runs from $-N_\omega$ to $N_\omega - 1$ (cf. also Sec. 3.3.9). For each combination of frequencies, the trace over a product of four operator matrices needs to be performed. This is the bottleneck of the calculations and requires optimization. Before evaluating the states in a given symmetry sector defined by a fixed total particle number and total spin, it is determined whether the trace is nonzero, as described in Sec. 2.5. In addition, the exponentials $\exp(-\beta E_i)$ are stored in memory. Terms where all energies are above a threshold defined by $\exp(-\beta E_i) < \epsilon$, where $\epsilon$ is the machine accuracy, are left out from the sum. Note that in contrast to the single-particle Green function, the two-particle Green function requires knowledge of the full spectrum, because even at low temperature transitions between nearly degenerate states at arbitrarily high energies contribute. The cutoff corresponds to neglecting transitions between states that are far apart.

Even with such optimization, the calculation of the two-particle Green function remains the computationally most expensive part. Computations for systems of up to five sites (including the impurity site) have been carried out in Ref. [39], which already requires parallelization. The computational cost is considerably reduced if the auxiliary self-energy is approximated on the Hartree-Fock level, i.e. by using diagram a) in Fig. 9.1 only. For this diagram, the vertex only needs to be evaluated on two instead of three independent frequencies. As shown below, this diagram yields by far the dominant contribution.

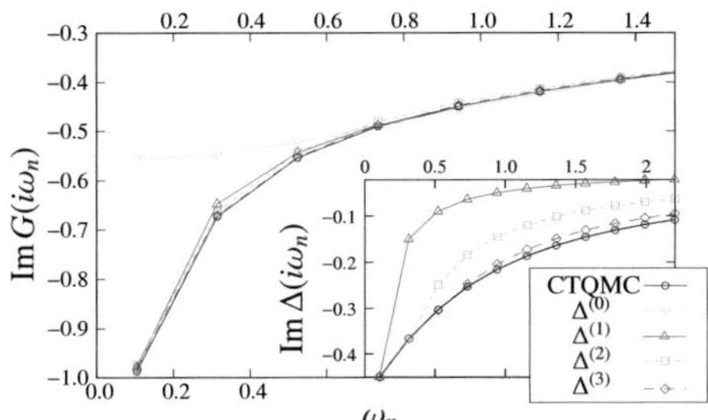

*Figure 9.2:* Imaginary part of the impurity Green function obtained by the superperturbation using different numbers of bath sites for $\beta = 30$ and $U = 3$. The representation of the exact hybridization (open circles) by $\Delta^{(N)}$ is shown in the inset for $N > 0$ ($\Delta^{(0)} \equiv 0$).

## 9.4  Results

In this section, representative results for the general behavior of the solver are presented. The calculations were performed using the diagrams a) to c) depicted in Fig. 9.1. If not otherwise stated, the self-energy was calculated in terms of bold-line diagrams, i.e. by making use of the Dyson iterations illustrated in the same figure. On the first iteration the self-energy was calculated using the bare auxiliary Green function, Eq. 9.13. Inserting the self-energy into the Dyson equation yields a new Green function which is subsequently used in the diagrams on the next iteration. This procedure is carried out until self-consistency is reached and converges typically in less than ten iterations.

In order to test the approach, calculations were carried out for up to $N = 3$ bath sites. The model was taken to be a single-orbital AIM with Hubbard interaction $H_{\mathrm{loc}} = U n_\uparrow n_\downarrow$, and the generic case of a semielliptical density of states of bandwidth $W = 4t$, with

$$\Delta(i\omega) = \frac{2t^2}{i\omega + i\sqrt{4t^2 - (i\omega)^2}} \tag{9.27}$$

(cf. chapter 4). The half-bandwidth is taken as the energy unit: $W/2 = 1$.

A central issue which is relevant for the practical applicability of the approach is the convergence with respect to the number of bath levels. It can only be competitive to other methods, such as conventional ED, if it produces accurate results for a small number of levels. This necessarily introduces some degree of arbitrariness to the determination of the bath parameters. Therefore, physical considerations should be used to impose certain conditions on how they should be determined.

There are basically two extremes, and one may switch between the two by adjusting the parameter $s$ and the frequency cutoff in the distance function (9.4). For $s$ chosen

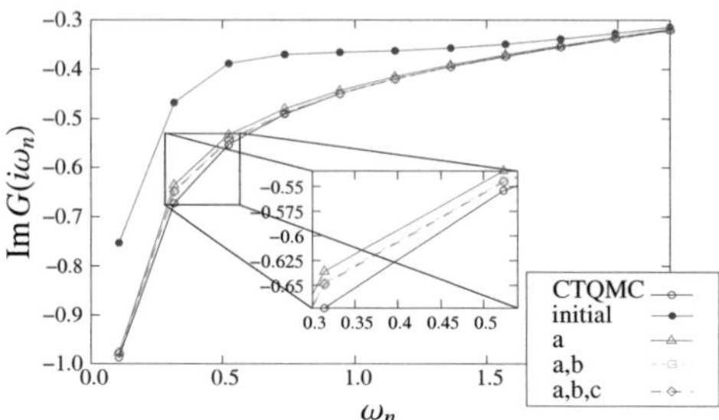


*Figure 9.3:* The contribution of different diagrams to the superperturbation result compared to the exact solution (open circles), for a single bath site. The parameters are otherwise the same as in Fig. 9.2. Diagram a) yields by far the largest correction (upward triangles) to the initial solution obtained from ED (filled circles).

large, the weight of the lower portion of the frequency spectrum is enhanced, so that the resulting discrete hybridization better represents the hybridization on the lowest Matsubara frequencies. This turns out to be a good starting point for the perturbation theory for many cases and seems to be a natural choice to describe metal-insulator transitions, because metallic or insulating behavior is most clearly reflected in the behavior at the few lowest Matsubara points. This choice however is not necessarily the best in general. A counterexample is given in Sec. 9.5.

The minimization of the distance function (9.4) is simplified by exploiting the particle-hole symmetry at half filling to eliminate some of the parameters. For example, in the case of a single bath site, the impurity level is located at the Fermi level, $\epsilon = 0$, and $V$ is the only free parameter. For two sites, the levels are symmetric about the Fermi level and the two hybridizations $V_k$ to both levels are equal. In general, this leaves $N$ out of $2N$ parameters $V_k$, $\epsilon_k$. By requiring that the discrete hybridization should equal (9.27) on the first $N$ Matsubara frequencies, one can obtain an analytical expression for the parameters for $N \leq 2$. For $N > 2$, or a general hybridization (9.35), one needs find the parameters numerically. Simulated annealing or conjugate gradient minimization can be employed for this purpose. Here we use the Fletcher-Reeves conjugate gradient algorithm provided by the GSL [158]. The routine often converges to a local minimum. We improve the procedure by performing the minimization repeatedly with varying starting parameters chosen randomly, within bounds determined by the bandwidth.

Fig. 9.2 shows results for $U = 3$, obtained for different number of bath sites up to $N = 3$. The bath parameters $V_k$ and $\epsilon_k$ have been obtained by minimizing the distance function, Eq. 9.4 for $s = 3$. Using a smaller $s$ leads to a $\Delta^{(N)}$ which better represents the tail of the hybridization function. The same is true for a large cutoff. The cutoff

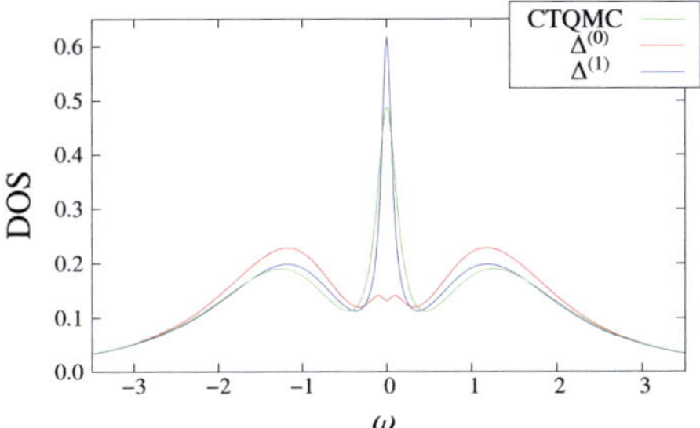



*Figure 9.4:* Comparison of the maximum entropy density of states generated from data obtained by superperturbation for no ($\Delta^{(0)}$) and one bath site ($\Delta^{(1)}$) and from continuous-time quantum Monte Carlo. While the superperturbation around the atomic limit ($\Delta^{(0)}$) does not reproduce the Kondo resonance, the perturbation around the solution for one bath site contains this physics.

is therefore chosen small, e.g. 6. In particular when the hybridization functions differ substantially on the first few Matsubara frequencies, by choosing $\Delta^{(N)}$ to emphasize the high frequency-tail, leads to worse results for the present case.

The quality of the representation of $\Delta$ (labeled 'CTQMC' in Fig. 9.2) by $\Delta^{(N)}$ is shown in the inset for $N > 0$. For $N = 0$, the auxiliary hybridization is zero, i.e. the expansion is performed around the atomic limit. For the particular choice of the distance function, $\Delta$ and $\Delta^{(N)}$ are equal on the first $N$ Matsubara frequencies and the approximation rapidly improves as the number of bath sites is increased. At low temperatures one should determine the discrete hybridization such that it resembles the original hybridization on a somewhat larger low-energy region instead of imposing this condition of the first $N$ successive points. In order to reflect the non-monotonous behavior of $\Delta$ in the insulating phase requires at least two bath sites. For an odd number of sites ($N > 1$), the peak at $\epsilon = 0$ in this case attains a lower weight than the ones which lie symmetrically around the Fermi level.

The superperturbation results are compared to numerically exact results obtained from CTQMC. One can see that while the expansion around the atomic limit (labeled $\Delta^{(0)}$) lacks accuracy, a drastic improvement occurs for a single bath site, although the difference between $\Delta^{(1)}$ and $\Delta$ is still significant. For $N > 1$, the results are essentially converged. Qualitatively the same behavior is found for a wide-range of $U$ values. The fast convergence with respect to the number of bath sites significantly reduces the computational effort compared to CTQMC calculations.

The results shown here have been obtained by summing up skeleton diagrams using the self-consistent renormalization procedure described in chapter 6. The use of skeleton diagrams is theoretically relevant since in this case the approximation is conserving in

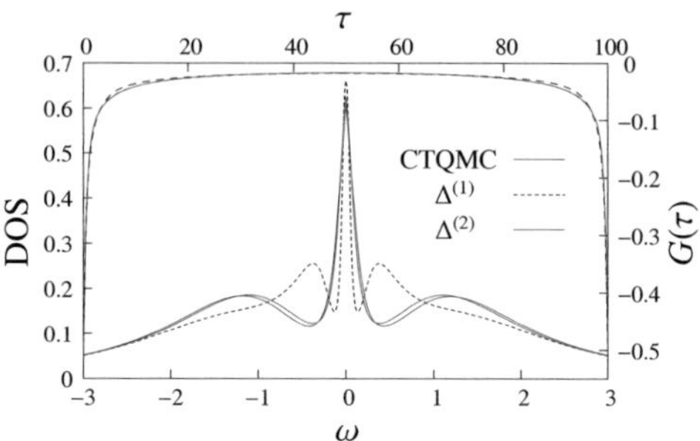



*Figure 9.5:* Low temperature results for $U = 3$ and $\beta = 100$. Shown are the imaginary time Green function (upper and right axis) and the corresponding maximum entropy density of states (lower and left axis). The result obtained by superperturbation with two bath sites requires considerably less computational effort compared to QMC, and is almost indistinguishable from the CTQMC result.

the Baym-Kadanoff sense [61]. Results obtained from the first Dyson iteration achieve a similar quality of approximation. The lowest-order approximation $G^{\mathrm{d}} \approx G_0^{\mathrm{d}} + G_0^{\mathrm{d}}\Sigma^{\mathrm{d}}G_0^{\mathrm{d}}$ to the dual Green function, in some cases, can differ substantially.

Another central point for the applicability of the perturbative solver regards the convergence of the perturbation series. Results representative for the role of different diagrams in the perturbation expansion are shown in Fig. 9.3. These are obtained for a single bath site, for which the exact solution (open circles) and the initial ED Green function (filled circles) differ significantly. The approximate solutions are shown for different combinations of the diagrams of Fig. 9.1, as indicated in the legend. The parameters are otherwise the same as in the previous figure.

One can clearly see that diagram a) yields by far the largest correction to the initial solution. The value on the first Matsubara frequency is almost exactly reproduced. Recall that $\Delta^{(1)}$ is identical to $\Delta$ on the first Matsubara frequency. The largest deviations occur for the second and third frequency, for which also the discrete hybridization differs most strongly from the input hybridization. Zooming into this region reveals that all diagrams give a correction in the right direction, whereby the correction by diagram c) is already negligible. The maximum entropy density of states in Fig. 9.4 shows that also spectral properties are correctly reproduced. The analytical continuation of the quantum Monte Carlo data (dashed-dotted line) exhibits the two Hubbard bands at $\omega = \pm U/2$ and shows the Kondo resonance at the Fermi level. The Kondo physics cannot be reproduced by perturbing around the atomic limit ($\Delta^{(0)}$, solid line). This is in accordance with the findings in Ref. [156]. However, perturbation around the ED solution for a single bath site already captures the Kondo resonance and yields good agreement compared to the exact solution. The different values at the Fermi level (the DOS should be pinned to its noninteracting value) are likely to be an artefact of the analytical continuation.

In order to demonstrate that the approach also works for lower temperatures, results for $T = 0.01$ are shown in Fig. 9.5. Although the expansion around the solution for a single bath site ($\Delta^{(1)}$) shows small deviations in the imaginary time Green function $G(\tau)$, the approximation appears insufficient as seen in the density of states. The superperturbation around the two bath-site solution however is almost exact. One could anticipate that more bath sites are required at lower temperatures. This point is elucidated in more detail in the next section.

## 9.5 Role of the Reference Problem

In the previous section it was shown that the superperturbation solver is able to reproduce the Kondo resonance of the Anderson impurity model. In agreement with Ref. [156], it is found that the Kondo physics cannot be recovered by perturbing around the atomic limit. Indeed, the Kondo physics is known to be nonperturbative [4]. The superperturbation appears to circumvent this problem as it allows to perturb a reference problem that already contains this physics. This recovers the Kondo resonance already for the perturbation around ED for a single bath site as seen in Fig. 9.4. However, the role of the reference problem remains obscure and deserves further analysis. The singlet formation in the Kondo regime is best analyzed using the local impurity susceptibility $\chi := \int_0^\beta d\tau \langle S_z(\tau)S_z(0)\rangle$, where $S_z$ is an impurity operator ($\chi_{dd}$ in the notation of chapter 5).

In analogy to the Green function, one could also evaluate the susceptibility in a perturbative manner. This can be done along the lines described in chapter 10, but requires additional approximations for the vertex. For the special case of the static response, it is also possible to start from the definition $\chi = \partial M/\partial h$ and evaluate it for small fields using the finite difference expression

$$\chi = \left.\frac{\partial M}{\partial h}\right|_{h=0} \approx \frac{M(h) - M(0)}{h} \, , \qquad (9.28)$$

where $M = (n_\uparrow - n_\downarrow)/2$ is the magnetization of the impurity. The average occupation in the superperturbation can be obtained from the imaginary time Green function $n_\sigma = -G_\sigma(\beta)$ within a spinpolarized calculation. For fields of the order of $h \sim 10^{-5}$ the dependence on the field is perfectly linear and allows an accurate evaluation of the susceptibility. In ED, the susceptibility can be obtained in the same way, but it is more natural to use the Lehmann representation

$$\chi(i\Omega_m) = \frac{1}{Z}\sum_{lm}\frac{\langle l|S_z|m\rangle\langle m|S_z|l\rangle}{i\Omega + E_l - E_m}\left(e^{-\beta E_m} - e^{-\beta E_l}\right) \, , \qquad (9.29)$$

where $\Omega_m = 2\pi m/\beta$ is a bosonic frequency. This result is obtained in the same manner as the Lehmann representation for the fermionic Green function (2.38). The static response is obtained by letting $i\Omega_m = 0$ (Eq. 3.79).

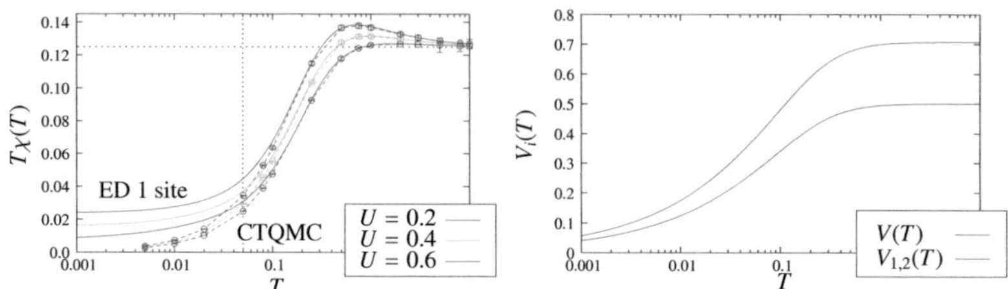


*Figure 9.6:* Left: Local susceptibility times temperature $T\chi(T)$ as a function of temperature obtained within CTQMC (dashed lines) and ED for a single bath site (solid lines) using the temperature dependent hybridization $V(T)$ shown in the right panel. In CTQMC the local moment is screened by the conduction electrons and $T\chi$ tends to zero for low temperatures (see chapter 5). In ED, the bath and impurity site become decoupled as the hybridization goes to zero. $T\chi$ approaches a finite constant value, i.e. $\chi$ diverges as $\sim 1/T$.

The superperturbation data presented in this section has been obtained using the implementation described in Sec. 9.7 below. The approximation to the dual self-energy therefore includes diagram a) of Fig. 9.1 only. The dual Green function was obtained by a single application of the Dyson equation, i.e. no self-consistent renormalization has been performed. The calculations have been carried out on Matsubara frequencies.

In Fig. 9.6, the impurity susceptibility times temperature (the effective local moment, see chapter 5) is shown as a function of temperature for different values of $U$ obtained from CTQMC according to Eq. 3.80 (left panel). The curves resemble those shown in Fig. 5.3. The increase of $T\chi$ at high temperatures is due to the formation of the local moment. At lower temperatures the moment is screened by the conduction electrons and $T\chi$ approaches zero. For comparison, the result from ED is plotted in the same figure. The bath parameters for $N = 1, 2$ were obtained according to the condition

$$\Delta^{(N)}(i\omega_n) = \Delta(i\omega_n) \, , n = 0, \ldots, N - 1 \tag{9.30}$$

as before. The parameter $V(T)$ for a single site and $V_{1,2}(T)$ for two sites as a function of temperature are shown in the right panel of Fig. 9.6. They strongly decrease as the temperature is lowered, while the parameters $\epsilon_1(T) = -\epsilon_2(T)$ for two sites have to approach the Fermi level (not shown). Note that as $\omega_0 = \pi T \to 0$, the hybridization (9.27) on the first Matsubara frequency approaches a constant, so that $V$ has to decrease in order for $|V|^2 /(i\omega_0)$ to remain finite. As a consequence, the impurity site and bath

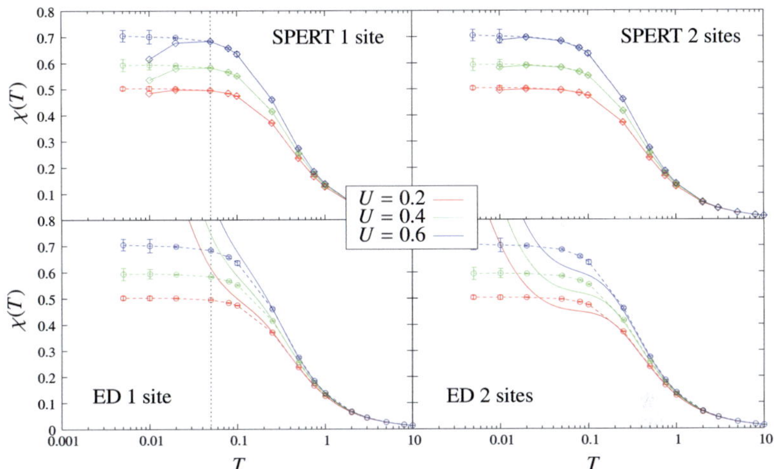

*Figure 9.7:* Local susceptibility of the impurity as a function of temperature for different values of $U$ obtained using the superperturbation (SPERT, top panel) and exact diagonalization (lower panel) for one (left) and two bath sites (right) in comparison with numerically exact continuous-time quantum Monte Carlo data (dashed lines). The hybridization $V$ was determined by condition (9.30) and is temperature dependent. $V(T)$ approaches zero so that at low temperatures the ED susceptibility approximately behaves as that of a free moment. The superperturbation cannot recover the (nonperturbative) singlet physics and has to break down at low temperature. The solution is systematically improved by increasing the number of bath sites.

site effectively become decoupled at low temperatures. This can be seen in the left panel of Fig. 9.6, where the ED curves approach a finite constant low temperature value and hence the susceptibility diverges as $\sim 1/T$.

The effect of this behavior on the superperturbation (SPERT) result is shown in Fig. 9.7. Here the susceptibility $\chi$ is plotted instead of $T\chi$, which better reveals the differences at low temperature. The CTQMC and ED data for one bath site in this figure is the same as in Fig. 9.6. The SPERT solution very closely follows the CTQMC result, even as the ED curves begin to depart from those obtained by CTQMC. For example, at $T = 0.05$ (vertical dotted lines) the SPERT and CTQMC results are virtually indistinguishable, although the ED curves differ from the latter substantially. Comparing to Fig. 9.6, one finds that at this temperature $T\chi$ obtained from ED is still far from being constant. At very low temperature, $T \sim 0.01$, the ED result for $T\chi$ is nearly constant, and the dual perturbation theory breaks down, reflected in a downturn of the suscepti-

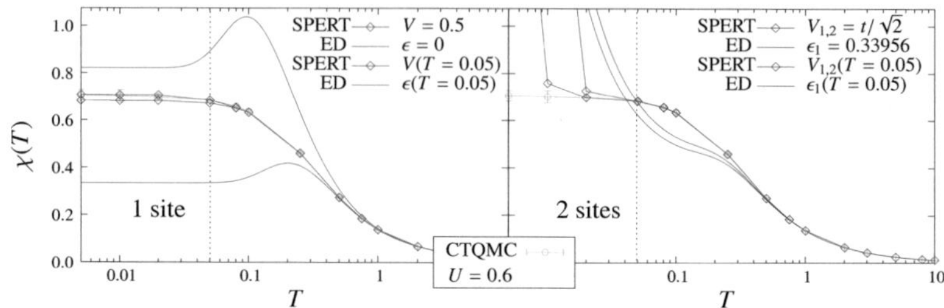

*Figure 9.8:* Local susceptibility $\chi(T)$ as a function of temperature for constant hybridization $V$ compared to ED for one (left) and two bath sites (right).

bility. The superperturbation cannot recover the correct behavior from the ED starting point, which exhibits local moment instead of singlet behavior. As can be seen from the figure, the solution is systematically improved down to lower temperatures by adding a bath site. The divergence of the ED result is shifted to lower temperatures, thus improving the superperturbation solution. These results are consistent with the observation made earlier, that a larger number of bath levels is required at low temperatures. For a given temperature, the deviation becomes larger as $U$ increases.

A divergent ED susceptibility can be avoided by using a hybridization independent of temperature. This in turn is achieved by choosing the discrete hybridization such that the high-frequency tail of $\Delta(i\omega_n)$ is well represented. In general, the hybridization sumrules given in Ref. [157] can be used for this purpose. For a Bethe lattice, these state that $\sum_k |V_k|^2 = t^2$. Inserting this into the discrete hybridization $\Delta^{(N)}(i\omega_n) = \sum_k |V_k|^2 / (i\omega_n - \epsilon_k) \to |V_k|^2 / (i\omega_n)$ in the limit $\omega_n \to \infty$, correctly reproduces the high energy asymptotics $t^2/(i\omega_n)$ of $\Delta(i\omega_n)$, which are obvious from (9.27). For a single bath site, this fixes $V = t = 1/2$. For two bath sites, $V_{1,2} = t/\sqrt{2} = 0.35355$. The free parameter $\epsilon_1 = -\epsilon_2$ is determined by conjugate gradient minimization of the distance function with a large cutoff $N_\omega = 600$ and $s = 0$ to emphasize the high frequency tail. This yields $\epsilon_1 = -\epsilon_2 = 0.33956$.

The result with thus obtained parameters is shown in Fig. 9.8. For a single bath site, the ED susceptibility does not diverge, as expected. Transitions to excited states are cut off for sufficiently low temperatures and the susceptibility becomes constant as expected for a ground state singlet. Remarkably, the superperturbation result recovers the proper low-temperature behavior and closely follows the CTQMC result. It therefore overcomes the problem of the discretization in ED, which otherwise necessitates the NRG treatment.

One can see that for $V = 1/2$, the result slightly differs from CTQMC for $T = 0.05$ (vertical dotted line), whereas the curves are virtually indistinguishable in Fig. 9.7.

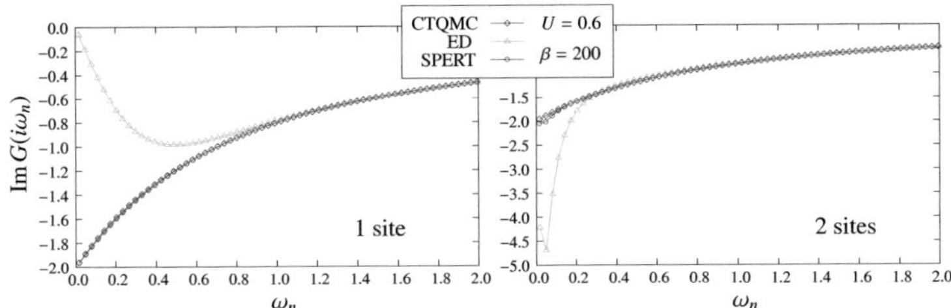



*Figure 9.9:* Imaginary part of the local Green function for $T = 0.005$. As for the susceptibility, the single-particle properties are reproduced only for one, but not for two bath sites.

Taking $V$ to be constant with its value at $T = 0.05$ according to Fig. 9.6, $V(T = 0.05) = 0.25916$, reproduces the result at this temperature, but also leads to better agreement for lower temperatures. This, of course, is an ad hoc condition, but it nevertheless illustrates that the superperturbation result for $V = 0.5$ can be improved by choosing the reference problem closer to the exact result.

In the right panel of Fig. 9.8, the susceptibility was calculated for a reference system with two bath sites. This shows that increasing the number of bath levels not necessarily leads to a systematic improvement of the results. The ED ground state is not a singlet as an unpaired spin remains at low temperature. Here the ED susceptibility again diverges as $1/T$. Correspondingly, the SPERT result departs from the CTQMC curve at low temperatures. Choosing $V_{1,2}(T = 0.05) = 0.36651$ and $\epsilon_1(T = 0.05) = -\epsilon_2 = 0.27207$ slightly improves the solution at lower temperatures, but does not avoid the breakdown of the perturbation theory.

It is crucial that the underlying ED reference problem contains the relevant physics of the problem. In renormalization group terminology, the ED reference problem with a single bath site contains the two relevant high-temperature local moment and low-temperature singlet fixed points. In this case, apart from the large energy scale $U$, the vertex also contains a small energy scale $J \sim |V|^2 /U$. The results look promising. However it is, at present, not clear if summing up the infinite partial series obtained by using Dyson's equation (no self-consistent renormalization has been performed here) is sufficient to recover an exponentially small Kondo scale $T_\mathrm{K} \sim \exp(-1/J)$. This point requires further clarification. In order to obtain sensible results in the Kondo regime, the number of bath sites should be chosen odd.

In Fig. 9.9, the Matsubara Green functions obtained in SPERT are plotted together with the corresponding ED Green function and the CTQMC result at low temperature $T = 0.005$ ($\beta = 200$). The bath parameters are taken from Fig. 9.8: $V = 0.5$, $\epsilon = 0$

and $V_{1,2} = t/\sqrt{2}$, $\epsilon_1 = 0.33956$ for one and two bath sites, respectively. For either reference system, the perturbation is nontrivial. For an expansion around the reference system with a single bath site, the Green function is very well reproduced. Although the difference for the case of two bath sites is small, the Green function clearly does not correctly reproduce the spectral intensity due to the Kondo peak at the Fermi level. This is consistent with the large deviation of the susceptibility.

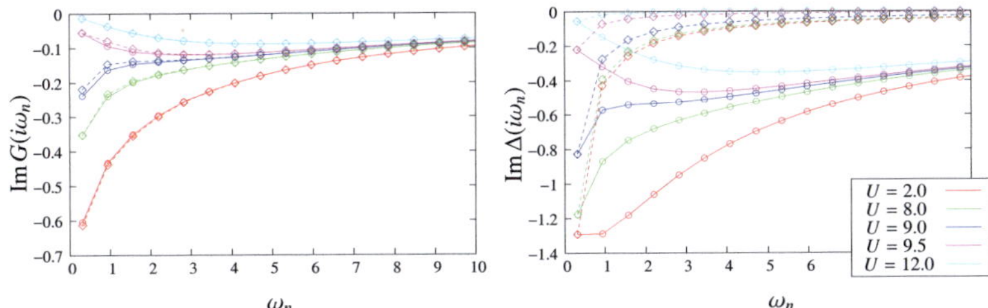

*Figure 9.10:* Left panel: Imaginary part of the selfconsistent solution for the local DMFT Green function $G(i\omega_n)$ obtained by superperturbation (dashed lines) in comparison with CTQMC results (solid lines), for $T/t = 0.1$. The largest deviations occur in the vicinity of the metal-insulator transition. Right panel: The representation of the self-consistent result for the hybridization function $\Delta(i\omega_n)$ (solid lines) by the discrete hybridization $\Delta^{(1)}$ (dashed lines). The curves coincide on the first Matsubara frequency by virtue of the condition (9.30).

## 9.6 Application within DMFT

Although it has been seen in the previous section that the solver works rather well for the Anderson impurity problem, it is a priori not clear how well it performs in the context of DMFT. Being an approximate solver, the error propagation through the iterations is possibly an issue, in particular for a small number of bath levels. As a test, the solver is applied to the two-dimensional Hubbard model at half-filling, in particular, to the Mott transition. In DMFT, the Mott transition in the paramagnetic phase has been shown to occur at a critical $U$ of $U_c \sim 9.35$ at $T/t \sim 0.1$ (see Ref. [149] and Sec. 6.8).

The implementation described in chapter 7 can be applied here. The self-consistency loop is depicted in Fig. 2.3. In the impurity solver step, the hybridization parameter is determined from the input hybridization $\Delta_{\text{old}}(i\omega_n)$ using the condition (9.30). The new hybridization function $\Delta_{\text{new}}(i\omega_n)$ is constructed from the impurity Green function in the usual way.

The calculations are carried out for a single bath site. Deviations are found to rapidly decrease as further bath sites are included, even close to the transition. The case of a small number of bath sites is also relevant for multiorbital problems, involving d- or f-systems. Including more bath sites for such systems is severely limited by computational resources. Results for the self-consistent local DMFT Green functions are shown in Fig. 9.10 (left panel) for various values of $U$. The solver works very well over the whole parameter range, even for intermediate coupling. The largest deviations occur for values of $U$ close to the critical value $U/t = 9.35$. The representation of the self-consistent DMFT hybridization through $\Delta^{(1)}$ is shown in the right panel of this figure. The deviation

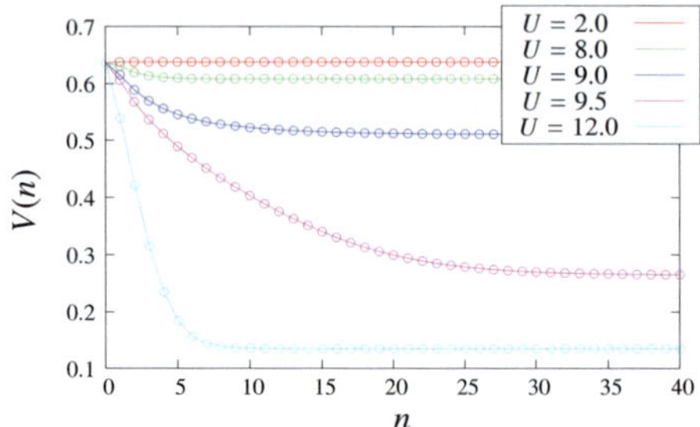


*Figure 9.11:* Hybridization parameter $V$ as a function of the number of DMFT iterations for different values of $U$. Close to the transition, the convergence is critically slow.

of the Green functions from the CTQMC result is not directly related to the degree of mismatch between $\Delta^{(1)}$ and $\Delta$. For illustration, the evolution of the bath parameter $V$ with the DMFT iterations is plotted in Fig. 9.11. For small to intermediate $U$, it converges rapidly to a constant value. For large $U$, the convergence is somewhat slower, since the same initial guess for the Weiss field, corresponding to $\Sigma(i\omega) \equiv 0$, was chosen for all calculations. For $U = 9.5$, right above the transition point, the convergence is critically slow. More than 50 iterations were required to reach a converged solution (this also occurs in DMFT with CTQMC). Remarkably, the solver very well captures the hysteresis at the Mott transition (not shown).

It should be noted that the computational cost is greatly reduced compared to CTQMC. A DMFT iteration for the single-site impurity problem with one bath site takes a few seconds on a standard PC. The gain in computational speed is of course relativized for problems with several orbitals. Nevertheless the solver can be expected to be useful as it becomes inherently difficult to obtain low noise data using QMC which would allow a reliable analytical continuation.

## 9.7   Analytical Continuation to the Real Axis

A particular strength of the superperturbation solver is that dynamical quantities can be analytically continued to the real axis. This is possible because analytical expressions are known for the building blocks of the perturbation expansion (the Lehmann representation of the single and two-particle Green function). The mathematically ill-posed problem of analytical continuation of the Matsubara Green function can hence be circumvented.

The formalism presented here was developed together with S. Brener and C. Jung. The major part of the implementation of the equations given in the remainder of this section was accomplished by C. Jung. The guidelines of the implementation are the same as described in chapter 7 and large parts of the structures could be reused. The container classes have been extended to work with functions of a continuous variable (on a discrete grid). This allowed a straightforward implementation of a general code, allowing multiorbital and spin-polarized calculations.

The analytical continuation is performed at Hartree-Fock level, i.e. using diagram a) of Fig. 9.1. This diagram was shown to give by far the dominant contribution to the self-energy. In principle, the treatment could be extended to higher orders, but this is an intricate task already for the subleading order and becomes computationally highly demanding. Inserting the expression for the vertex (9.12) into the appropriate expression for the dual self-energy (9.18) yields

$$
\begin{aligned}
(\Sigma^{\mathrm{d}}_{(a)})_{12} &= -\gamma_{1234}(G^{\mathrm{d}}_0)_{43} = -g^{-1}_{11'}g^{-1}_{33'}(\chi_{1'2'3'4'} - \chi^0_{1'2'3'4'})g^{-1}_{2'2}g^{-1}_{4'4}(G^{\mathrm{d}}_0)_{43} \\
&= -g^{-1}_{11'}g^{-1}_{33'}\,\chi_{1'2'3'4'}\,g^{-1}_{2'2}g^{-1}_{4'4}(G^{\mathrm{d}}_0)_{43} + g^{-1}_{11'}g^{-1}_{33'}\,\chi^0_{1'2'3'4'}\,g^{-1}_{2'2}g^{-1}_{4'4}(G^{\mathrm{d}}_0)_{43} \\
&=: \Sigma^{\mathrm{d}\,1}_{12} + \Sigma^{\mathrm{d}\,0}_{12}\ .
\end{aligned}
\tag{9.31}
$$

$\Sigma^{\mathrm{d}\,0}_{12}(\Sigma^{\mathrm{d}\,1}_{12})$ denote the (non-)trivial part of the dual self-energy. For notational convenience and to establish a connection to Ref. [156], it is useful to introduce the function

$$
\Delta^{\mathrm{d}}_\omega := \left[D^{-1}_\omega + g_\omega\right]^{-1}\ ,
\tag{9.32}
$$

which will be referred to as the dual "hybridization". In terms of this quantity, the auxiliary Green function (9.13) can be written

$$
G^{\mathrm{d}}_{0\,\omega} = -g_\omega\Delta^{\mathrm{d}}_\omega g_\omega\ .
\tag{9.33}
$$

The trivial part of the self-energy is readily found to be

$$
\begin{aligned}
\Sigma^{\mathrm{d}\,0}_{12} &= -g^{-1}_{11'}g^{-1}_{33'}\beta(g_{1'2'}g_{3'4'} - g_{1'4'}g_{3'2'})g^{-1}_{2'2}g^{-1}_{4'4}g_{45}\Delta^{\mathrm{d}}_{56}g_{63} \\
&= -g^{-1}_{11'}g^{-1}_{33'}\beta(\delta_{1'2}\delta_{3'4} - \delta_{1'4}\delta_{3'2})g_{45}\Delta^{\mathrm{d}}_{56}g_{63} \\
&= -g^{-1}_{12}\beta\,\mathrm{Tr}[\Delta^{\mathrm{d}}\,g] + \Delta^{\mathrm{d}}_{12}\ .
\end{aligned}
\tag{9.34}
$$

This expression can straightforwardly be evaluated on the real axis provided that the ED Green function $g$ and $\Delta^{\mathrm{d}}$ are known on the real axis. The Green function is analytically continued by letting $i\omega \to E + i0^+$ in the Lehmann representation, Eq. 2.38. $\Delta^{\mathrm{d}}$ additionally contains $\Delta$ and $\Delta^{(N)}$. Both are known for a complex argument according to Eqs. 9.3 and 9.27. For realistic calculations, an analytical form of the hybridization $\Delta(i\omega)$ can be constructed from the LDA Hamiltonian $h_{\mathbf{k}}$ as

$$
\Delta(i\omega) = (i\omega + \mu)\mathbb{1} - \left\{\frac{1}{N}\sum_{\mathbf{k}}\left[(i\omega + \mu)\mathbb{1} - h_{\mathbf{k}}\right]^{-1}\right\}^{-1}
\tag{9.35}
$$

In the following, the semielliptical form (9.27) was used. When analytically continuing this expression, some care must be taken for the branch cut. The square root has a branch cut discontinuity along the negative real axis. Correspondingly, $\Delta(z)$ inherits a branch cut for $-2t < \omega < 2t$. The real part is continuous across the real axis, but the imaginary part changes sign, so that $\Delta(\omega + i0^+) - \Delta(\omega + i0^-) = 2i \, \mathrm{Im} \, \Delta(\omega)$.

The analytical continuation of the nontrivial part of the dual self-energy requires more effort. First, write it in the form

$$S_{1'2'} := g_{1'1} \Sigma_{12}^{\mathrm{d}\,1} g_{22'} = g_{33'}^{-1} \chi_{1'2'3'4'} g_{4'4}^{-1} g_{45} \Delta_{56}^{\mathrm{d}} g_{63} = \chi_{1'2'3'4'} \Delta_{4'3'}^{\mathrm{d}} \,. \tag{9.36}$$

The same expression appears in Eq. 9.16 of Ref. [156] with $\Delta^{\mathrm{d}}$ replaced by $\Delta$. In this reference, it was shown how to perform an analytical continuation of this expression for the particular case of the hybridization function (9.27), but no calculations were performed. The relevant formula can be derived in a similar fashion as the derivation of the Lehmann representation of the two-particle Green function in ED. In the notation of Ref. [156], the result is

$$S_{\alpha\beta}(i\omega_n) = \frac{1}{\beta} \sum_{\omega_n'} \chi_{\alpha\beta\gamma\delta}(i\omega_n, i\omega_n, i\omega_n', i\omega_n') \Delta_{\delta\gamma}^{\mathrm{d}}(i\omega_n') = \sum_{ijkl} \sum_{\delta\gamma} \times$$

$$\times (C_\alpha)_{ij}(C_\gamma^\dagger)_{jk}(C_\delta)_{kl}(C_\beta^\dagger)_{li} \left[ \frac{\mathcal{R}_{\gamma\delta}(E_j, E_k)}{E_{jl}(i\omega_n - E_{ji})} + \frac{\mathcal{R}_{\gamma\delta}(E_l, E_k)}{E_{lj}(i\omega_n - E_{li})} + \frac{Q_{\gamma\delta}(i\omega_n, E_i, E_k)}{(i\omega_n - E_{li})(i\omega_n - E_{ji})} \right]$$

$$+ (C_\alpha)_{ij}(C_\delta)_{jk}(C_\gamma^\dagger)_{kl}(C_\beta^\dagger)_{li} \left[ \frac{\mathcal{R}_{\gamma\delta}(E_k, E_j)}{E_{jl}(i\omega_n - E_{ji})} + \frac{\mathcal{R}_{\gamma\delta}(E_k, E_l)}{E_{lj}(i\omega_n - E_{li})} - \frac{Q_{\gamma\delta}(-i\omega_n, E_k, E_i)}{(i\omega_n - E_{li})(i\omega_n - E_{ji})} \right]$$

$$+ (C_\gamma^\dagger)_{ij}(C_\delta)_{jk}(C_\alpha)_{kl}(C_\beta^\dagger)_{li} \left[ \frac{\mathcal{R}_{\gamma\delta}(E_k, E_j)}{E_{ki}(i\omega_n - E_{lk})} + \frac{\mathcal{R}_{\gamma\delta}(E_i, E_j)}{E_{ik}(i\omega_n - E_{li})} - \frac{Q_{\gamma\delta}(-i\omega_n, E_l, E_j)}{(i\omega_n - E_{li})(i\omega_n - E_{lk})} \right]$$

$$+ (C_\delta)_{ij}(C_\gamma^\dagger)_{jk}(C_\alpha)_{kl}(C_\beta^\dagger)_{li} \left[ \frac{\mathcal{R}_{\gamma\delta}(E_j, E_k)}{E_{ki}(i\omega_n - E_{lk})} + \frac{\mathcal{R}_{\gamma\delta}(E_j, E_i)}{E_{ik}(i\omega_n - E_{li})} + \frac{Q_{\gamma\delta}(i\omega_n, E_j, E_l)}{(i\omega_n - E_{li})(i\omega_n - E_{lk})} \right]$$

$$+ (C_\delta)_{ij}(C_\alpha)_{jk}(C_\gamma^\dagger)_{kl}(C_\beta^\dagger)_{li} \frac{1}{(i\omega_n - E_{kj})(i\omega_n - E_{li})} \times$$

$$\times \left[ \mathcal{R}_{\gamma\delta}(E_k, E_l) - \mathcal{R}_{\gamma\delta}(E_j, E_i) + Q_{\gamma\delta}(i\omega_n, E_j, E_l) - Q_{\gamma\delta}(-i\omega_n, E_k, E_i) \right]$$

$$+ (C_\gamma^\dagger)_{ij}(C_\alpha)_{jk}(C_\delta)_{kl}(C_\beta^\dagger)_{li} \frac{1}{(i\omega_n - E_{kj})(i\omega_n - E_{li})} \times$$

$$\times \left[ \mathcal{R}_{\gamma\delta}(E_l, E_k) - \mathcal{R}_{\gamma\delta}(E_i, E_j) + Q_{\gamma\delta}(i\omega_n, E_i, E_k) - Q_{\gamma\delta}(-i\omega_n, E_l, E_j) \right] , \tag{9.37}$$

where the abbreviation $E_{ij} := E_i - E_j$ and the functions

$$\mathcal{R}_{\gamma\delta}(E_i, E_j) := (X_i + X_j) \frac{1}{\beta} \sum_{i\omega_n'} \frac{\Delta_{\delta\gamma}^{\mathrm{d}}(i\omega_n')}{i\omega_n' - E_{ij}} \,, \tag{9.38}$$

$$Q_{\gamma\delta}(i\omega_n, E_i, E_j) := (X_i - X_j) \frac{1}{\beta} \sum_{i\omega_n'} \frac{\Delta_{\gamma\delta}^{\mathrm{d}}(i\omega_n')}{i\omega_n' - i\omega_n - E_{ij}} \tag{9.39}$$

have been introduced. Latin indices label the eigenstates. The quantities $X_i$ denote the probabilities $X_i = (1/\mathcal{Z})\exp(-\beta E_i)$ of the eigenstates. Using Eqs. 9.36, 9.37 together with (9.34) in (9.31) is equivalent to the previous implementation which uses Eq. 9.18 with the vertex expressed in terms of the two-particle Green function according to (9.12). We have verified that the two implementations numerically (i.e. within the effective machine precision $\epsilon \sim 10^{-16}$) yield the same results when evaluated on Matsubara frequencies. Some care needs to be taken in the numerical evaluation of (9.37). The terms containing $\mathcal{R}(E_i, E_j)$ in angular brackets in the second to fifth line are singular for degenerate levels. The proper limit $E_i \rightarrow E_j$ is evaluated using l'Hôpital's rule. Note further that division by zero occurs in (9.39) for $i\omega_n' = i\omega_n$ and $E_i = E_j$. Simultaneously the prefactor $(X_i - X_j)$ vanishes. Thus the value of the sum in this case stems only from the term $i\omega_n' = i\omega_n$. The result is obtained by using the identity

$$X_i \pm X_j = \frac{1}{\mathcal{Z}}\left(e^{-\beta E_i} \pm e^{-\beta E_j}\right) = \frac{\pm 1}{\mathcal{Z}}e^{-\beta E_i}\left(e^{\beta E_{ij}} \pm 1\right) = \pm X_i\left(e^{\beta E_{ij}} \pm 1\right) \tag{9.40}$$

for the negative sign and l'Hôpital's rule to get

$$Q(i\omega_n, E_i, E_i) = X_i \Delta^{\mathrm{d}}(i\omega_n') . \tag{9.41}$$

With (9.37), the analytical continuation to the real axis is effectively reduced to analytical continuation of the function $Q$. The quantity $\mathcal{R}$ can in principle be evaluated on Matsubara frequencies, or using the analytical continuation of $\Delta^{\mathrm{d}}(i\omega)$. Since (9.37) depends on these quantities in a complicated manner and either method involves numerical errors, we evaluate $\mathcal{R}$ in the same manner as $Q$. $\mathcal{R}$ is straightforwardly rewritten in terms of a contour integral using the standard procedure depicted in Fig. 9.12. The sum over Matsubara frequencies is expressed through a contour integral by means of the residue theorem

$$\frac{1}{2\pi i}\oint_C f(z)dz = \sum_k \mathrm{Res}(f, a_k) , \tag{9.42}$$

where $f$ is analytic in the complex plane except for a discrete set of points $a_k$, which are the poles of $f$. The function $f$ is chosen such that the sum of residues reproduces the sum over Matsubara frequencies. This is accomplished by the Fermi function, which has simple poles at the location of the Matsubara frequencies. Expanding $\exp(\beta\omega)$ about $i\omega_n$ yields $\exp(i\beta\omega_n)[1 + \beta(\omega - i\omega_n) + \ldots] = -1[1 + \beta(\omega - i\omega_n) + \ldots]$, by the definition of the Matsubara frequencies. Hence the residue of the Fermi function $f(\omega) = 1/[1 + \exp(\beta\omega)] = -1/[\beta(\omega - i\omega_n) + \ldots]$ is read off to be $-1/\beta$. The summation of a function $F(i\omega_n)$ over Matsubara frequencies is hence expressed in terms of its unique analytical continuation $F(z)$ through a contour integral as

$$\frac{1}{\beta}\sum_{\omega_n} F(i\omega_n) = -\sum_n \mathrm{Res}\left(\frac{F(z)}{e^{\beta z} + 1}, \omega_n\right) = -\frac{1}{2\pi i}\oint_C f(z)F(z)dz . \tag{9.43}$$

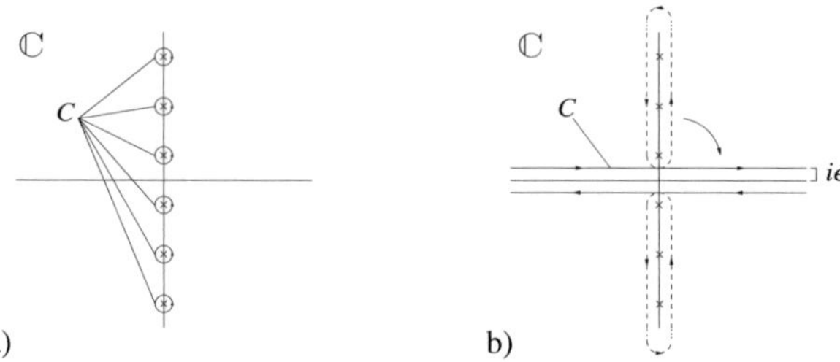

*Figure 9.12:* Contour integration for $\mathcal{R}(E_i, E_j)$ (9.38). The sum over Matsubara frequencies located at $z = i\omega_n = i(2n+1)\pi/\beta$ (crosses) is transformed into a contour integral over the Fermi function, which has simple poles at these frequencies (a). The residues produce the individual terms in the sum. The contour is deformed to be parallel to the imaginary axis (dashed lines) and closed at infinity (b). The final contour runs parallel to the real axis and is offset by $\pm i\epsilon$ (solid lines)

The contour is deformed to be parallel to the real axis as shown in Fig. 9.12 b). Application to (9.38) yields the following expression:

$$\mathcal{R}_{\alpha\beta}(E_i, E_j) = -\frac{X_i + X_j}{2\pi i}\left[\int_{-\infty}^{\infty} \frac{f(z^+)\Delta_{\alpha\beta}^{\mathrm{d}}(z^+)}{z^+ - E_{ij}}dz - \int_{-\infty}^{\infty} \frac{f(z^-)\Delta_{\alpha\beta}^{\mathrm{d}}(z^-)}{z^- - E_{ij}}dz\right], \qquad (9.44)$$

where $z^\pm = z \pm i\epsilon$. Here $\epsilon$ is a small finite number, which is restricted to be smaller than the first Matsuara frequency $\pi/\beta$ (otherwise one needs to account for the residue of the pole(s) excluded from the contour). For the hybridization function $\Delta$ in (9.27), this result can be further simplified by taking $\epsilon \to 0$. In this limit, $f(x + i\epsilon)\Delta(x + i\epsilon) - f(x - i\epsilon)\Delta(x - i\epsilon) = 2if(x)\,\mathrm{Im}\,\Delta(x)$. Defining the real principal value integrals

$$\Delta^\pm(z) = -\frac{1}{\pi}\mathcal{P}\int_{-\infty}^{\infty} \frac{f(\pm x)\,\mathrm{Im}\,\Delta(x)dx}{z - x}, \qquad (9.45)$$

(9.44) can be expressed as

$$\mathcal{R}(E_i, E_j) = (X_i + X_j)[-\Delta^+(E_{ij}) + f(E_{ij})\,\mathrm{Re}\,\Delta(E_{ij})] . \qquad (9.46)$$

The second term stems from the pole at $x = E_{ij}$ which for $\epsilon \to 0$ lies on the real axis. It has to be omitted from the contour by encircling it clockwise on a circle with the radius $r$ taken to zero. Using the parameterization for the circle, $z(\varphi) = E_{ij} + re^{i\varphi}$, the corresponding contour integral is

$$-\frac{(X_i + X_j)}{2\pi i}\int_0^{2\pi} i\,d\varphi f(E_{ij} + re^{i\varphi})\Delta(E_{ij} + re^{i\varphi}) \underset{r \to 0}{=} -(X_i + X_j)f(E_{ij})\,\mathrm{Re}\,\Delta(E_{ij}) , \qquad (9.47)$$

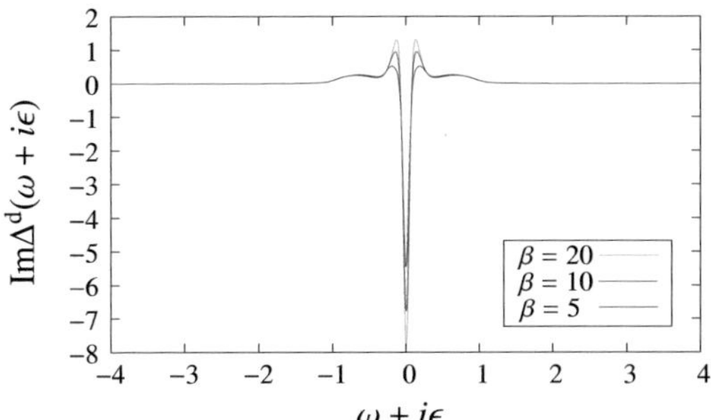

*Figure 9.13:* Dual hybridization function $\Delta^{\mathrm{d}}(\omega+i\epsilon)$ on the integration contour parallel to the real axis with offset $i\epsilon$ for different temperatures. The parameters are $U = 4$, $V = 0.5$ and $\epsilon = 0.1$.

where it was used that $\Delta(x \pm i0^+) = \operatorname{Re}\Delta(x) \pm \operatorname{Im}\Delta(x)$. An additional sign enters to account for the converse sense of rotation. With $f(x) + f(-x) = 1$, $\Delta(E_{ij}) = \Delta^+(E_{ij}) + \Delta^-(E_{ij})$ and (9.40), Eq. 9.46 can be written

$$\mathcal{R}(E_i, E_j) = \operatorname{Re}[X_i\Delta^-(E_{ij}) - X_j\Delta^+(E_{ij})] , \tag{9.48}$$

which is the result of Ref. [156]. This expression is not applicable here, due to the structure of the "hybridization" $\Delta^{\mathrm{d}}$. Take the case of a single bath site as a generic example. In this case the discrete hybridization is given by $\Delta^{(N)} = |V|^2 /(i\omega)$ (the impurity level is located at zero for symmetry reasons). $\Delta^{(N)}$ has a pole on the real axis at real frequency $\omega = 0$. As a consequence, since $\Delta$ remains finite, $D^{-1} = (\Delta^{(N)} - \Delta)^{-1}$ is negligible and the dual hybridization according to (9.32) behaves as $\Delta^{\mathrm{d}} \sim g^{-1}$. This can in turn be written $g^{-1} = i\omega - \Delta^{(N)} - \Sigma(i\omega)$. As $\Sigma$ remains finite, $\Delta^{\mathrm{d}}$ has a pole of first order, and the integrand has a pole of second order at zero. In order to shift the contour onto the real axis would hence require the knowledge of the residues of $\Delta^{\mathrm{d}}$ on the real axis, which are not directly accessible. We therefore use the integration contour in Fig. 9.12 b). The dual hybridization is plotted on the contour in Fig. 9.13. Due to the pole at $\omega = 0$, $\Delta(\omega + i\epsilon)$ is sharply peaked. The range of numerical integration is determined by the behavior of the integrand. It contains a product of $\Delta$ with the Fermi function and therefore falls off exponentially at high energy.

Having the quantities $\mathcal{R}$ at hand, the first term in the trivial part of the dual self-energy, Eq. 9.34, is determined according to

$$\operatorname{Tr}[\Delta^{\mathrm{d}}g] = \sum_{ij} \sum_{\delta\gamma}(C_\delta)_{ij}(C_\gamma^\dagger)_{ji}\mathcal{R}_{\gamma\delta}(E_j, E_i) . \tag{9.49}$$

The analytical continuation of the function $Q$ requires some care. For fixed argument $\omega_n$, the Matsubara sum is replaced by a contour integral

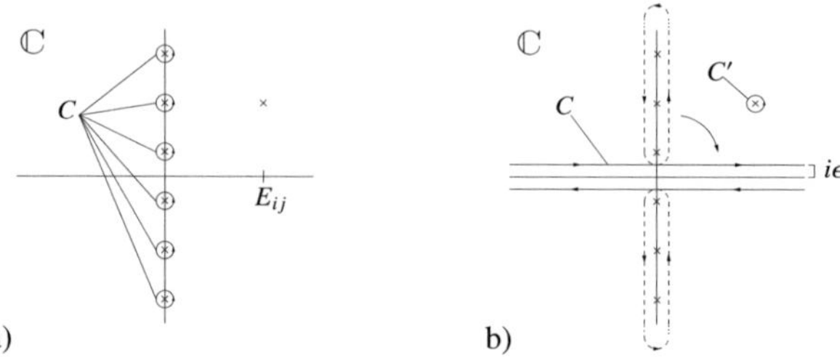



*Figure 9.14:* Contour integration for the analytical continuation of $Q(i\omega_n, E_i, E_j)$ (9.39). The integrand in (9.50) has an additional pole at $z = E_{ij} + i\omega_n$ (a). Deforming the contour as shown in b) crosses the pole and the residue of the pole, given by the integral over the contour $C'$, has to be subtracted.

$$Q(i\omega_n, E_i, E_j) = -\frac{1}{2\pi i} \oint_C \frac{f(z)\Delta^{\mathrm{d}}(z)}{z - i\omega_n - E_{ij}} dz \,, \tag{9.50}$$

where the contour is parallel to and infinitesimally shifted from the imaginary axis. The integrand obviously has simple poles for $E_{ij} \neq 0$ at $z = i\omega_n + E_{ij}$. Deforming the contour as depicted in Fig. 9.14 requires to pass the pole and hence the residue must be subtracted. The residue is

$$\frac{-1}{2\pi i} \oint_{C'} \frac{f(z)\Delta^{\mathrm{d}}(z)}{z - i\omega_n - E_{ij}} = -\frac{\Delta^{\mathrm{d}}(i\omega_n + E_{ij})}{e^{\beta(i\omega_n + E_{ij})} + 1} = \frac{\Delta^{\mathrm{d}}(i\omega_n + E_{ij})}{e^{\beta E_{ij}} - 1}\,. \tag{9.51}$$

The definition of the Matsubara frequencies is used to set $\exp(\beta\omega_n) = -1$ before analytical continuation. The same situation is encountered when evaluating certain Matsubara frequency sums using contour integration [98]. Deforming the contour, one arrives at

$$Q(\tilde{z}, E_i, E_j) = (X_i - X_j)\left[\frac{-1}{2\pi i}\left(\int_{-\infty}^{\infty} \frac{f(z^+)\Delta^{\mathrm{d}}(z^+)}{\tilde{z} - z^+ - E_{ij}} dz - \int_{-\infty}^{\infty} \frac{f(z^-)\Delta^{\mathrm{d}}(z^-)}{\tilde{z} - z^- - E_{ij}} dz\right) - \frac{\Delta^{\mathrm{d}}(\tilde{z} + E_{ij})}{e^{\beta E_{ij}} - 1}\right] \tag{9.52}$$

with $z^{\pm} = z \pm i\epsilon$, as before. Note that in the limit $E_i \to E_j$, one has $X_i - X_j = 0$ so that the first term in the angular brackets vanishes. The singular contribution from the residue (9.51) is canceled by $(X_i - X_j)$ according to (9.40), which leaves $X_1 \Delta^{\mathrm{d}}_{\alpha\beta}(\tilde{z})$, thus recovering (9.41). The resulting function is the analytic continuation for all points in the complex plane with $|\mathrm{Im}\, z| > \epsilon$, i.e. inside the contour.

Eq. 9.37 and ultimately the Green function of the impurity model are evaluated parallel to the real axis for $z = \omega + i\delta$ with some small offset $\delta$. The shift leads to a

Lorentzian broadening of the $\delta$-peaks on the real axis. This should not be confused with the displacement $\epsilon$ of the contour from the real axis. Note that $\delta$ is restricted to values larger than $\epsilon$.

The computational effort to evaluate (9.37) needed to obtain the dual self-energy is considerable: For a given point $\omega + i\delta$ near the real axis, for which the final Green function or spectral function is to be calculated, one needs to evaluate $Q(\omega + i\delta, E_i, E_j)$ by contour integration for all possible energy differences $E_i - E_j$. It is advisable to temporarily store $\mathcal{R}(E_i, E_j)$ and $Q(\omega + i\delta, E_i, E_j)$ in memory, since these quantities are repeatedly used in the four-dimensional sum over the states in the Hilbert space. Each contour integral in turn requires the evaluation of $\Delta^d$ and therefore the ED Green function on a grid determined by the integration routine.

Due to the structure of $\Delta^d(i\omega)$ close to the real axis (Fig. 9.13), it is difficult to accurately evaluate the integrals (9.44) and (9.52) using standard quadrature rules (e.g. Simpson's rule) on a regular grid. We use an adaptive integration routine `gsl_integration_qag` from the GNU Scientific Library [158], which yields very accurate results. The stability of the integration is confirmed by the fact that for $\epsilon$ not too small, the result is independent of $\epsilon$ as it should be. The accuracy can further be checked by evaluating $\mathcal{R}(E_i, E_j)$ and $Q(z, E_i, E_j)$ on Matsubara frequencies and comparing the results obtained by contour integration (Eqs. 9.44 and 9.52) and summation over frequencies according to (9.38) and (9.39).

The disadvantage of adaptive integration is that the positions at which the dual hybridization and consequently the ED Green function are evaluated are not known a priori. As a consequence, the ED Green function cannot be stored in memory, but has to be evaluated repeatedly, possibly at the same points. The computational cost to evaluate this function increases rapidly with the size of the Hilbert space.

Using the above equations for the dual self-energy, the dual Green function can be obtained on the real axis. Rather than using (9.17) for the evaluation of the dual propagator, Dyson's equation is used,

$$(G^{d})^{-1} = (G_0^{d})^{-1} - \Sigma_{(a)}^{d} \, . \tag{9.53}$$

With the equations in the form given here, it is not directly possible to perform the Dyson iterations, as they use the specific structure of $G_0^d$. They can however straightforwardly be written in a suitable form. For example, Eq. 9.36 has to be written in the form $g_{1'1}\Sigma_{12}^{d\,1}g_{22'} = \chi_{1'2'3'4'}g_{4'4}^{-1}(G_0^d)_{43}g_{33'}^{-1}$. Note that in order to perform the Dyson iterations, the dual Green function would have to be evaluated on the integration contour, i.e. for $i\delta = i\epsilon$. In this case the integrand in (9.52) has poles on the contour, which have to be accounted for in a similar fashion as in (9.46). Once an approximation for the dual Green function is obtained, the impurity Green function is calculated using the identity (9.19). For illustration, the density of states (DOS) $A(\omega + i\delta) := (-1/\pi)\,\text{Im}\,G(i\omega + i\delta)$ obtained by direct analytical continuation in comparison with the analytical continuation

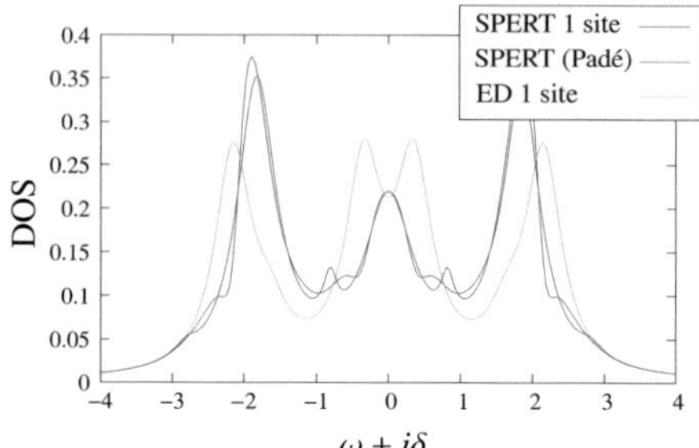


*Figure 9.15:* Density of states $-(1/\pi)\operatorname{Im}G(\omega + i\delta)$ obtained by direct analytical continuation of the superperturbation equations. The result is compared to analytical continuation of the Matsubara Green function using Padé approximants. In addition, the density of states of the ED reference problem is shown. Parameters are $U = 3, \beta = 10, t = 0.5, V = 0.5$ and $\delta = 0.3$.

using Padé approximants and the DOS of the reference problem with a single bath site is shown in Fig. 9.15. The Padé algorithm was implemented according to Ref. [108]. The stability of this routine sensitively depends on the numerical accuracy. Long double (24 Byte) precision arithmetic is used for this algorithm. The analytical continuation using Padé resembles the overall behavior of the direct analytical continuation rather well. The best agreement is found in the vicinity of the Fermi level, as one would expect. Deviations occur at higher energies. For complicated multiorbital systems, the direct analytical continuation may become superior.

The direct analytical continuation indicates causality problems at very low temperatures. The DOS becomes negative for certain energies. This is related to the fact that the perturbation $\Delta^{(N)} - \Delta$ diverges at the energy of the bath levels for $T = 0$ and the perturbation theory breaks down for low temperatures. Temperature plays the role of a low-energy cutoff. This is reminiscent of the breakdown of the non-crossing approximation. Increasing the number of bath sites alleviates this problem.

## 9.8 Conclusions and Outlook

The superperturbation solver presented here has been shown to be an efficient approximate solver for the quantum impurity problem. It produces results with good agreement compared to numerically exact quantum Monte Carlo results already for a single bath site and over a wide range of parameters. It has further been shown to be applicable as an impurity solver in the context of DMFT. For not too large system sizes, the computational effort for such calculations is considerably reduced compared to QMC.

The perturbation series of the dual self-energy has the favorable property that the lowest-order diagram represents by far the dominant contribution. In addition, the results can be systematically improved by increasing the number of bath sites. However, it was found that the improvement is not necessarily monotonous, but may depend crucially on the physics of the underlying reference problem. The choice of the bath geometry and the bath parameters should therefore be guided by physical considerations. If the reference problem contains the physics of the singlet formation, the Kondo resonance of the model is well reproduced.

The results make the solver a promising tool for the solution of multiorbital quantum impurity problems in the context of LDA+DMFT. The formulation given here is suitable for multiorbital systems. For the application as a solver in the context of realistic LDA+DMFT calculations, it is important that it can handle general, complex local interactions. For such applications, it is vital to test the behavior of the solver away from half-filling. Since the choice of the underlying reference problem has proven significant, it needs to be elucidated how bath sites should be connected to the impurity levels for a multiorbital problem. A priori, different topologies are possible.

In particular for multiorbital problems, analytical continuation of imaginary time quantum Monte Carlo data afflicted with significant statistical errors becomes inherently difficult. The present solver circumvents this problem as the Green function can be obtained without statistical errors or directly analytically continued to the real frequency axis. These properties make the solver predestined for investigating multiplet effects in solids.

For such applications the solver has to be parallelized, since the computational effort in the ED part of the calculation increases rapidly with the number of orbitals. In Ref. [39], calculations of the full two-particle Green function on three independent frequencies for up to five sites (including impurity and bath sites) were performed with a parallelized code. Here, using only the first diagram in the perturbation expansion, the evaluation of the two-particle Green function on two independent frequencies is sufficient. More significantly, formula (9.37) allows to obtain the self-energy on a given Matsubara frequency by a single evaluation of the trace ($Q$ and $\mathcal{R}$ require the evaluation of the ED Green function and should be stored in memory). This way, calculations on Matsubara frequencies for at least d-systems become tractable with a parallelized code. However as discussed in the previous section, the direct analytical continuation is computationally more costly. A further interesting perspective is the application to nonequilibrium systems, for which the approach is reformulated within the Keldysh formalism. Work on these issues is currently in progress.

# Chapter 10

# Dual Fermion Approach to Susceptibility of Correlated Lattice Fermions

S. Brener, H. Hafermann, A. N. Rubtsov,
M. I. Katsnelson and A. I. Lichtenstein
Phys. Rev. B **77**, 195105 (2008)

Susceptibilities are important to determine the response of a system to an external perturbation. They are related to measurable quantities via linear response theory (see. Sec. 3.3.10) and provide a means to study, e.g., transport properties. Susceptibilities also contain information of the system in absence of a relevant perturbation, as they are indicative of phase transitions. For example, the instability of a system to a magnetically ordered state is reflected in the divergence of the susceptibility in the corresponding channel: An infinitesimal external field will trigger a finite response, the system has a finite (spontaneous) magnetization.

The standard approach to this problem is to start from the Bethe-Salpeter equation (BSE) for the two-particle Green function. To simplify the notation, it is convenient to introduce the four-vector representation $k \equiv \{\mathbf{k}, i\omega_n\}$ where $\sum_k$ denotes the summation $(T/N) \sum_\mathbf{k} \sum_\omega$. In the following, $q \equiv \{\mathbf{q}, \Omega_m\}$ is used to label the momentum and frequency of collective two-particle excitations, so that $\Omega_m$ is a bosonic Matsubara frequency. In this notation, the Bethe-Salpeter equations in the spin and charge channels can be written (see, e.g., Refs. [15, 40])

$$\chi^{\text{sp/ch}}(k, k'q) = \chi_0(k, q)\delta_{kk'} \pm \chi^0(k, q) \sum_{k''} \Gamma_{\text{irr}}^{\text{sp/ch}}(k, k'', q)\chi^{\text{sp/ch}}(k'', k', q) , \qquad (10.1)$$

which defines the irreducible vertex in the respective channel and the upper (positive) sign corresponds to the spin channel. The disconnected part of the two-particle Green

*Figure 10.1:* Bethe-Salpeter equation for the two-particle Green function. $\Gamma_{(\mathrm{irr})}$ stands for the (irreducible) vertex and lines denote dressed Green functions.

function which contributes to the susceptibility is given by $\chi_0(k, q) = -G(k)G(k + q)$. Using supermatrix inversion, the equations are solved according to

$$[\chi^{\mathrm{sp/ch}}(k, k', q)]^{-1} = [\chi_0^{\mathrm{sp/ch}}(k, k', q)]^{-1} \mp \Gamma_{\mathrm{irr}}^{\mathrm{sp/ch}}(k, k', q) ,\qquad (10.2)$$

where all quantities are matrices in $k$, $k'$. Here the sign of the irreducible vertex is chosen such that taking $\Gamma_{\mathrm{irr}}^{\mathrm{sp/ch}} = U$ recovers the familiar result for the random-phase approximation (RPA) result for the spin (charge) susceptibility, $\chi_{\mathrm{RPA}} = \chi_0/(1 \mp U\chi_0)$. The BSEs are exact in the sense that, given the exact single-particle Green function and the exact irreducible vertex in one of the possible channels, one can extract the exact two-particle Green function. In general, neither quantity is known exactly, and the two-particle Green function is evaluated based on an approximation to the irreducible vertex (and the single-particle Green function). Once an approximate solution for the two-particle Green function is found, the corresponding susceptibility is obtained as

$$\chi^{\mathrm{sp/ch}}(q) = \sum_{kk'} \chi^{\mathrm{sp/ch}}(k, k', q) .\qquad (10.3)$$

Diagrammatically, this operation corresponds to tapering the ends as depicted in Fig. 10.3, which fixes the endpoints on either side of the diagram to the same (imaginary) time and position. The calculation of susceptibilities has been performed within DMFT [15] and cluster approaches [35, 159]. In the following, it is shown how susceptibilities are obtained within the dual fermion approach.

## 10.1   Bethe-Salpeter Equations for Dual Fermions

For dual fermions, the simplest approximation to the irreducible vertex is the bare interaction of dual fermions, $\gamma_{\omega\omega'\Omega}^{(4)\sigma_1\sigma_2\sigma_3\sigma_4}$ (the index '(4)' is omitted in the following). Note that this quantity is the full (reducible) vertex of the impurity, but plays the role of a bare (irreducible) interaction for dual fermions. Instead of working with the BSE for the two-particle Green function, we will work with the BSEs for the vertex function [160, 161]. The two approaches are equivalent, as can straightforwardly be seen by expanding the corresponding Bethe-Salpeter equations. Because the irreducible vertex is approximated, the solution of the BSE will depend on the particular channel,

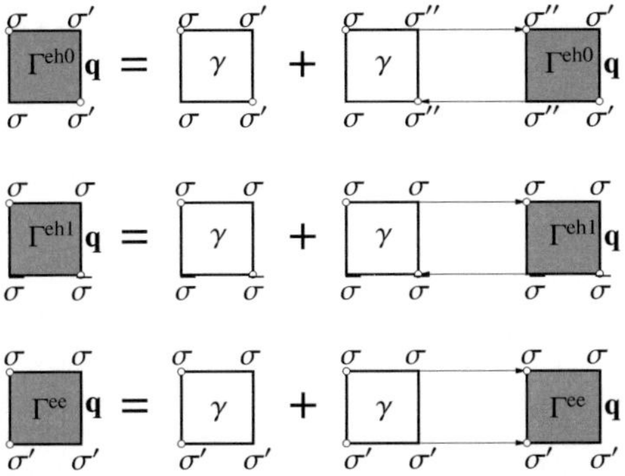

*Figure 10.2:* Bethe-Salpeter equations in the three different channels, with a local approximation $\Gamma_{\text{irr}} = \gamma$ to the irreducible vertex. The first two equations correspond to two different spin channels of the electron-hole channel and the last equation is for the particle-particle channel.

which should therefore be chosen based on physical considerations. The BSE in the spin channel is suitable to describe magnetic phenomena, while the particle-particle (electron-electron) channel should be chosen to describe pairing in a superconductor[1].

The BSE in the three different channels is depicted in Fig. 10.2. The first two equations correspond to two different spin channels of the (horizontal) electron-hole channel and the third one to the electron-electron channel. Using two different horizontal channels instead of a horizontal and a vertical channel is related to the use of an antisymmetrized theory. A detailed discussion of the Bethe-Salpeter equations is postponed to chapter 11. The relation to the spin and charge susceptibilities is established in Sec. 10.1.1 below.

In the following, the BSEs in the electron-hole channels and in the paramagnetic state are considered (all quantities are invariant under exchange of the direction of all spins). A possible orbital dependence is also omitted; the general BSEs are given in appendix A.8.1. In real space, these BSEs read

$$\Gamma^{\text{eh0}\,\sigma\sigma'}_{\omega\omega'\Omega}(\mathbf{x}) = \gamma^{\sigma\sigma'}_{\omega\omega'\Omega}\delta_{\mathbf{x},0} - T \sum_{\omega''\sigma''} \sum_{\mathbf{x}'} \gamma^{\sigma\sigma''}_{\omega\omega''\Omega} G^{\sigma''}_{\omega''}(\mathbf{x}')G^{\sigma''}_{\omega''}(-\mathbf{x}')\Gamma^{\text{eh0}\,\sigma''\sigma'}_{\omega''\omega'\Omega}(\mathbf{x} - \mathbf{x}') , \quad (10.4)$$

$$\Gamma^{\text{eh1}\,\sigma\bar{\sigma}}_{\omega\omega'\Omega}(\mathbf{x}) = \gamma^{\sigma\bar{\sigma}}_{\omega\omega'\Omega}\delta_{\mathbf{x},0} - T \sum_{\omega''} \sum_{\mathbf{x}'} \gamma^{\sigma\bar{\sigma}}_{\omega\omega''\Omega} G^{\sigma}_{\omega''}(\mathbf{x}')G^{\bar{\sigma}}_{\omega''}(-\mathbf{x}')\Gamma^{\text{eh1}\,\sigma\bar{\sigma}}_{\omega''\omega'\Omega}(\mathbf{x} - \mathbf{x}') . \quad (10.5)$$

---

[1] The BSE in the electron-electron channel with the local approximation to the irreducible vertex does not provide a proper description of the pairing instability. Nonlocal spin fluctuations are expected to be important in this case. An approximation which includes these fluctuations is given in Sec. 10.4.

The spin on Green functions is written for completeness. Here and in the following, all quantities are dual by default and the index 'd' is omitted. The above form is not well suited for numerical calculations, because the vertex is not diagonal in position labels. Instead, consider the equations in reciprocal space,

$$\Gamma^{\text{eh}0\,\sigma\sigma'}_{\omega\omega'\Omega}(\mathbf{q}) = \gamma^{\sigma\sigma'}_{\omega\omega'\Omega} - \frac{T}{N}\sum_{\omega''\sigma''}\sum_{\mathbf{k}} \gamma^{\sigma\sigma''}_{\omega\omega''\Omega} G^{\sigma''}_{\omega''}(\mathbf{k}) G^{\sigma''}_{\omega''+\Omega}(\mathbf{k}+\mathbf{q})\, \Gamma^{\text{eh}0\,\sigma''\sigma'}_{\omega''\omega'\Omega}(\mathbf{q})\,, \qquad (10.6)$$

$$\Gamma^{\text{eh}1\,\sigma\bar{\sigma}}_{\omega\omega'\Omega}(\mathbf{q}) = \gamma^{\sigma\bar{\sigma}}_{\omega\omega'\Omega} - \frac{T}{N}\sum_{\omega''}\sum_{\mathbf{k}} \gamma^{\sigma\bar{\sigma}}_{\omega\omega''\Omega} G^{\bar{\sigma}}_{\omega''}(\mathbf{k}) G^{\sigma}_{\omega''+\Omega}(\mathbf{k}+\mathbf{q})\, \Gamma^{\text{eh}1\,\sigma\bar{\sigma}}_{\omega''\omega'\Omega}(\mathbf{q})\,. \qquad (10.7)$$

The physical content of these BSEs is the multiple scattering of particle-hole pairs. The vector $\mathbf{q}$ is the center of mass momentum and $\Omega$ is the (bosonic) frequency transferred in scattering processes. Due to the particular approximation to the irreducible vertex, the BSEs depend on transferred momentum and frequency only. Since these are conserved, the matrices are blockdiagonal and can be solved for each $\mathbf{q}$ and $\Omega$ separately. The designation 0 and 1 indicates the $z$-component of the spin of the scattered particle-hole pair. The first equation describes the scattering of a particle and a hole with the same spin $\sigma$ and hence $S_z = 0$.

### 10.1.1   Spin Diagonalization

The Bethe-Salpeter equations (10.6,10.7) differ by the $z$-component of the spin of the particle-hole pair, which is conserved in scattering processes. The second equation also corresponds to a definite value of the total spin $S$, as can be seen from the fact that it is spin diagonal, i.e. it only involves the vertex for a single spin component (there is no sum over spins). It corresponds to the $S_z = \pm 1$ components of the triplet channel (depending on the sign of $\sigma$). For the first equation, the total spin is not conserved and hence the two spin components $\Gamma^{\text{eh}0\sigma\sigma}$ and $\Gamma^{\text{eh}0\sigma\bar{\sigma}}$ mix. The BSE can still be solved by matrix inversion, but in this case the vertices have to be viewed as matrices in the superindex $1 \equiv \{\sigma\omega\}$. The two matrix equations can be decoupled by introducing the linear combinations

$$\Gamma^{\text{s}} = \Gamma^{\text{eh}0\uparrow\uparrow} + \Gamma^{\text{eh}0\uparrow\downarrow}\,, \qquad \Gamma^{\text{t}} = \Gamma^{\text{eh}0\uparrow\uparrow} - \Gamma^{\text{eh}0\uparrow\downarrow}\,. \qquad (10.8)$$

Note that both have $S_z = 0$. Here 's' and 't' label the singlet and triplet channels of the vertex. Note that the vertex in the singlet (triplet) channel enters the calculation of the charge (spin) susceptibility, so that these designations of the channels are also customary. In order to distinguish the singlet and triplet channels from the corresponding channels of the particle-particle vertex, the designations 'd' and 'm' for "density" and "magnetic" will be used in the following. In the paramagnetic state, the values of the vertex for different projections of the spin should not differ due to rotational invariance in spin space. Indeed, it is straightforward to show that [161]

$$\Gamma^{eh1} = \Gamma^{eh0\uparrow\uparrow} - \Gamma^{eh0\uparrow\downarrow} \tag{10.9}$$

holds in the paramagnetic state, i.e. $\Gamma^{t}(\Gamma^{m})$ corresponds to the $S_z = 0$ projection of the $S = 1$ triplet channel. $\Gamma^{s}(\Gamma^{d})$ on the other hand denotes the vertex in the $S_z = 0$, $S = 0$ singlet channel. The positive sign for the singlet channel seems counterintuitive. Note however, that this is the spin of a particle-hole pair. For the vertex in the particle-particle channel, the negative sign corresponds to the singlet (see Sec. 10.4). To summarize, from the nonvanishing spin combinations of the vertices, one obtains the following spin diagonalized linear combinations [162]

$$\begin{aligned}
\Gamma^{d} &= \Gamma^{\uparrow\uparrow\uparrow\uparrow} + \Gamma^{\uparrow\uparrow\downarrow\downarrow} \,, \\
\Gamma^{m0} &= \Gamma^{\uparrow\uparrow\uparrow\uparrow} - \Gamma^{\uparrow\uparrow\downarrow\downarrow} \,, \\
\Gamma^{m+1} &= \Gamma^{\uparrow\downarrow\downarrow\uparrow} \,, \\
\Gamma^{m-1} &= \Gamma^{\downarrow\uparrow\uparrow\downarrow} \,,
\end{aligned} \tag{10.10}$$

where $0, \pm1$ are the different projections of the $S = 1$ channel. The $S = 1$ vertices must be equal in the paramagnetic state. After spin diagonalization, instead of (10.6), one obtains the two decoupled Bethe-Salpeter equations

$$\Gamma^{d/m}_{\omega\omega'\Omega}(\mathbf{q}) = \gamma^{d/m}_{\omega\omega'\Omega} - \frac{T}{N} \sum_{\omega''} \sum_{\mathbf{k}} \gamma^{d/m}_{\omega\omega''\Omega} G_{\omega''}(\mathbf{k}) G_{\omega''+\Omega}(\mathbf{k}+\mathbf{q}) \Gamma^{d/m}_{\omega''\omega'\Omega}(\mathbf{q}) \,. \tag{10.11}$$

After the equations are solved, the vertex $\Gamma^{eh0\sigma\sigma'}$ may simply be obtained as

$$\Gamma^{eh0\uparrow\uparrow} = \frac{1}{2}(\Gamma^{d} + \Gamma^{m}) \,, \qquad \Gamma^{eh0\uparrow\downarrow} = \frac{1}{2}(\Gamma^{d} - \Gamma^{m}) \,. \tag{10.12}$$

Note that in contrast to (10.1), the BSEs for the charge (density) and spin (magnetic) vertex appear with the same sign. This is related to the fact that the dual interaction is fully antisymmetric and both equations correspond to the same (horizontal) electron-hole channel. For details, see chapter 11 and appendix A.8.1.

## 10.1.2 Numerical Solution

The Bethe-Salpeter equation may be solved straightforwardly using iteration either in the non-diagonal or spin diagonalized form. As a starting guess, the irreducible vertex is inserted in place of the reducible one on the right-hand side of BSE: $\Gamma^{(0)} = \gamma$. Evaluation of the diagram on the right-hand side yields a new approximation $\Gamma^{(1)}$ for the reducible vertex, which is in turn inserted into the BSE. This successively generates a sum of ladder diagrams which contains all ladders with $1, \ldots, n + 1$ irreducible rungs in the

$n$-th order approximation $\Gamma^{(n)}$. This procedure is analogous to successively summing the terms of an infinite geometric series. In the vicinity of an instability at which the susceptibility has to diverge, the number of iterations to obtain a converged solution diverges. In general, if one is interested only in the position of the instability instead of the actual value of the susceptibility itself, it is sufficient to determine the leading eigenvalue of the matrix

$$-\frac{T}{N}\Gamma^{\alpha}_{\text{irr}}(k,k',q)\,G(k')\,G(k'+q)\,,\tag{10.13}$$

which is the building block of the particle-hole ladder. Here $\alpha$ denotes the respective channel. The instability is indicated by the leading eigenvalue being equal to one, $\lambda_{\max} = 1$. The iterative solution becomes inefficient in the vicinity of an instability, but allows to sum the ladder up to a given finite order. This will be used in chapter 11. For eigenvalues $\lambda \neq 1$, the BSE is efficiently solved by matrix inversion according to

$$[\Gamma^{\alpha}_{\omega\omega'\Omega}(\mathbf{q})]^{-1} = (\gamma^{\alpha}_{\omega\omega'\Omega})^{-1} + \frac{T}{N}\sum_{\mathbf{k}} G_{\omega}(\mathbf{k})G_{\omega+\Omega}(\mathbf{k}+\mathbf{q})\delta_{\omega\omega'}\,,\tag{10.14}$$

for the spin diagonal channels $\alpha = \text{d}, \text{m}$. A consistent frequency cutoff (see Sec. 3.3.9) is imposed, because extending the frequency sums to infinity may introduce large errors when matrix inverses are performed [163].

The convolution in Eq. 10.14 is evaluated using FFT as described in Sec. 6.6. Due to the frequency cutoff for the three fermionic Matsubara frequencies, the size $N$ of the matrix for a fixed bosonic frequency $\Omega_m = 2\pi m T$ depends on $m$: $N = 2N_\omega - m$. For the vertices, basically two different storage schemes are possible. One can either store them as function of either three independent fermionic frequencies, or alternatively one bosonic and two fermionic Matsubara frequencies. While the local vertex is stored using the former scheme, here the latter is preferable, because the susceptibility is usually only calculated for the first bosonic frequency and only for positive frequencies if an analytical continuation is desired.

## 10.2   Calculation of Lattice Susceptibility

The (longitudinal) spin susceptibility and the charge susceptibility are given by

$$\langle S_z S_z\rangle(\mathbf{q},\Omega) = \frac{1}{4}\langle(n_\uparrow - n_\downarrow)^2\rangle(\mathbf{q},\Omega) = \frac{1}{2}\left(\langle n_\uparrow n_\downarrow\rangle - \langle n_\uparrow n_\downarrow\rangle\right)(\mathbf{q},\Omega)\,,$$

$$\langle n\,n\rangle(\mathbf{q},\Omega) = \langle(n_\uparrow + n_\downarrow)^2\rangle(\mathbf{q},\Omega) = 2\left(\langle n_\uparrow n_\uparrow\rangle + \langle n_\uparrow n_\downarrow\rangle\right)(\mathbf{q},\Omega)\,.\tag{10.15}$$

The averages are expressed according to

$$\langle n_\sigma n_{\sigma'}\rangle(\mathbf{q},\Omega) = \chi_0^{\sigma\sigma'}(\mathbf{q},\Omega) + \chi^{\sigma\sigma'}(\mathbf{q},\Omega)\,,\tag{10.16}$$

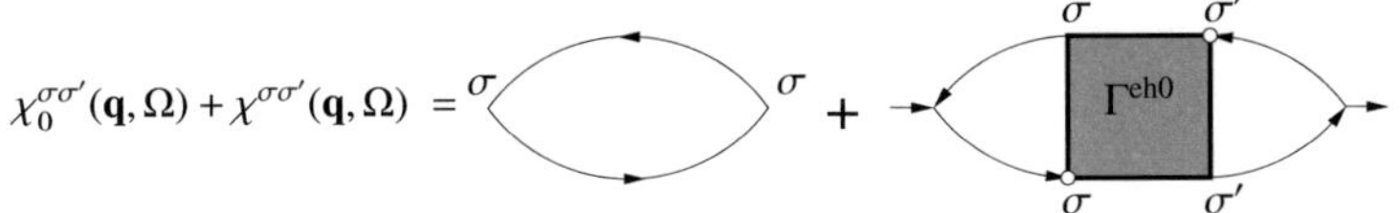


*Figure 10.3:* Diagrammatic representation of $\langle n_\sigma\, n_{\sigma'}\rangle = \chi_0^{\sigma\sigma'}(\mathbf{q}, \Omega) + \chi^{\sigma\sigma'}(\mathbf{q}, \Omega)$.

where the shorthand notation $\chi^{\sigma\sigma'}$ stands for $\chi^{\sigma\sigma\sigma'\sigma'}$. This is depicted diagrammatically in Fig. 10.3. The first term is given by the bubble diagram and reads

$$\chi_0^{\sigma\sigma'}(\mathbf{q}, \Omega) = -\frac{T}{N} \sum_\omega \sum_{\mathbf{k}} G_\omega(\mathbf{k}) G_{\omega+\Omega}(\mathbf{k}+\mathbf{q})\, \delta_{\sigma\sigma'} . \tag{10.17}$$

The nontrivial (connected) part of the diagram is given by

$$\chi^{\sigma\sigma'}(\mathbf{q}, \Omega) = \frac{T^2}{N^2} \sum_{\omega\omega'} \sum_{\mathbf{k}\mathbf{k}'} G_\omega(\mathbf{k}) G_{\omega+\Omega}(\mathbf{k}+\mathbf{q}) \Gamma^{\text{eh0}\,\sigma\sigma'}_{\omega\omega'\Omega} G_{\omega'}(\mathbf{k}') G_{\omega'+\Omega}(\mathbf{k}'+\mathbf{q}) . \tag{10.18}$$

So far, these relations are valid for dual and lattice fermions. From the Bethe-Salpeter equations, one obtains an approximation to the dual vertex, which allows to calculate the dual susceptibility. The corresponding quantity in terms of lattice fermions is obtained using the exact relations of Sec. 6.3. The bubble diagram for lattice fermions is computed by expressing the lattice fermion Green function through the identity (6.26). Using the relation (6.35) between the dual vertex and the dual two-particle Green function and the transformation (6.33), it is readily seen that the nontrivial part of the dual susceptibility is transformed by simply replacing the dual Green functions to the left and right of the vertex by the propagator-like functions

$$\mathcal{G}^{\text{L}}_\omega(\mathbf{k}) = (\Delta_\omega - h_{\mathbf{k}})^{-1} g_\omega^{-1} G_\omega^{\text{d}}(\mathbf{k}) ,$$
$$\mathcal{G}^{\text{R}}_\omega(\mathbf{k}) = G_\omega^{\text{d}}(\mathbf{k}) g_\omega^{-1} (\Delta_\omega - h_{\mathbf{k}})^{-1} , \tag{10.19}$$

which are identical for a diagonal basis.

## 10.2.1   Relation to DMFT

In DMFT, the irreducible lattice vertex is approximated by the local irreducible vertex of the impurity problem [15]. The latter can then be extracted from the reducible impurity vertex $\gamma$ by inverting a local Bethe-Salpeter equation

$$(\gamma^\alpha_{\omega\omega'\Omega})^{-1} = (\gamma^{\text{irr}\,\alpha}_{\omega\omega'\Omega})^{-1} - \chi^0_{\omega\Omega}\delta_{\omega\omega'}, \qquad \alpha = \text{d}, \text{m} . \tag{10.20}$$

The approximation to the lattice vertex is obtained from the irreducible vertex by inverting the BSE

$$[\Gamma^{\alpha}_{\omega\omega'\Omega}(\mathbf{q})]^{-1} = (\gamma^{\text{irr}\,\alpha}_{\omega\omega'\Omega})^{-1} - \chi^{0}_{\omega\Omega}(\mathbf{q})\delta_{\omega\omega'}. \tag{10.21}$$

The disconnected part in the two equations is defined as

$$\chi^{0}_{\omega\Omega} = -TG^{\text{loc,DMFT}}_{\omega}G^{\text{loc,DMFT}}_{\omega+\Omega},$$

$$\chi^{0}_{\omega\Omega}(\mathbf{q}) = -\frac{T}{N}\sum_{\mathbf{k}} G^{\text{DMFT}}_{0\,\omega}(\mathbf{k})G^{\text{DMFT}}_{0\,\omega+\Omega}(\mathbf{k}+\mathbf{q}), \tag{10.22}$$

where the local DMFT Green function is given by $G^{\text{loc,DMFT}}_{\omega} = (1/N)\sum_{\mathbf{k}} G^{\text{DMFT}}_{\omega}(\mathbf{k})$. Recall that within DMFT, the latter is equal to the Green function $g_{\omega}$ of the impurity model. The susceptibilities are obtained by the same expressions as in the previous section, with the vertex and Green functions replaced by the respective DMFT quantities.

In Sec. 6.4 it was established that for the selfconsistency condition (6.44), DMFT is recovered if no diagrammatic corrections to the dual self-energy are taken into account. In that case, the bare dual Green function transforms to the DMFT Green function when inserted into the exact relation (6.26) between the dual and lattice Green functions. It is worthwhile to analyze the analogous statement for the two-particle quantities. The disconnected part of the dual two-particle Green function contains bare dual Green functions and hence transforms into the corresponding DMFT result. The connected part of the two-particle Green function in DMFT is given by $G^{\text{DMFT}}_{11'}G^{\text{DMFT}}_{33'}\Gamma^{\text{DMFT}}_{1'2'3'4'}G^{\text{DMFT}}_{2'2}G^{\text{DMFT}}_{4'4}$, where $\Gamma^{\text{DMFT}}$ is obtained according to Eqs. 10.20, 10.21. In the dual fermion approach, the connected part transformed back to lattice fermions is

$$L_{11'}L_{33'}\left[\chi^{\text{d}} - G^{\text{d}}\otimes G^{\text{d}}\right]_{1'2'3'4'} R_{2'2}R_{4'4} = L_{11'}L_{33'}\, G^{\text{d}}_{0\,1'\bar{1}}G^{\text{d}}_{0\,3'\bar{3}}\Gamma^{\text{d}}_{\bar{1}\bar{2}\bar{3}\bar{4}}G^{\text{d}}_{0\,\bar{2}2'}G^{\text{d}}_{0\,\bar{4}4'}\, R_{2'2}R_{4'4}$$

$$= G^{\text{DMFT}}_{11'}G^{\text{DMFT}}_{33'}\Gamma^{\text{d}}_{1'2'3'4'}G^{\text{DMFT}}_{2'2}G^{\text{DMFT}}_{4'4}. \tag{10.23}$$

Here e.g. $G^{\text{DMFT}}_{12} = -[(\Delta - h)^{-1}g^{-1}]_{11'}G^{\text{d}}_{0\,1'2}$ has been used, which follows from (A.39). The dual vertex $\Gamma^{\text{d}}$ has to be transformed back to the lattice vertex $\Gamma$ using the relation (6.37): $\Gamma = L'_{11'}L'_{33'}\Gamma^{\text{d}}_{1'2'3'4'}R'_{2'2}R'_{4'4}$, where $L'_{12} = -[\mathbb{1} + \Sigma^{\text{d}}g]^{-1}_{12}$ and similarly for $R'$ (Eqs. A.136, A.137). In DMFT, one has $\Sigma^{\text{d}} = 0$ and it follows that $\Gamma = \Gamma^{\text{d}}$. The right hand side of (10.23) hence reads $G^{\text{DMFT}}_{11'}G^{\text{DMFT}}_{33'}\Gamma_{1'2'3'4'}G^{\text{DMFT}}_{2'2}G^{\text{DMFT}}_{4'4}$. As expected, the same result is obtained by first transforming the dual vertex and then connecting the DMFT Green functions. Note that $\Gamma = \Gamma^{\text{d}}$ is the rule how to transform the dual vertex to the corresponding quantity for lattice fermions. It does not imply that the two approaches to calculate the susceptibility are equivalent. Rather, since the transformation from dual to lattice fermions in case of DMFT is the identity, it remains to be shown that the dual vertex calculated according to (10.14) is exactly equivalent to $\Gamma^{\text{DMFT}}$. Combining Eqs. 10.20 and 10.21 to eliminate the irreducible vertex, one obtains

$$[\Gamma^{\alpha}_{\omega\omega'\Omega}(\mathbf{q})]^{-1} = (\gamma^{\alpha}_{\omega\omega'\Omega})^{-1} - \left[\chi^{0}_{\omega\Omega}(\mathbf{q}) - \chi^{0}_{\omega\Omega}\right]\delta_{\omega\omega'}. \tag{10.24}$$

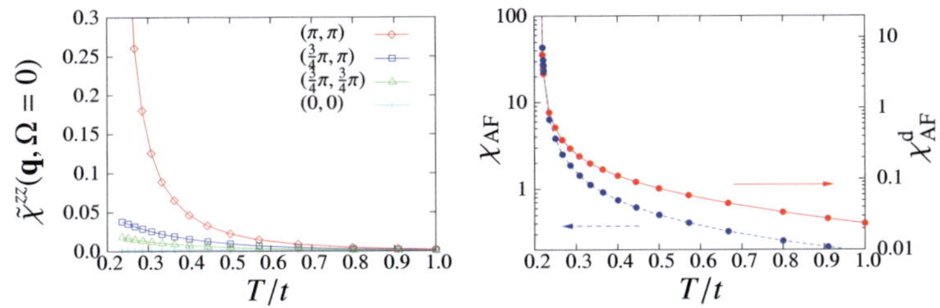

*Figure 10.4:* Left: nontrivial part $\tilde{\chi}^{zz}(\mathbf{q}, \Omega = 0)$ for different wave vectors at $U/t = 4$. The susceptibility diverges for $\mathbf{q} = (\pi, \pi)$, indicating the antiferromagnetic instability. Right: comparison of the antiferromagnetic dual and lattice fermion susceptibility for the same parameters. The two susceptibilities differ in absolute value, but diverge at the same temperature: Two-particle excitations and corresponding instabilities are the same for dual and lattice fermions.

Comparing this to the BSE (10.14) for dual fermions,

$$[\Gamma^{\alpha}_{\omega\omega'\Omega}(\mathbf{q})]^{-1} = (\gamma^{\alpha}_{\omega\omega'\Omega})^{-1} - \chi^{0\,\mathrm{d}}_{\omega\Omega}(\mathbf{q})\delta_{\omega\omega'}, \tag{10.25}$$

one sees that the equations differ only through the last term. For the case of DMFT, (10.25) can be further rewritten. Using the relation between the DMFT and bare dual Green functions $G^{\mathrm{DMFT}}_{\omega}(\mathbf{k}) - g_{\omega} = G^{\mathrm{d}}_{\omega}(\mathbf{k})$ (A.39) and $(1/N)\sum_{\mathbf{k}} G_{\omega}(\mathbf{k}) = (1/N)\sum_{\mathbf{k}} G_{\omega}(\mathbf{k} + \mathbf{q}) = g_{\omega}$ yields

$$\chi^{0\,\mathrm{d}}_{\omega\Omega}(\mathbf{q}) = -\frac{T}{N} \sum_{\mathbf{k}} \left[ G^{\mathrm{DMFT}}_{\omega}(\mathbf{k}) - g_{\omega} \right] \left[ G^{\mathrm{DMFT}}_{\omega+\Omega}(\mathbf{k} + \mathbf{q}) - g_{\omega+\Omega} \right]$$

$$= -\frac{T}{N} \sum_{\mathbf{k}} G^{\mathrm{DMFT}}_{\omega}(\mathbf{k})G^{\mathrm{DMFT}}_{\omega+\Omega}(\mathbf{k} + \mathbf{q}) + T\, g_{\omega}g_{\omega+\Omega}, \tag{10.26}$$

which exactly recovers the term in angular brackets in (10.24). This is a nontrivial and remarkable result: in the DMFT limit, the dual fermion approach to calculate the two-particle Green function and susceptibilities is exactly equivalent to the standard procedure in DMFT.

## 10.3 Numerical Results

In the following, the approach is tested on the 2D Hubbard model. The nontrivial part of the dual spin susceptibility is plotted as a function of temperature for different values of $\mathbf{q}$ in the left panel of Fig. 10.4. As the temperature is lowered, the susceptibility diverges in the $\mathbf{q} = (\pi, \pi)$ momentum channel, while it remains finite for other wave vectors. For

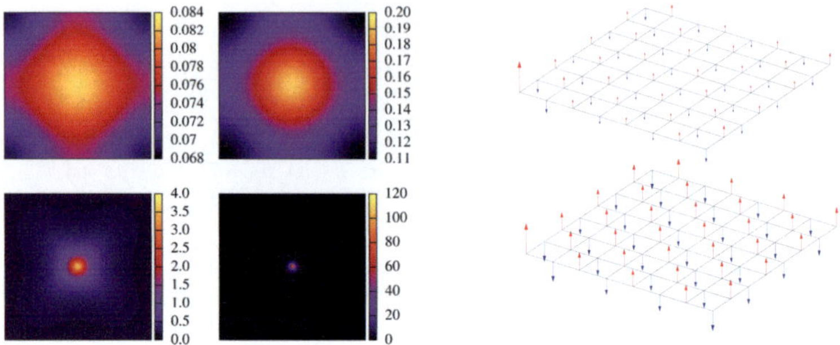

*Figure 10.5:* Left panel: Momentum dependence of the static susceptibility (centered at the $M$-point) for $U/t = 4$ and different temperatures. From left to right and top to bottom: $T/t = 2.0, 1.0, 0.25, 0.215$. Right panel: Static susceptibility in real space for $T/t = 0.235$ (top) and $0.222$ (bottom). For higher temperature, the correlations decay with distance (note that periodic boundary conditions are imposed). At the critical temperature, the correlation length exceeds the lattice size.

the lattice fermions, one would expect this to happen at the same wave vector, due to the antiferromagnetic instability in the Hubbard model, which in turn is a consequence of the perfect nesting of the Fermi surface.

In the paramagnetic state, the relation $(1/2)(\langle S^+S^-\rangle + \langle S^-S^+\rangle) = 2\langle S^zS^z\rangle$ holds. The transverse susceptibilities $\langle S^\pm S^\mp\rangle$ are obtained from an expression similar to (10.18) with $\Gamma^{\mathrm{eh}0}$ replaced by $\Gamma^{\mathrm{eh}1}$. The above relation follows from the symmetry (10.9), which is fulfilled up to small numerical error. The BSEs preserve this symmetry exactly, provided the same holds for the irreducible vertex $\gamma$. The numerical error is therefore traced back to Monte Carlo averaging. The two-particle Green functions are obtained from the measurements $\tilde{g}_{12}$ (see chapter 3) as $\chi_{1234}^{\sigma\sigma\sigma'\sigma'} = \langle \tilde{g}_{12}^\sigma \tilde{g}_{34}^{\sigma'}\rangle_{\mathrm{MC}} - \delta_{\sigma\sigma'}\langle \tilde{g}_{14}^\sigma \tilde{g}_{32}^\sigma\rangle_{\mathrm{MC}}$ and $\chi_{1234}^{\sigma\bar\sigma\bar\sigma\sigma} = -\langle \tilde{g}_{14}^\sigma \tilde{g}_{32}^{\bar\sigma}\rangle_{\mathrm{MC}}$. It follows that

$$\chi_{1234}^{\sigma\sigma\sigma\sigma} - \chi_{1234}^{\sigma\sigma\bar\sigma\bar\sigma} = (\langle \tilde{g}_{12}^\sigma \tilde{g}_{34}^\sigma\rangle_{\mathrm{MC}} - \langle \tilde{g}_{14}^\sigma \tilde{g}_{32}^\sigma\rangle_{\mathrm{MC}}) - \langle \tilde{g}_{12}^\sigma \tilde{g}_{34}^{\bar\sigma}\rangle_{\mathrm{MC}} \neq \chi_{1234}^{\sigma\bar\sigma\bar\sigma}, \qquad (10.27)$$

because the quantities $\tilde{g}_{12}^\sigma$ and $\tilde{g}_{12}^{\bar\sigma}$ differ even for paramagnetic systems. The CTQMC measurements do not preserve rotational invariance in spin space and this symmetry is fulfilled up to a statistical error.

In the right panel of Fig. 10.4, the complete dual susceptibility (including the trivial part) is compared to the corresponding quantity in terms of lattice fermions for $\mathbf{q} = (\pi, \pi)$. The latter was obtained from the same calculation, using the exact transformation described earlier. The susceptibilities differ in absolute value, but diverge at the same temperature. This is a numerical illustration of the fact that two-particle excitations and the corresponding instabilities are the same for dual and lattice fermions.

In Fig. 10.5, one can see how the divergence occurs in reciprocal space. The left panel shows a surface plot of the lattice susceptibility centered at the $M$-point for different temperatures. For high temperatures, the susceptibility clearly reflects the symmetry of the lattice. This becomes much less pronounced as the temperature is lowered and the susceptibility diverges at the $M$-point. The right panel shows how the spin-correlations evolve in real space: The length of the arrows at a given distance is proportional to the static spin-spin correlation with the spin to the left. The antiferromagnetic correlations are clearly visible. At higher temperatures, these decay quickly as a function of distance. As the temperature is lowered, the correlations become essentially independent of distance. At the instability, the correlation length exceeds the lattice size.

In general, one would expect that a transition to the ordered state for the two-dimensional model cannot occur, as it requires to break a continuous symmetry and is hence forbidden by the Mermin-Wagner theorem. It should be noted, however, that eventhough the thermodynamic limit is built into this theory as in DMFT, the Mermin Wagner theorem is not strictly applicable here. The theory is essentially based on a diagrammatic expansion around the mean-field solution. Long-range fluctuations, which are essential to the proof of the Mermin-Wagner theorem, are not fully accounted for in any approximation to the self-energy. Hence, eventhough nonlocal corrections are included, the approximation partially retains its mean-field character. It is known that theories with a residual mean-field character yield finite transition temperatures [35]. In the current implementation, the use of a discrete Fourier transform involves a discretization of reciprocal space. This corresponds to large but finite lattices with periodic boundary conditions in real space. The results may be compared to lattice QMC calculations.

The inverse of the static antiferromagnetic susceptibility $\chi_{\mathrm{AF}} := \chi(\mathbf{q}_{\mathrm{AF}}, \Omega = 0)$ for $\mathbf{q}_{\mathrm{AF}} = (\pi, \pi)$ obtained within the dual fermion approach is compared to the inverse DMFT susceptibility in Fig. 10.6. The critical temperature in the respective approximation is given by the point for which the curves extrapolate to zero. One can see that the Néel temperature in the DF calculation is reduced compared to DMFT. Because the DF approach according to Sec. 10.2.1 yields the DMFT susceptibility if no diagrams are taken into account, the reduction of the critical temperature can be attributed to the effect of spatial fluctuations, included through diagram b) of Fig. 6.3. These suppress the antiferromagnetism and are absent in the DMFT. At high temperatures, the DF susceptibility approaches the DMFT result as thermal fluctuations tend to wash out the correlations. One can further see that finite size effects are very small for $U/t = 4$ and are virtually absent for $U/t = 8$. This sets a scale for the range of the correlations included. There is no clear departure of the DF results from mean-field behavior, for which the susceptibility diverges as $(T - T_{\mathrm{N}})^{-\gamma}$ with critical exponent $\gamma \sim 1$ (this has not

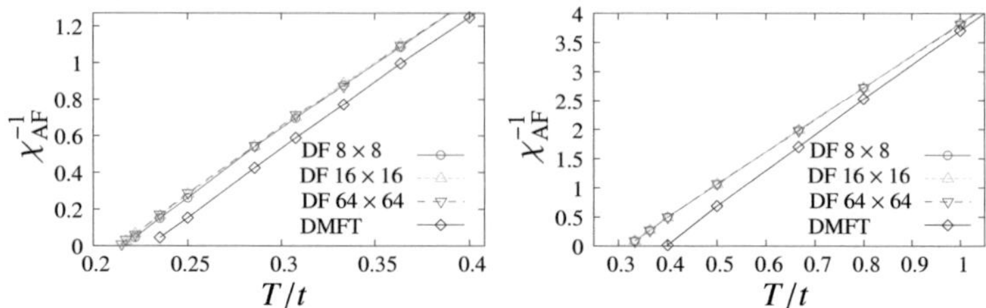

*Figure 10.6:* Inverse of the static antiferromagnetic susceptibility $\chi_{\mathrm{AF}}$ as a function of temperature as obtained within DMFT and the dual fermion approach for $U/t = 4$ (left) and $U/t = 8$ (right). The label indicates the number of k-points or lattice size, respectively. The DMFT calculation was carried out for a Brillouin zone discretization of $64 \times 64$ points. The extrapolation of $\chi_{\mathrm{AF}}^{-1}$ to zero yields the critical temperature.

been analyzed quantitatively).

Eventhough the critical temperature is reduced compared to DMFT, the results are in disagreement with lattice QMC simulations [74, 164]. Close to the antiferromagnetic instability it becomes crucial to take spin fluctuation (ladder) diagrams into account, which is facilitated by the approximation presented in chapter 11.

## 10.3.1   Comparison to CTQMC

It is instructive to compare two-particle quantities obtained within this approach to known results. In Fig. 10.7 they are compared to results from CTQMC calculations for the $2 \times 2$ plaquette, which were extracted from the same calculation as the results of Fig. 6.13. The CTQMC is employed as a lattice solver (periodic boundary conditions for the $2 \times 2$ plaquette are enforced by taking $t \to 2t$). For a direct comparison, the dual single- and two-particle Green functions have been transformed to lattice fermions. The vertices are taken in the magnetic spin channel and the index is omitted in the following. The full vertex of the cluster is measured in real space and transformed to reciprocal space using a multidimensional discrete Fourier transform. In four-momentum notation, it is denoted as $\Gamma(k_1, k_2, k_3, k_4)\delta_{k_1+k_3,k_2+k_4}$ and with $k = k_2$, $k' = k_3$ and $q = k_1 - k_2 = k_4 - k_3$ abbreviated as $\Gamma(k, k', q)$. The irreducible lattice vertex in the electron-hole channel is defined through the BSE

$$\Gamma(k, k'q) = \Gamma^{\mathrm{irr}}(k, k'q) - \sum_{k''} \Gamma^{\mathrm{irr}}(k, k'', q)G(k'')G(k'' + q)\Gamma(k'', k', q), \qquad (10.28)$$

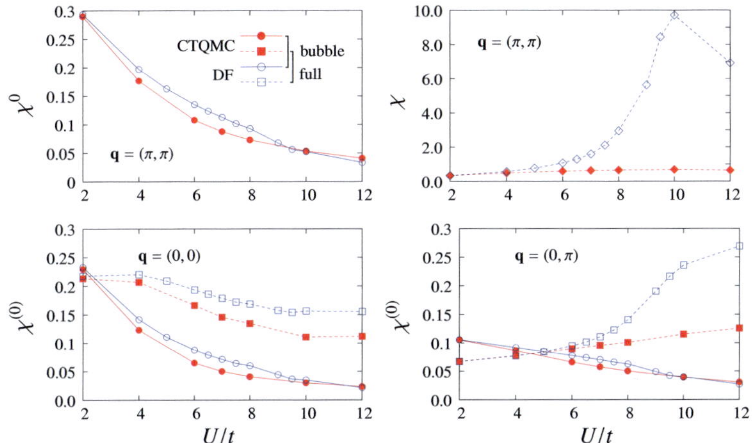

*Figure 10.7:* Comparison of the the DF susceptibility to numerically exact CTQMC results for the two-dimensional Hubbard model on the $2 \times 2$ plaquette with periodic boundary conditions for different wave vectors at $T/t = 0.4$. The particle-hole bubble $\chi^0$ (solid lines) agrees well in particular for weak and strong coupling. The full susceptibility (dashed lines) strongly deviates for large $U$.

where $\sum_k$ stands for $(T/N) \sum_{\mathbf{k}} \sum_\omega$. The exact irreducible vertex is obtained by matrix inversion according to

$$(\Gamma^{\mathrm{irr}})^{-1}(k, k'q) = \Gamma^{-1}(k, k'q) - \frac{T}{N} G(k)G(k + q)\delta_{kk'} . \tag{10.29}$$

The Green functions $G(k) = G(\mathbf{k}, \omega_n)$ are the exact lattice Green functions. No approximations are involved here. Eigenvalues are obtained from the solution of the eigenvalue problem

$$-\sum_{k'} \Gamma^{\mathrm{irr}}(k, k', q)G(k')G(k' + q)\phi(k') = \lambda\phi(k) . \tag{10.30}$$

Due to the ladder approximation, the dual fermion vertex depends on transferred momentum $\mathbf{q}$ only. For better comparability, the $\mathbf{k},\mathbf{k}'$-dependence of the full QMC vertex is therefore traced out according to

$$\Gamma_{\omega\omega'\Omega}(\mathbf{q}) := \frac{1}{N^2} \sum_{\mathbf{k}\mathbf{k}'} \Gamma_{\omega\omega'\Omega}(\mathbf{k}, \mathbf{k}', \mathbf{q}) \tag{10.31}$$

and susceptibilities are calculated with the reduced vertex as described earlier. Comparison of the spectra of the full and the reduced vertex reveals that the $\mathbf{k},\mathbf{k}'$-dependence becomes significant in particular for strong coupling. Neglecting it, however, is reasonable for small to intermediate coupling.

One can see that the particle-hole bubble, which is constructed from the single-particle Green functions, is in good agreement to the CTQMC result. This is true in particular for weak and strong coupling and rather independent of the wave vector. In agreement with the data shown in Fig. 6.13, single-particle quantities are confirmed to be well approximated. This, however, is clearly not the case for two-particle quantities. The static spin-spin susceptibility is significantly overestimated for strong coupling, in particular for the antiferromagnetic wave vector.

From the solution of the eigenvalue problem (10.30) it is found that for weak coupling, the dual fermion approximation to the irreducible lattice vertex very well reproduces the entire spectrum of the irreducible QMC vertex (10.31). For strong coupling, in particular the leading and subleading eigenvalues are overestimated. An overestimation of the effective two-particle interaction and hence the critical temperature indicate that critical fluctuations are not properly taken into account. Indeed, the single-particle Green functions in the ladder are renormalized only by the two lowest order diagrams a) and b), for which no significant contribution to such fluctuations was found. It is conceivable that two-particle quantities can be sensitive to small details of the Green function, because it enters the ladder at every order.

In that case, the feedback of two-particle excitations onto the single-particle Green functions must be taken into account. At this temperature, neglecting such diagrams appears to have a small effect on the single-particle Green functions. This changes as the temperature is lowered and the strength of the antiferromagnetic fluctuations increases (note that the calculations were performed at a rather high temperature $T/t = 0.4$).

The deviations are larger for strong coupling because the model is closer to the antiferromagnetic instability than for weak coupling at the same temperature. The deviations could also partially originate from a significant $\mathbf{k},\mathbf{k}'$-dependence of the vertex.


## 10.4    Application to Superconductivity

Superconductivity and related phenomena have been the subject of several studies on the two-dimensional Hubbard model [165, 166, 145], as it is believed to capture the low-energy physics of the cuprates [167]. However, the pairing mechanism in this model remains controversial [168, 169]. Large scale numerical studies at the limit of todays computational feasibility are able to separate finite size effects and have provided evidence that this model indeed exhibits a superconducting ground state. While numerical studies of finite lattices in the relevant part of the phase diagram are hampered by the fermionic sign problem [37], cluster generalizations of DMFT allow to access the superconducting phase while operating in the thermodynamic limit [32]. Correlations are included up to the extension of the cluster. Recently, the combination of numerical and

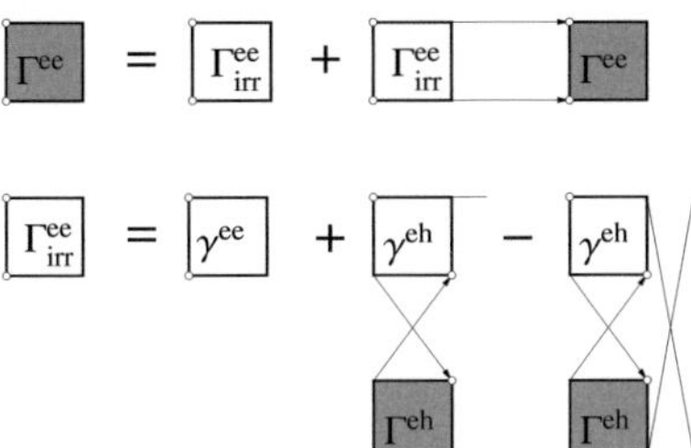

*Figure 10.8:* Diagrammatic representation of the Bethe-Salpeter equation for the particle-particle channel and the approximation to the irreducible particle-particle vertex.

diagrammatic techniques has been used to gain insight into the pairing mechanism of this model [165]. The need to perform calculations on large clusters makes these simulations highly demanding. Therefore, the phase diagram has not been calculated in this study.

The approach developed in the preceding sections allows to address this problem via the calculation of generalized susceptibilities. For example, the phase boundary to the superconducting state can be determined by locating the parameters at which the pairing susceptibility diverges. As for the case of antiferromagnetism, there cannot be a finite temperature phase transition in the two-dimensional model, but rather a Kosterlitz-Thouless transition to a phase with long-range order, with a large but finite correlation length. A small coupling to a third dimension may induce long-range order, which may also be expected to occur due to the residual mean-field character of the approximation. Being primarily interested in the phase boundary, it is sufficient to consider the eigenvalue problem derived from the BSE. The transition is located from the condition that the leading eigenvalue of the Bethe-Salpeter kernel is equal to one. One may evaluate the dual kernel instead, because the leading eigenvalue must reach unity for the same parameters.

The first step is to construct a suitable approximation for the dual irreducible particle-particle vertex. The simplest approximation is the bare interaction of the dual fermions, i.e. the vertex $\gamma$:

$$\Gamma^{\text{ee,irr}}_{\text{p}12;34} \approx \gamma_{\text{p}12;34} = -\gamma_{13;24} \, . \tag{10.32}$$

The label 'p' indicates that the vertex is defined to describe scattering of electron-electron (particle-particle) pairs (the points where an incoming line terminates have labels 1,2, cf. Fig. 10.9), while $\gamma$ describes particle-hole scattering (ingoing endpoints have index 1,3). From a physical point of view, this approximation is expected to be insufficient as it does not contain spin fluctuation diagrams. The corresponding particle-hole ladder diagrams are particle-particle irreducible and can be added to the irreducible particle-particle vertex as depicted in Fig. 10.8. This approximation can be motivated more formally by systematically improving the approximation to the irreducible vertex (see Sec. 11.2.3); instead a purely diagrammatic construction is given here. The ladder diagrams are obtained from the solution of the Bethe-Salpeter equation $\Gamma^{\mathrm{eh}}$ in the electron-hole channel:

$$\tilde{\Gamma}^{\mathrm{eh}}_{12;34} := \Gamma^{\mathrm{eh}}_{12;34} - \gamma_{12;34} \, . \tag{10.33}$$

Adding the diagrams with and without the outgoing endpoints crossed can be written as

$$\Gamma^{\mathrm{ee,irr}}_{\mathrm{p}12;34} = \gamma_{\mathrm{p}12;34} + \tilde{\Gamma}^{\mathrm{eh}}_{\mathrm{p}12;34} - \tilde{\Gamma}^{\mathrm{eh}}_{\mathrm{pc}12;34} \, . \tag{10.34}$$

The index 'c' indicates crossing of endpoints. Expressing this in terms of the corresponding original quantities in the electron-hole channel gives

$$\begin{aligned}
\Gamma^{\mathrm{ee,irr}}_{\mathrm{p}12;34} &= -\gamma_{13;24} - \tilde{\Gamma}^{\mathrm{eh}}_{13;24} + \tilde{\Gamma}^{\mathrm{eh}}_{14;23} \\
&= -\Gamma^{\mathrm{eh}}_{13;24} + \Gamma^{\mathrm{eh}}_{14;23} - \gamma_{14;23} \, .
\end{aligned} \tag{10.35}$$

Subtracting $\gamma$ avoids double counting of diagrams. In order to separate singlet and triplet contributions to pairing, write spins explicitly and define the vertices $\Gamma^{\mathrm{ee}+}$ and $\Gamma^{\mathrm{ee}-}$ as

$$\Gamma^{\mathrm{ee}+}_{\mathrm{p}12;34} = \Gamma^{\mathrm{ee}\,\sigma\bar{\sigma}\bar{\sigma}\sigma}_{\mathrm{p}12;34} = -\Gamma^{\mathrm{eh}1\,\sigma\bar{\sigma}\bar{\sigma}\sigma}_{13;24} \qquad \Gamma^{\mathrm{ee}-}_{\mathrm{p}12;34} = \Gamma^{\mathrm{ee}\,\sigma\bar{\sigma}\sigma\bar{\sigma}}_{\mathrm{p}12;34} = -\Gamma^{\mathrm{eh}0\,\sigma\sigma\bar{\sigma}\bar{\sigma}}_{13;24} \, . \tag{10.36}$$

These have spin projection $S_z = 0$ of the particle-particle pair. The vertices in the singlet (s) and triplet (t) electron-electron channels are hence given by

$$\Gamma^{\mathrm{ee}\,s} = \Gamma^{\mathrm{ee}+} - \Gamma^{\mathrm{ee}-} \, , \qquad \Gamma^{\mathrm{ee}\,t} = \Gamma^{\mathrm{ee}+} + \Gamma^{\mathrm{ee}-} \, . \tag{10.37}$$

It is readily verified that e.g. the singlet vertex is antisymmetric under exchange of spin variables in the initial (12) or final (34) states and symmetric under the exchange of frequencies and momenta. The spin configurations of the crossed channels transform to those of the particle-hole vertices as

$$\Gamma^{\mathrm{ee}+}_{\mathrm{pc}12;34} = \Gamma^{\mathrm{ee}\,\sigma\bar{\sigma}\bar{\sigma}\sigma}_{\mathrm{pc}12;34} = -\Gamma^{\mathrm{eh}0\,\sigma\sigma\bar{\sigma}\bar{\sigma}}_{14;23} \qquad \Gamma^{\mathrm{ee}-}_{\mathrm{pc}12;34} = \Gamma^{\mathrm{ee}\,\sigma\bar{\sigma}\sigma\bar{\sigma}}_{\mathrm{pc}12;34} = -\Gamma^{\mathrm{eh}1\,\sigma\bar{\sigma}\bar{\sigma}\sigma}_{14;23} \, . \tag{10.38}$$

Therefore, writing spin indices in Eq. 10.35 explicitly, yields the two vertices

$$\begin{aligned}
\Gamma^{\mathrm{ee,irr}+}_{\mathrm{p}12;34} &= \Gamma^{\mathrm{ee,irr}\,\sigma\bar{\sigma}\bar{\sigma}\sigma}_{\mathrm{p}12;34} = -\Gamma^{\mathrm{eh}1\,\sigma\bar{\sigma}\bar{\sigma}\sigma}_{13;24} + \Gamma^{\mathrm{eh}0\,\sigma\sigma\bar{\sigma}\bar{\sigma}}_{14;23} - \gamma^{\sigma\sigma\bar{\sigma}\bar{\sigma}}_{14;23} \\
\Gamma^{\mathrm{ee,irr}-}_{\mathrm{p}12;34} &= \Gamma^{\mathrm{ee,irr}\,\sigma\bar{\sigma}\sigma\bar{\sigma}}_{\mathrm{p}12;34} = -\Gamma^{\mathrm{eh}0\,\sigma\sigma\bar{\sigma}\bar{\sigma}}_{13;24} + \Gamma^{\mathrm{eh}1\,\sigma\bar{\sigma}\bar{\sigma}\sigma}_{14;23} - \gamma^{\sigma\bar{\sigma}\bar{\sigma}\sigma}_{14;23}
\end{aligned} \tag{10.39}$$

From this, one readily obtains the irreducible vertex in the singlet channel as

$$\begin{aligned}
\Gamma^{\mathrm{ee,irr}\,s}_{\mathrm{p}12;34} &= \Gamma^{\mathrm{irr},+\,\sigma\bar{\sigma}\bar{\sigma}\sigma}_{\mathrm{p}12;34} - \Gamma^{\mathrm{irr},-\,\sigma\bar{\sigma}\sigma\bar{\sigma}}_{\mathrm{p}12;34} \\
&= \Gamma^{\mathrm{eh}0\,\sigma\sigma\bar{\sigma}\bar{\sigma}}_{13;24} - \Gamma^{\mathrm{eh}1\,\sigma\bar{\sigma}\bar{\sigma}\sigma}_{13;24} + \Gamma^{\mathrm{eh}0\,\sigma\sigma\bar{\sigma}\bar{\sigma}}_{14;23} - \Gamma^{\mathrm{eh}1\,\sigma\bar{\sigma}\bar{\sigma}\sigma}_{14;23} - \gamma^{\sigma\sigma\bar{\sigma}\bar{\sigma}}_{14;23} + \gamma^{\sigma\bar{\sigma}\bar{\sigma}\sigma}_{14;23} \, .
\end{aligned} \tag{10.40}$$

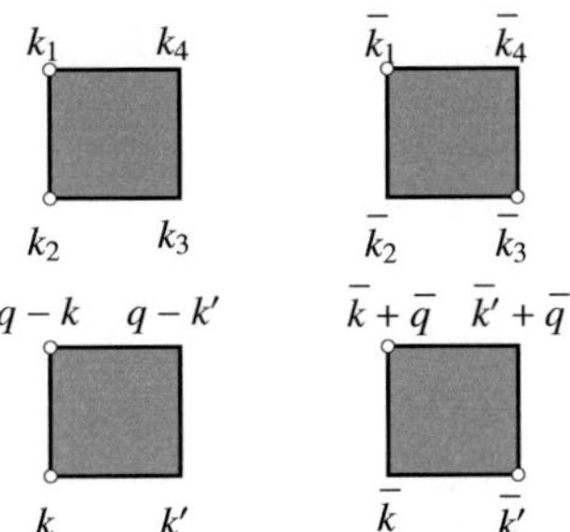



*Figure 10.9:* Frequency labels for the particle-particle (left) and particle-hole vertices (right). $k$ stands for the combined index $(\mathbf{k}, \omega_n)$ and $q = (\mathbf{q}, \Omega_m)$.

Using rotational invariance in spin-space to eliminate $\Gamma^{\mathrm{eh1}}$ yields

$$\Gamma^{\mathrm{ee,irr\,s}}_{\mathrm{p}12;34} = 2\Gamma^{\mathrm{eh0}\sigma\sigma\bar\sigma\bar\sigma}_{13;24} - \Gamma^{\mathrm{eh0}\sigma\sigma\sigma\sigma}_{13;24} + 2\Gamma^{\mathrm{eh0}\sigma\sigma\bar\sigma\bar\sigma}_{14;23} - \Gamma^{\mathrm{eh0}\sigma\sigma\sigma\sigma}_{14;23} - 2\gamma^{\sigma\sigma\bar\sigma\bar\sigma}_{14;23} + \gamma^{\sigma\sigma\sigma\sigma}_{14;23} \ . \tag{10.41}$$

This can be expressed in terms of the particle-hole density (d) and magnetic (m) channels (10.10) in the form

$$\Gamma^{\mathrm{ee,irr\,s}}_{12;34} = \left(\frac{3}{2}\tilde\Gamma^{\mathrm{eh\,m}}_{13;24} - \frac{1}{2}\tilde\Gamma^{\mathrm{eh\,d}}_{13;24}\right) + \left(\frac{3}{2}\tilde\Gamma^{\mathrm{eh\,m}}_{14;23} - \frac{1}{2}\tilde\Gamma^{\mathrm{eh\,d}}_{14;23}\right) - \left(\frac{3}{2}\tilde\gamma^{\mathrm{m}}_{14;23} - \frac{1}{2}\tilde\gamma^{\mathrm{d}}_{14;23}\right) \ . \tag{10.42}$$

Writing frequencies and momenta explicitly (labels are defined as in Fig. 10.9) and accounting for the fact that the vertices in the electron-hole channel depend on transferred momentum only, one finds that

$$\begin{aligned}
\Gamma^{\mathrm{ee,irr\,s}}_{\omega,\omega',\Omega}(\mathbf{k},\mathbf{k}',\mathbf{q}) = {}& 2\Gamma^{\mathrm{eh0}\,\sigma\sigma\bar\sigma\bar\sigma}_{\omega',\omega,\Omega-\omega-\omega'}(\mathbf{q}-\mathbf{k}-\mathbf{k}') - \Gamma^{\mathrm{eh0}\,\sigma\sigma\sigma\sigma}_{\omega',\omega,\Omega-\omega-\omega'}(\mathbf{q}-\mathbf{k}-\mathbf{k}') \\
& + 2\Gamma^{\mathrm{eh0}\,\sigma\sigma\bar\sigma\bar\sigma}_{\Omega-\omega',\omega,\omega'-\omega}(\mathbf{k}'-\mathbf{k}) - \Gamma^{\mathrm{eh0}\,\sigma\sigma\sigma\sigma}_{\Omega-\omega',\omega,\omega'-\omega}(\mathbf{k}'-\mathbf{k}) \\
& - 2\gamma^{\sigma\sigma\bar\sigma\bar\sigma}_{\Omega-\omega',\omega,\omega'-\omega} + \gamma^{\sigma\sigma\sigma\sigma}_{\Omega-\omega',\omega,\omega'-\omega} \ .
\end{aligned} \tag{10.43}$$

Likewise, for the triplet channel one obtains

$$\Gamma^{\mathrm{ee,irr\,t}}_{\omega,\omega',\Omega}(\mathbf{k},\mathbf{k}',\mathbf{q}) = \Gamma^{\mathrm{eh0}\,\sigma\sigma\sigma\sigma}_{\Omega-\omega',\omega,\omega'-\omega}(\mathbf{k}'-\mathbf{k}) - \Gamma^{\mathrm{eh0}\,\sigma\sigma\sigma\sigma}_{\omega',\omega,\Omega-\omega-\omega'}(\mathbf{q}-\mathbf{k}-\mathbf{k}') - \gamma^{\sigma\sigma\sigma\sigma}_{\Omega-\omega',\omega,\omega'-\omega} \ . \tag{10.44}$$

Finally, the Bethe-Salpeter equation for the particle-particle channel is given by

$$\begin{aligned}
\Gamma^{\mathrm{ee\,s/t}}_{\omega\omega'\Omega}(\mathbf{k},\mathbf{k}',\mathbf{q}) = {}& \Gamma^{\mathrm{ee,irr\,s/t}}_{\omega\omega'\Omega}(\mathbf{k},\mathbf{k}',\mathbf{q}) + \frac{T}{2N}\sum_{\omega''\mathbf{k}''}\Gamma^{\mathrm{irr},s/t}_{\mathrm{p}\omega\omega''\Omega}(\mathbf{k},\mathbf{k}'',\mathbf{q})G^{\mathrm{d}}_{\Omega-\omega''}(\mathbf{q}-\mathbf{k}'')\times \\
& \times G^{\mathrm{d}}_{\omega''}(\mathbf{k}'')\,\Gamma^{\mathrm{ee\,s/t}}_{\omega''\omega'\Omega}(\mathbf{k}'',\mathbf{k}',\mathbf{q}) \ .
\end{aligned} \tag{10.45}$$

At present, the solution of this equation retaining the full frequency and momentum dependence is not possible, mainly due to large memory requirements. However, it is possible to solve the corresponding eigenvalue problem. For transferred frequency $\Omega = 0$ and momentum $\mathbf{q} = 0$ it takes the form

$$\frac{T}{2N} \sum_{\omega' \mathbf{k}'} \Gamma^{\text{ee,irr s/t}}_{\omega\omega'\Omega=0}(\mathbf{k}, \mathbf{k}', \mathbf{q} = 0)\, G^{\text{d}}_{-\omega'}(-\mathbf{k}')\, G^{\text{d}}_{\omega'}(\mathbf{k}')\phi_{\omega'}(\mathbf{k}') = \lambda\phi_\omega(\mathbf{k}) . \tag{10.46}$$

This differs from Eq. 4 in Ref. [165] by a sign due to a different definition of $\Gamma^{\text{ee}}$ and a factor of $1/2$, which here avoids double counting of diagrams. With the four-momentum index $k = (\mathbf{k}, \omega)$ one needs to calculate eigenvalues and eigenfunctions of the supermatrix

$$A(k, k') := \frac{T}{2N}\Gamma^{\text{ee,irr s/t}}_{\omega\omega'\Omega=0}(\mathbf{k}, \mathbf{k}', \mathbf{q} = 0)\, G^{\text{d}}_{-\omega'}(-\mathbf{k}')\, G^{\text{d}}_{\omega'}(\mathbf{k}') . \tag{10.47}$$

### 10.4.1    Implementation Notes

The linear dimension of the matrix in the eigenvalue problem is $N \cdot 2N_\omega$. Calculations have been carried out for frequency cutoffs up to $N_\omega = 50$ for the vertex and lattice sizes up to $N = 64 \times 64$. The matrices are too large to be stored in memory. Keeping instead only the vertex from the particle-hole channel and the Green functions in memory allows a fast computation of the matrix elements of the irreducible particle-particle vertex. Since only the leading eigenvalue is needed, an iterative solver for the eigenvalue problem is employed (the implicitly restarted Arnoldi algorithm from the ARPACK library [170]). The matrix elements are recalculated at each iteration.

### 10.4.2    Results

The Hamiltonian of the Hubbard model is given by (6.48). Including nearest and next-nearest neighbor hopping the bare dispersion reads

$$h_{\mathbf{k}} = -2t(\cos k_x + \cos k_y) - 4t' \cos k_x \cos k_y . \tag{10.48}$$

Often the next-nearest-neighbor hopping amplitude $t'/t = -0.3$ is included for a more realistic description of the experimental situation in the cuprates. Here, in order to compare with Ref. [165], the calculations were performed with $t' = 0$. Fig. 10.10 shows the leading dual eigenvalues of the Bethe-Salpeter equation in various channels as a function of temperature. An eigenvalue of $\lambda^{\text{d}}_{\text{max}} = 1$ would indicate a transition to an ordered state. Clearly, the dominant contribution to pairing stems from the $\mathbf{q} = (0, 0)$ singlet pairing channel. As shown in the inset, the momentum dependence of the eigenfunction (this is essentially the gap function [165]) exhibits d-wave symmetry (this symmetry is not present at very high temperatures, e.g. $T/t = 0.33$). It naturally emerges when

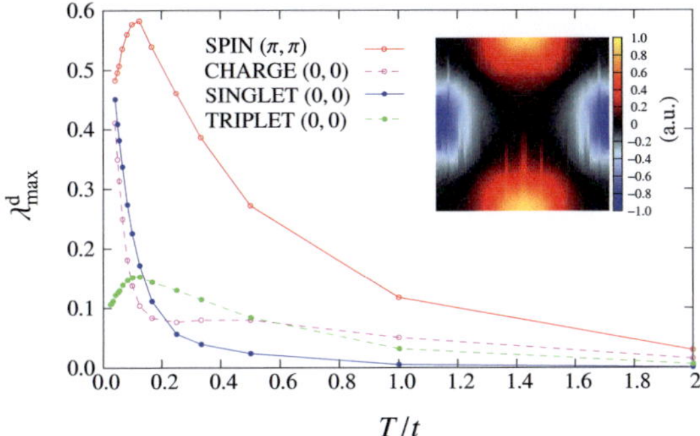

*Figure 10.10:* Leading dual eigenvalue in various channels for $U/t = 4$ and $\delta = 15\%$ hole doping as a function of temperature. The lattice size is $16 \times 16$. The dominant contribution to pairing originates from the singlet pairing channel. The inset shows the momentum dependence of the corresponding eigenfunction ($64 \times 64$ points, $T/t = 0.042$). The d-wave symmetry is apparent.

the spin-fluctuation diagrams are taken into account. These diagrams dominate the tendency to pairing. The frequency structure of the singlet channel shows even-frequency pairing, while the triplet channel exhibits odd-frequency pairing. Further decomposition into density (singlet) and magnetic (triplet) particle-hole channels according to (10.42) shows that the pairing fluctuations are dominated by the $S = 1$ triplet particle-hole channel. These findings are in accordance to those of Ref. [165]. Note that the approximations are rather different. In the DCA study of this reference, the irreducible particle-particle vertex $\Gamma^{\text{ee,irr}}$ was extracted from the full vertex of the cluster. Here an approximation for $\Gamma^{\text{ee,irr}}$ is constructed diagrammatically, starting from the local irreducible vertex.

The eigenvalue in the $\mathbf{q}_{\text{AF}} = (\pi, \pi)$ particle-hole channel is found to first increase and then drop as the temperature is lowered. The reason is that the peak in the response at $\mathbf{q}_{\text{AF}}$ shown in Fig. 10.5 is splitted into four parts. As the temperature is lowered, the peaks become narrower, leading to the observed drop. The splitting is due to the deformation of the Fermi surface upon doping, which relaxes the nesting condition. Being a Fermi surface effect, this is already contained in DMFT. The steep drop was not observed in Ref. [165], due to the coarse graining in the DCA. Here a similar behavior is obtained when focusing on the maximal response, which is slightly shifted from $(\pi, \pi)$. At this value of $U$, the charge eigenvalue shows the tendency towards a homogeneous charge ordered state (this is not observed for $U/t = 8$). For some values of the doping, the leading eigenvalue is found at $(\pi, 0)$ showing a tendency to a stripe-like modulation of the charge order. At these parameters, no transition to the superconducting state is

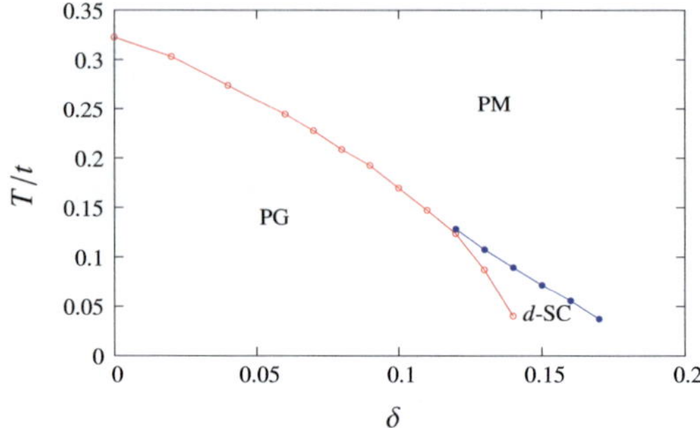


*Figure 10.11:* Tentative phasediagram of the 2D Hubbard model for $U/t = 8$. The region labeled PG is characterized by strong antiferromagnetic fluctuations. As the phase boundary is approached, the DOS of the ladder approximation (see chapter 11) develops a pseudogap. The merging of the d-wave phase boundary into the PG boundary is an artefact of the present approximation. A d-wave instability exists independently of the AF instability for $\delta \sim 15\%$.

observed for the accessible temperatures. For a bandwidth of 2eV [145], the lowest temperatures accessed here ($T/t \sim 0.02$) correspond to $\sim 60$K. For lower temperatures, convergence of the results with respect to the number of frequencies in the vertex cannot be guaranteed.

Fig. 10.11 shows the phasediagram obtained from the leading eigenvalues in the $\mathbf{q} = (\pi, \pi)$ particle-hole and $\mathbf{q} = (0, 0)$ particle-particle channels for $U/t = 8$. The calculations for the phasediagram, and the refinement and convergence checks of the results in the previous figure, were carried out by M. Kecker during the course of his diploma thesis using the implementation developed here. Anticipating the results of chapter 11, the region enclosed by the boundary at which the particle-hole ladder diverges is characterized by strong antiferromagnetic fluctuations. In the ladder approximation, they lead to the formation of a pseudogap in the density of states. This region can hence be interpreted as the pseudogap region, which resembles the situation in the underdoped cuprates [171].

A transition to the d-wave superconducting state is found independently of the pseudogap phase boundary for $\delta \sim 15\%$. For smaller doping the boundary turns upwards an merges into the pseudogap phase boundary. This is clearly an artefact of the present approximation. The divergence of the ladder sum in the particle-hole channel triggers a divergence in the particle-particle channel. The divergence of the ladder sum in turn is due to the fact that the antiferromagnetic fluctuations are not sufficiently accounted for. This problem is resolved within the ladder approximation (it has not yet been applied in this context).

To conclude, the behavior of the dual eigenvalues is in qualitative agreement with previous QMC studies. The results look promising, but the results for the phasediagram are still unsatisfactory and the method needs further improvement. The antiferromagnetic fluctuations should be properly taken into account through the ladder approximation. In order to access the part of the phasediagram deep inside the pseudogap region, where the ladder approximation does not converge, one may instead start in the symmetry broken state as described in Sec. 6.10.1. The magnetization is expected to be reduced when fluctuations are included through the ladder approximation. One can expect the phase boundary to the superconducting state to be significantly affected in this case. Short-range singlet correlations can be treated explicitly within cluster dual fermion calculations. A Moriya-$\lambda$ correction to the vertex has been proposed in Ref. [172] to enforce $T_N = 0$ as required by the Mermin-Wagner theorem.

A self-consistent treatment, which accounts for the feedback effects of particle-particle fluctuations on the particle-hole excitations, requires a parquet-like construction and is computationally highly demanding. The present approach may nevertheless provide insight into the nature of the pairing fluctuations near $T_c$. In contrast to cluster approaches, it does not break translational invariance and allows to account for anomalies associated with a narrow region of k-space, such as van-Hove singularities. The diagrammatic approach has the advantage that different contributions can be unraveled by selectively including or excluding them from the calculation.

# Chapter 11

# The Ladder Dual Fermion Approach

H. Hafermann, G. Li, A. N. Rubtsov, M. I. Katsnelson,
A. I. Lichtenstein and H. Monien
Phys. Rev. Lett. **102**, 206401 (2009)

The ladder dual fermion approach (LDFA) has been introduced in the above listed publication entitled *Efficient Perturbation Theory for Quantum Lattice Models*. It has been developed as part of this work and will be presented in detail below. The paper further contains results on the convergence properties of the approach, which were obtained in collaboration. The contribution made here is to refine the eigenvalue analysis by including higher-order contributions in the perturbation expansion of the self-energy. It is shown that the dual fermion approach has superior convergence properties compared to straightforward diagrammatic expansions around DMFT. The convergence properties are addressed in Sec. 11.3.

## 11.1   Motivation

In the preceding chapters, there have been indications that the dual fermion (DF) approach does not correctly capture the physics of the 2D Hubbard model at low temperatures. In particular, the Néel temperature is not significantly reduced compared to DMFT calculations (for a discussion regarding the applicability of the Mermin-Wagner theorem, see Sec. 10.3). Moreover, two-particle quantities are also not accurately approximated (Sec. 10.3.1). This suggests that antiferromagnetic fluctuations are not properly taken into account. On the other hand, the recently proposed dynamical vertex approximation (DΓA) [39] and a related approach [40], have been shown to include such contributions beyond DMFT, leading to, e.g., pseudogap formation.

Due to the abstract nature of the transformation to dual fermions, it is a priori not clear how physically meaningful approximations can be constructed in this approach. The fact that dual fermions couple to the lattice degrees of freedom *locally* (Eq. 6.8), suggests that in order to account for long-range correlations, e.g. through diagrams with loops, one should take into account topologically equivalent diagrams in terms of dual fermions. Moreover, it is known that in order to capture the effects of nonlocal antiferromagnetic fluctuations in the vicinity of the antiferromagnetic instability of the Hubbard model, it is essential to take into account the paramagnon contributions to the self-energy. In Refs. [39, 40], the feedback of two-particle excitations onto the self-energy is included by means of a ladder approximation.

The equivalence between two-particle excitations for dual and lattice fermions is therefore crucial to the approach. It points to the need to include ladder diagrams to dual fermions to correctly describe these effects. In general, this equivalence guarantees that approximations can be constructed much as in the usual way, i.e. based on physical considerations. In the case considered here, one is led to include diagrams from the particle-hole fluctuation channels in the vicinity of the magnetic instability, while the particle-particle contribution is expected to be negligible. On the same basis, one may construct more complex approximations, such as an analog to the parquet approach. The relation between $n$-particle correlation functions for dual and lattice fermions (Sec. 6.3.2) means that an approximation on the two-particle level corresponds to the two-particle level in terms of dual fermions. The truncation of the dual potential after the first-order term is hence justified by the same reasoning as for lattice fermions, namely by a phase space argument [93] (the available phase space for, e.g., three-particle scattering processes is much smaller than for two-particle scattering). Note that the fluctuation-exchange approximation (FLEX) or the parquet approach are restricted to the two-particle level.

For weak coupling, similar results can be expected from the LDFA, DΓA and related approaches (and possibly FLEX), where $\gamma^{(4)} \sim U$ (no quantitative analysis has been performed so far). For strong coupling however, it was already pointed out that the dual fermion approach has superior convergence properties and that the adjustment of the hybridization function is essential for a consistent description of the full interaction range. It is hence desirable to extend the dual fermion approach to include long-range correlations. This leads to a versatile approach for a realistic description of materials beyond mean-field.

In order to illustrate that the dual two-particle Green function indeed contains the relevant physics, the dynamical susceptibility $\chi(\mathbf{q}, \omega)$ is plotted in Fig. 11.1. It has been obtained using the methods of chapter 10, i.e. by transforming the dual susceptibility back to lattice fermions and subsequent analytical continuation using Padé approximants. The bubble diagram was supplemented with the high-energy tails of the Green functions as described in appendix. B, prior to the analytical continuation. The dynamical susceptibility clearly displays the magnon spectrum in the paramagnetic state. The dispersion from spin wave theory is shown for comparison. It is given by the expression [173]

$$\epsilon(\mathbf{k}) = 2zJS \sqrt{1 - \gamma(\mathbf{k})^2} \, , \qquad (11.1)$$

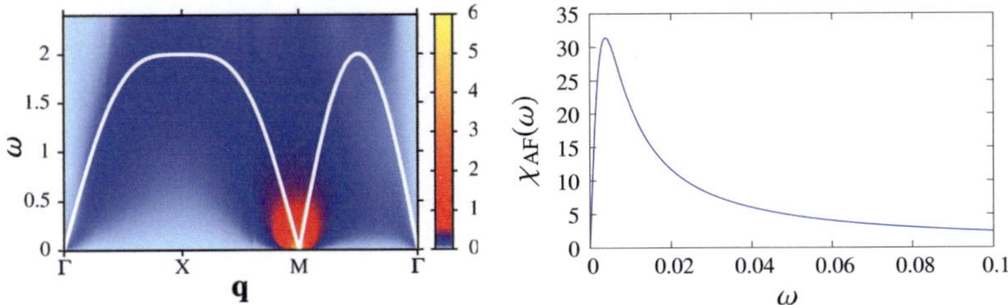

*Figure 11.1:* Left: Dynamical susceptibility $\chi(\mathbf{q}, \omega)$ for $U/t = 4$ and $T/t = 0.19$, obtained from a dual fermion calculation. It shows the magnon spectrum in the paramagnetic state. The dispersion from spin wave theory with effective exchange coupling $J = 4t^2/U$ is shown for comparison. Values for $\chi > 6$ are excluded from the colormap to improve the contrast (the spectrum is less visible in print). Right: Cross-section through the peak at the $M$-point. The displacement from zero is consistent with a small energy scale $J/\xi$, where $\xi$ is the correlation length (in units of the lattice constant).

where $z$ is the coordination number, $S = 1/2$ is the spin of the fermions and

$$\gamma(\mathbf{k}) = \frac{1}{z} \sum_{\text{NN}} e^{i\mathbf{k}\mathbf{r}_{\text{NN}}} = \frac{1}{2}\left(\cos k_x + \cos k_y\right) \tag{11.2}$$

for the square lattice. Here the sum is over the $z = 4$ nearest neighbors at positions $\mathbf{r}_{\text{NN}}$. The effective AF exchange coupling is given by $J = 4t^2/U$. The right panel of Fig. 11.1 shows a cross-section for the antiferromagnetic wave vector $\mathbf{q}_{\text{AF}} = (\pi, \pi)$ ($M$-point). The peak is broadened and slightly shifted from zero. Such a behavior is reminiscent of a 2D Heisenberg model at finite temperature, where long-range order with a correlation length $\xi \gg a$ takes place ($a$ is the lattice constant) and a corresponding small energy scale of order $Ja/\xi$ arises [151]. The correlation length determined from the peak displacement $\Delta\omega$ from zero is of the same order as the lattice size. This is consistent with the fact that the temperature is very close to the DF Néel temperature.

## 11.2 Derivation of the LDFA Equations

The LDFA is a $\Phi$-derivable approximation. Similarly to FLEX [174, 175], the (dual) Luttinger-Ward functional contains the ring diagrams $\delta\Phi_{\mathrm{d}}^{(n)}$ up to all orders $n$. The contributions to this functional are depicted in Fig. 11.2. The dual self-energy of the ladder approximation is obtained by functional derivative according to $\Sigma^{\mathrm{LDFA}} = \delta\Phi^{\mathrm{d}}/\delta G^{\mathrm{d}}$. Note that the functional derivative with respect to any of the $2n$ lines in a ring diagram yields

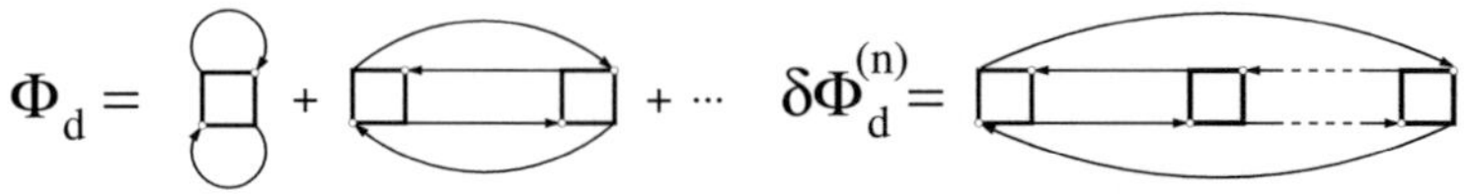

Figure 11.2: The contributions to the dual Luttinger-Ward functional for the ladder approximation. The generic $n$-th order ring diagrams are included up to all orders $n$.

the same contribution to the self-energy, so that the symmetry factors $S = 2n$ of the diagrams exactly cancel (see appendix A.4.3). This construction is theoretically relevant, as it ensures a conserving and thermodynamically consistent approximation [61, 143]. The resulting theory is also conserving in terms of lattice fermions [41, 139]. It is however not suitable for numerical calculations. These are instead based on the solution of the Bethe-Salpeter equations, which are used to construct an approximation to the full vertex, similar to Refs. [39, 40]. From the vertex, the self-energy is obtained through the Schwinger-Dyson equation.

The DΓA, in its actual implementation as a ladder approximation, has been motivated through the parquet equations. In general, the full vertex of the problem is constructed from the fully irreducible $n$-particle vertices and full Green functions as building blocks. In DΓA, only the fully irreducible two-particle vertex is considered and it is assumed to be equal to the DMFT solution for the corresponding quantity of the impurity model. Hence the DΓA is restricted to the two-particle level. This is also the case for the LDFA. In principle, it is possible to construct a parquet approximation to dual fermions. Not only for this reason, but also because it is instructive for the derivation of the ladder approximation, the parquet equations are considered first.

## 11.2.1 Parquet Equations

The parquet equations are motivated by the goal to iteratively construct an approximation to the full vertex of the lattice, which has full crossing symmetry (see below). This is done starting from the lowest-order approximation $\Gamma \approx \Lambda_{\text{firr}}$. $\Lambda_{\text{firr}}$ is the fully irreducible vertex, which is local to good approximation. The next order already is a plaquette built from four (reducible) vertices connected by full Green functions [176]. In what is commonly understood as the parquet approach, $\Lambda_{\text{firr}}$ is taken to be the bare interaction $U$. In DΓA, it is approximated by the local irreducible vertex of the impurity problem. In the LDFA, it is taken to be the bare interaction of the dual fermions, $\Lambda_{\text{firr}} = \gamma^{(4)}$.

For a heuristic derivation, first decompose any diagram that contributes to the full vertex $\Gamma$ into an irreducible part and the reducible rest. In symbolic notation:

$$\Gamma = \Lambda_{\text{irr}}^{\alpha} + \Lambda_{\text{red}}^{\alpha} . \tag{11.3}$$

The difference to the single-particle case is that this decomposition is not unique. Reducibility or irreducibility needs to be defined with respect to one of the three possible channels: $\alpha$ =eh,v or ee, where 'eh' stands for the electron-hole, 'ee' for the electron-electron and 'v' for the vertical channel, respectively. A diagram is said to be irreducible e.g. in the electron-hole channel if it cannot be cut in two by cutting a pair of antiparallel (horizontal) lines. The sum of all reducible contributions to the vertex for a specific channel is obtained as $\langle \Lambda_{irr}^{\alpha} * G * G * \Gamma \rangle_{\alpha}$, where the symbol '$*$' indicates summation over all internal degrees of freedom and $\alpha$ is the respective channel. Together with Eq. 11.3 one obtains three different equations for the vertex:

$$\Gamma = \Lambda_{irr}^{eh} + \langle \Lambda_{irr}^{eh} * G * G * \Gamma \rangle_{eh} \, , \tag{11.4}$$

$$\Gamma = \Lambda_{irr}^{ee} + \langle \Lambda_{irr}^{ee} * G * G * \Gamma \rangle_{ee} \, , \tag{11.5}$$

$$\Gamma = \Lambda_{irr}^{v} + \langle \Lambda_{irr}^{v} * G * G * \Gamma \rangle_{v} \, . \tag{11.6}$$

These are the Bethe-Salpeter equations. Given the exact irreducible vertex in the respective channel, these equations yield the exact vertex. To proceed, note that any diagram which contributes to the irreducible vertex falls into one of three different categories: either it is fully irreducible, or reducible in one of the two other channels. This yields the decompositions

$$\Lambda_{irr}^{eh} = \Lambda_{firr} + \Lambda_{red}^{ee} + \Lambda_{red}^{v} \, , \tag{11.7}$$

$$\Lambda_{irr}^{ee} = \Lambda_{firr} + \Lambda_{red}^{eh} + \Lambda_{red}^{v} \, , \tag{11.8}$$

$$\Lambda_{irr}^{v} = \Lambda_{firr} + \Lambda_{red}^{eh} + \Lambda_{red}^{ee} \, . \tag{11.9}$$

Using (11.3), the vertex can therefore be written

$$\Gamma = \Lambda_{firr} + \Lambda_{red}^{eh} + \Lambda_{red}^{ee} + \Lambda_{red}^{v} \, . \tag{11.10}$$

An alternative representation for $\Gamma$ is obtained by adding the three equations (11.7-11.9), which yields

$$\Lambda_{irr}^{eh} + \Lambda_{irr}^{ee} + \Lambda_{irr}^{v} = 3\Lambda_{firr} + 2\left(\Lambda_{red}^{eh} + \Lambda_{red}^{ee} + \Lambda_{red}^{v}\right) \, . \tag{11.11}$$

Comparing with Eq. 11.10, the full vertex can hence be expressed in terms of the irreducible vertices as

$$\Gamma = \frac{1}{2}\left(\Lambda_{irr}^{eh} + \Lambda_{irr}^{ee} + \Lambda_{irr}^{v} - \Lambda_{firr}\right) \, . \tag{11.12}$$

To arrive at the parquet equations, the reducible vertex in channel $\alpha$ is expressed as $\Lambda_{red}^{\alpha} = \langle \Lambda_{irr}^{\alpha} * G * G * \Gamma \rangle_{\alpha}$ as before. Inserting here the alternative expressions for the irreducible vertex, Eqs. 11.7-11.9, together with (11.10) one arrives at a closed set of equations,

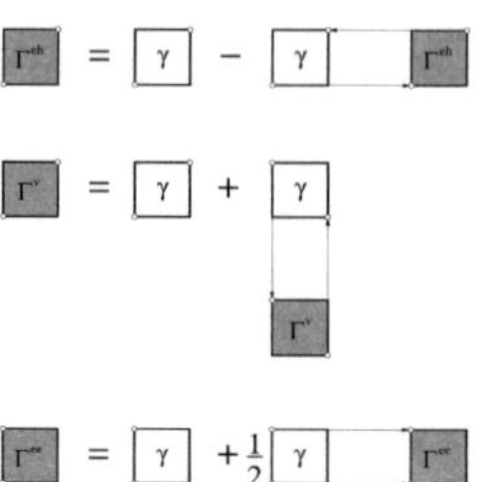

*Figure 11.3:* Diagrammatic representation of the Bethe-Salpeter equations in the three different channels: Horizontal (electron-hole) channel, vertical channel and electron-electron channel. In LDFA, the irreducible vertex is approximated by the bare dual interaction $\gamma$. The solution for the vertex $\Gamma^\alpha$ depends on the channel and thus acquires an index.

$$\Gamma = \Lambda_{\text{firr}} + \Lambda_{\text{red}}^{\text{eh}} + \Lambda_{\text{red}}^{\text{ee}} + \Lambda_{\text{red}}^{\text{v}} , \tag{11.13}$$

$$\Lambda_{\text{red}}^{\text{eh}} = \langle (\Lambda_{\text{firr}} + \Lambda_{\text{red}}^{\text{ee}} + \Lambda_{\text{red}}^{\text{v}}) * G * G * \Gamma \rangle_{\text{eh}} , \tag{11.14}$$

$$\Lambda_{\text{red}}^{\text{ee}} = \langle (\Lambda_{\text{firr}} + \Lambda_{\text{red}}^{\text{eh}} + \Lambda_{\text{red}}^{\text{v}}) * G * G * \Gamma \rangle_{\text{ee}} , \tag{11.15}$$

$$\Lambda_{\text{red}}^{\text{v}} = \langle (\Lambda_{\text{firr}} + \Lambda_{\text{red}}^{\text{ee}} + \Lambda_{\text{red}}^{\text{eh}}) * G * G * \Gamma \rangle_{\text{v}} . \tag{11.16}$$

which are known as the parquet equations[1]. The iterative solution of these equations yields a vertex with full crossing symmetry. However, the complexity often requires additional approximations [163, 176]. The solution to these equations is not guaranteed to be conserving in the Baym-Kadanoff sense.

## 11.2.2  Approximation to the Vertex: DΓA

In the DΓA, the exact irreducible vertex is approximated by the irreducible vertex of the impurity problem. According to Ref. [39], the vertex that generates the ladder approximation is obtained by assuming the locality of the reducible vertex parts on the right-hand side of Eqs. 11.14-11.16. In doing so, one obtains

$$\Lambda_{\text{red}}^{\text{eh}} = \langle (\Lambda_{\text{firr,loc}} + \Lambda_{\text{red,loc}}^{\text{ee}} + \Lambda_{\text{red,loc}}^{\text{v}}) * G * G * \Gamma \rangle_{\text{eh}} \tag{11.17}$$

and likewise for the two other channels. Using (11.7), this becomes

$$\Lambda_{\text{red}}^{\text{eh}} = \langle \Lambda_{\text{irr,loc}}^{\text{eh}} * G * G * \Gamma \rangle_{\text{eh}} . \tag{11.18}$$

---

[1]When the vertices are depicted by horizontal and vertical rectangles indicating the channels (see e.g. Ref. [39]), the diagrammatic representation of these equations is reminiscent of a parquet floor and hence the name.

Using $\Lambda_{\text{red}}^{\text{eh}} = \Gamma - \Lambda_{\text{irr,loc}}^{\text{eh}}$, this is equivalent to the Bethe-Salpeter equation, Eq. 11.4, for a local irreducible vertex. Since the solution $\Gamma$ of the BSE in this approximation depends on the type of channel, the vertex acquires an index, e.g. $\Gamma^{\text{eh}}$ for the electron-hole channel. Inserting the result for $\Lambda_{\text{red}}^{\alpha}$ for the three channels into Eq. 11.13, one obtains

$$\Gamma = \Lambda_{\text{firr,loc}} + (\Gamma^{\text{eh}} - \Lambda_{\text{irr,loc}}^{\text{eh}}) + (\Gamma^{\text{ee}} - \Lambda_{\text{irr,loc}}^{\text{ee}}) + (\Gamma^{\text{v}} - \Lambda_{\text{irr,loc}}^{\text{v}}) . \tag{11.19}$$

In order to simplify this equation further, one may rewrite Eq. 11.12 for the corresponding local quantities as

$$\Lambda_{\text{firr,loc}} = \Lambda_{\text{irr,loc}}^{\text{eh}} + \Lambda_{\text{irr,loc}}^{\text{ee}} + \Lambda_{\text{irr,loc}}^{\text{v}} - 2\Gamma_{\text{loc}} . \tag{11.20}$$

Inserting this into the approximation for $\Gamma$ yields the final expression

$$\Gamma = \Gamma^{\text{eh}} + \Gamma^{\text{v}} + \Gamma^{\text{ee}} - 2\Gamma_{\text{loc}} , \tag{11.21}$$

where $\Gamma_{\text{loc}}$ stands for the local part of the full vertex, which in the D$\Gamma$A is taken to be the *reducible* vertex $\gamma^{(4)}$ of the impurity. The above relation avoids an explicit calculation of the fully irreducible vertex. The irreducible vertices in the respective channels, which enter the BSEs, have to be obtained from the full vertex of the impurity by inverting a corresponding local Bethe-Salpeter equation. The vertices $\Gamma^{\alpha}$ are then obtained from the standard BSEs and inserted into (11.21) in order to obtain an approximation to the full vertex.

### 11.2.3  Approximation to the Vertex: LDFA

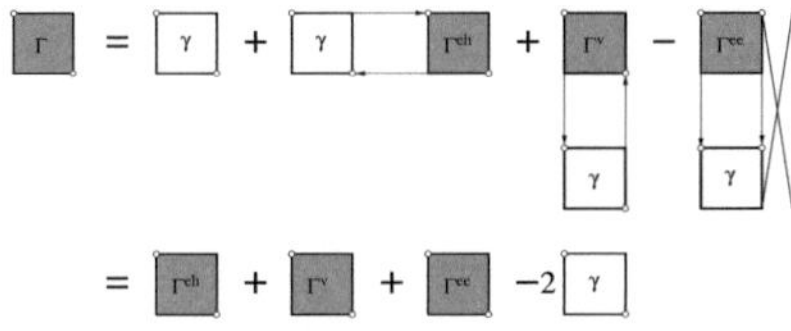

*Figure 11.4:* Diagrammatic construction of an approximation to the full vertex in the ladder dual fermion approximation. The ladder sums are obtained as the solution of the Bethe-Salpeter equations in the tree different channels (Fig. 11.3).

In the LDFA, the fully irreducible vertex is naturally approximated by the bare interaction of dual fermions, i.e. $\Lambda_{\text{firr}} = \gamma^{(4)}$ (the index on $\gamma$ will be omitted in what follows). In the first iteration of the parquet equations, the reducible vertices are taken to be zero,

$\Lambda_{\text{red}}^{\text{eh}} = \Lambda_{\text{red}}^{\text{v}} = \Lambda_{\text{red}}^{\text{ee}} = 0$. In this approximation, the irreducible vertices are equal to the fully irreducible one by virtue of Eqs. 11.7-11.9: $\Lambda_{\text{firr}} = \Lambda_{\text{irr}}^{\alpha} = \gamma$. A first approximation to the reducible vertices is obtained from the BSEs (11.4-11.6) and are thus given by $\Lambda_{\text{red}}^{\alpha} = \Gamma^{\alpha} - \gamma$. The full vertex is hence given by (cf. Fig. 11.4)

$$\Gamma = \gamma + \Lambda_{\text{red}}^{\text{eh}} + \Lambda_{\text{red}}^{\text{v}} + \Lambda_{\text{red}}^{\text{ee}} = \Gamma^{\text{eh}} + \Gamma^{\text{v}} + \Gamma^{\text{ee}} - 2\gamma \, . \tag{11.22}$$

Subtracting $\gamma$ appears as a double counting correction. In contrast to (11.20), here it is given by the bare interaction.

In general, the approximation to the vertex should contain the contributions from all fluctuation channels. In the following, only the contributions from the two particle-hole channels will be considered. These are expected to dominate near a magnetic instability. Neglecting $\Lambda_{\text{red}}^{\text{ee}}$ in (11.22), the approximation to the vertex gives

$$\Gamma = \gamma + \Lambda_{\text{red}}^{\text{eh}} + \Lambda_{\text{red}}^{\text{v}} = \Gamma^{\text{eh}} + \Gamma^{\text{v}} - \gamma \, . \tag{11.23}$$

Note that the approximation to electron-electron vertex in Sec. 10.4, which has been motivated diagrammatically, can be obtained in a similar fashion: Eq. 11.8 states that $\Lambda_{\text{irr}}^{\text{ee}} = \Lambda_{\text{firr}} + \Lambda_{\text{red}}^{\text{eh}} + \Lambda_{\text{red}}^{\text{v}}$. Using $\Lambda_{\text{firr}} = \Lambda_{\text{irr}}^{\alpha} = \gamma$ as above and expressing the reducible vertex through the solution of the BSEs yields $\Lambda_{\text{irr}}^{\text{ee}} = \Gamma^{\text{eh}} + \Gamma^{\text{v}} - \gamma$.

### 11.2.4  Bethe-Salpeter Equations

As a next step, the Bethe-Salpeter equations are considered in detail. For the (horizontal) electron-hole channel, the BSE in the $S_z = 0$ channel is

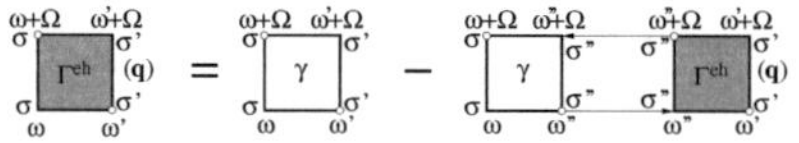

$$\Gamma_{\omega\omega'\Omega}^{\text{eh}\,\sigma\sigma\sigma'\sigma'}(\mathbf{q}) = \gamma_{\omega\omega'\Omega}^{\sigma\sigma\sigma'\sigma'} - \frac{T}{N} \sum_{\omega''\mathbf{k}''} \sum_{\sigma''} \gamma_{\omega\omega''\Omega}^{\sigma\sigma\sigma''\sigma''} G_{\omega''}(\mathbf{k}'') G_{\omega''+\Omega}(\mathbf{k}'' + \mathbf{q}) \Gamma_{\omega''\omega'\Omega}^{\text{eh}\,\sigma''\sigma''\sigma'\sigma'}(\mathbf{q})$$

$$\tag{11.24}$$

and the equation for the $S_z = 1$ channel is given by

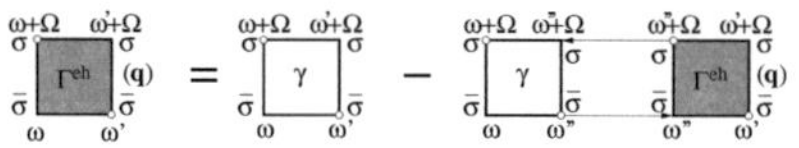

$$\Gamma_{\omega\omega'\Omega}^{\text{eh}\,\sigma\bar{\sigma}\bar{\sigma}\sigma}(\mathbf{q}) = \gamma_{\omega\omega'\Omega}^{\sigma\bar{\sigma}\bar{\sigma}\sigma} - \frac{T}{N} \sum_{\omega''\mathbf{k}''} \gamma_{\omega\omega''\Omega}^{\sigma\bar{\sigma}\bar{\sigma}\sigma} G_{\omega''}(\mathbf{k}'') G_{\omega''+\Omega}(\mathbf{k}'' + \mathbf{q}) \Gamma_{\omega''\omega'\Omega}^{\text{eh}\,\sigma\bar{\sigma}\bar{\sigma}\sigma}(\mathbf{q}) \, . \tag{11.25}$$

For the vertical channel, the equations for the two spin channels are

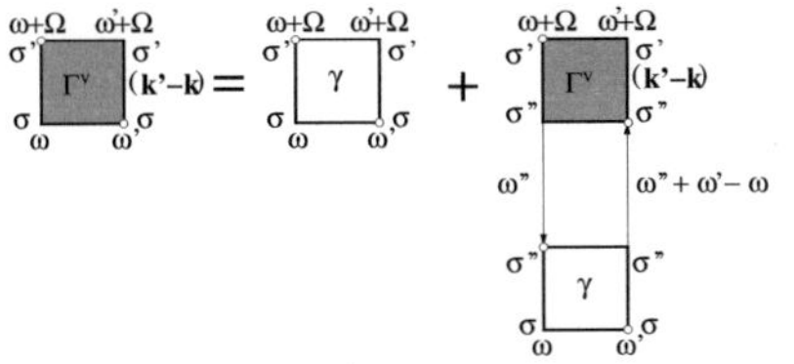

$$\Gamma^{\mathrm{v}\,\sigma'\sigma\sigma\sigma'}_{\omega\,\omega+\Omega\,\omega'-\omega}(\mathbf{k}'-\mathbf{k}) = \gamma^{\sigma'\sigma\sigma\sigma'}_{\omega\omega'\Omega} + \frac{T}{N}\sum_{\omega''\mathbf{k}''}\sum_{\sigma''}\gamma^{\sigma''\sigma\sigma\sigma''}_{\omega\omega'\omega''-\omega}G_{\omega''}(\mathbf{k}'')G_{\omega''+\omega'-\omega}(\mathbf{k}''+\mathbf{k}'-\mathbf{k})\times$$

$$\times\,\Gamma^{\mathrm{v}\,\sigma'\sigma''\sigma''\sigma'}_{\omega''\,\omega+\Omega\,\omega'-\omega}(\mathbf{k}'-\mathbf{k})\,,\qquad(11.26)$$

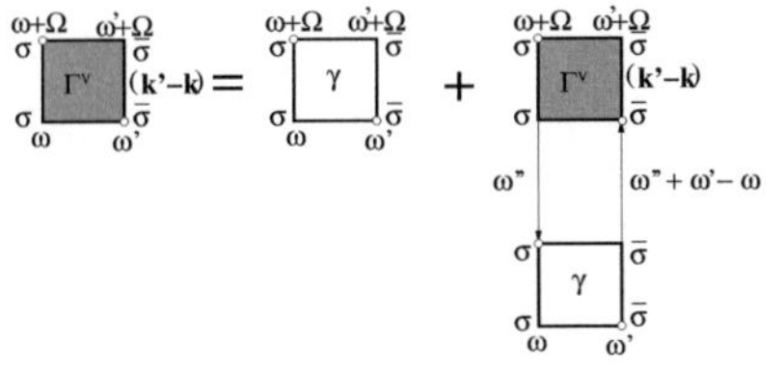

$$\Gamma^{\mathrm{v}\,\sigma\sigma\bar\sigma\bar\sigma}_{\omega\,\omega+\Omega\,\omega'-\omega}(\mathbf{k}'-\mathbf{k}) = \gamma^{\sigma\sigma\bar\sigma\bar\sigma}_{\omega\omega'\Omega} + \frac{T}{N}\sum_{\omega''\mathbf{k}''}\gamma^{\sigma\sigma\bar\sigma\bar\sigma}_{\omega\omega'\omega''-\omega}G_{\omega''}(\mathbf{k}'')G_{\omega''+\omega'-\omega}(\mathbf{k}''+\mathbf{k}'-\mathbf{k})\times$$

$$\times\,\Gamma^{\mathrm{v}\,\sigma\sigma\bar\sigma\bar\sigma}_{\omega''\,\omega+\Omega\,\omega'-\omega}(\mathbf{k}'-\mathbf{k})\,.\qquad(11.27)$$

From the structure of the diagrams it is evident that the horizontal and vertical channels are equivalent. The horizontal channel, when viewed as to describe scattering processes from left to right, looks like the vertical channel viewed from bottom to top (note that apart from a sign, the equations in the vertical channel take the same form as the ones in the horizontal channel by introducing the wave vector $\mathbf{q} = \mathbf{k}' - \mathbf{k}$). Therefore, the vertex in the vertical channel can be expressed in terms of the electron-hole vertex. This can be done either by exchanging the two outgoing or the two incoming endpoints (depicted by open circles). For reasons outlined below, the latter is done here. Taking care of the sign due to exchange of channels (for details, see appendix A.8.1), one finds that

$$\begin{aligned}
\Gamma^{\mathrm{v}\,\sigma'\sigma\sigma\sigma'}_{\omega\omega'\Omega}(\mathbf{q}) &= -\Gamma^{\mathrm{eh}\,\sigma\sigma\sigma'\sigma'}_{\omega\,\omega+\Omega\,\omega'-\omega}(\mathbf{k}'-\mathbf{k})\,,\\
\Gamma^{\mathrm{v}\,\sigma\sigma\bar\sigma\bar\sigma}_{\omega\omega'\Omega}(\mathbf{q}) &= -\Gamma^{\mathrm{eh}\,\bar\sigma\sigma\sigma\bar\sigma}_{\omega\,\omega+\Omega\,\omega'-\omega}(\mathbf{k}'-\mathbf{k})\,.
\end{aligned}\qquad(11.28)$$

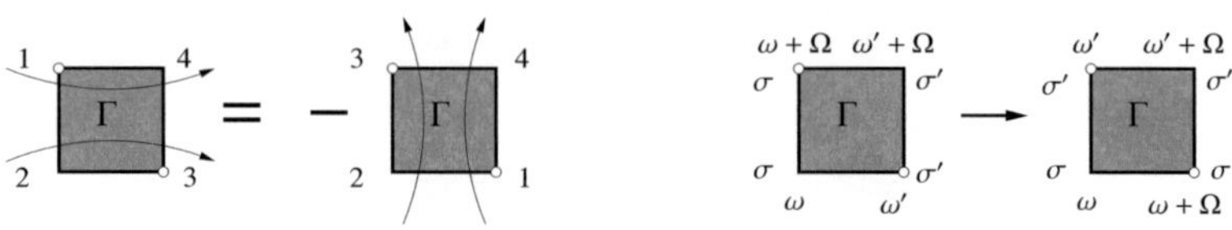

*Figure 11.5:* Left: crossing symmetry of the full vertex. Exchanging two endpoints gives a minus sign by the anticommutation relations and corresponds to a change of channel. Right: by exchanging labels on (incoming) endpoints, the vertex transforms according to $\Gamma^{\sigma\sigma\sigma'\sigma'}_{\omega\omega'\Omega} \rightarrow \Gamma^{\sigma'\sigma\sigma\sigma'}_{\omega\,\omega+\Omega\,\omega'-\omega}$.

Using these relations, the two spin channels of the full vertex (11.23) can be expressed as

$$\Gamma^{\sigma\sigma\sigma'\sigma'}_{\omega\omega'\Omega}(\mathbf{k},\mathbf{k}',\mathbf{q}) = \Gamma^{\mathrm{eh}\,\sigma\sigma\sigma'\sigma'}_{\omega\omega'\Omega}(\mathbf{q}) - \Gamma^{\mathrm{eh}\,\sigma'\sigma\sigma\sigma'}_{\omega\,\omega+\Omega\,\omega'-\omega}(\mathbf{k}'-\mathbf{k}) - \gamma^{\sigma\sigma\sigma'\sigma'}_{\omega\omega'\Omega}\,,$$

$$\Gamma^{\bar\sigma\sigma\sigma\bar\sigma}_{\omega\omega'\Omega}(\mathbf{k},\mathbf{k}',\mathbf{q}) = \Gamma^{\mathrm{eh}\,\bar\sigma\sigma\sigma\bar\sigma}_{\omega\omega'\Omega}(\mathbf{q}) - \Gamma^{\mathrm{eh}\,\sigma\sigma\bar\sigma\bar\sigma}_{\omega\,\omega+\Omega\,\omega'-\omega}(\mathbf{k}'-\mathbf{k}) - \gamma^{\bar\sigma\sigma\sigma\bar\sigma}_{\omega\omega'\Omega}\,. \tag{11.29}$$

It depends on all three momenta and has the property that it obeys partial crossing symmetry. For the exact vertex, this symmetry is illustrated in Fig. 11.5. Due to the anticommutation relations of fermions, i.e. as a consequence of the Pauli principle, the vertex changes sign by exchanging two of its endpoints, e.g. by crossing the lines connected to 1 and 3. As indicated by the arrows, this changes the channel from horizontal to vertical. While neither of the vertices $\Gamma^{\mathrm{eh}}$ and $\Gamma^{\mathrm{v}}$ obey this symmetry, exchanging the incoming endpoints of the approximate vertex (11.29), one finds that

$$\Gamma^{\sigma'\sigma\sigma\sigma'}_{\omega\,\omega+\Omega\,\omega'-\omega}(\mathbf{k},\mathbf{k}+\mathbf{q},\mathbf{k}'-\mathbf{k}) = \Gamma^{\mathrm{eh}\,\sigma'\sigma\sigma\sigma'}_{\omega\,\omega+\Omega\,\omega'-\omega}(\mathbf{k}'-\mathbf{k}) - \Gamma^{\mathrm{eh}\,\sigma\sigma\sigma'\sigma'}_{\omega\omega'\Omega}(\mathbf{q}) + \gamma^{\sigma\sigma\sigma'\sigma'}_{\omega\omega'\Omega}$$

$$= -\Gamma^{\sigma\sigma\sigma'\sigma'}_{\omega\omega'\Omega}(\mathbf{k},\mathbf{k}',\mathbf{q})\,,$$

$$\Gamma^{\sigma\sigma\bar\sigma\bar\sigma}_{\omega\,\omega+\Omega\,\omega'-\omega}(\mathbf{k},\mathbf{k}+\mathbf{q},\mathbf{k}'-\mathbf{k}) = \Gamma^{\mathrm{eh}\,\sigma\sigma\bar\sigma\bar\sigma}_{\omega\,\omega+\Omega\,\omega'-\omega}(\mathbf{k}'-\mathbf{k}) - \Gamma^{\mathrm{eh}\,\bar\sigma\sigma\sigma\bar\sigma}_{\omega\omega'\Omega}(\mathbf{q}) + \gamma^{\bar\sigma\sigma\sigma\bar\sigma}_{\omega\omega'\Omega}$$

$$= -\Gamma^{\bar\sigma\sigma\sigma\bar\sigma}_{\omega\omega'\Omega}(\mathbf{k},\mathbf{k}',\mathbf{q})\,, \tag{11.30}$$

where the antisymmetry property of the local impurity vertex was used, e.g. $\gamma^{\sigma'\sigma\sigma\sigma'}_{\omega\,\omega+\Omega\,\omega'-\omega} = -\gamma^{\sigma\sigma\sigma'\sigma'}_{\omega\omega'\Omega}$. Exchanging endpoints involves a transformation between spin channels. The vertex is not crossing symmetric with respect to the outgoing endpoints. In the present approximation, only partial crossing symmetry can be maintained [175]. A vertex with full crossing symmetry requires the solution of the parquet equations [163].

### 11.2.5 Schwinger-Dyson Equation

From the equation of motion, one can establish an exact relation between the exact vertex and the self-energy. According to the rules of the perturbation theory (Sec. 6.2), the Schwinger-Dyson equation [176] for the present case reads

$$
\begin{aligned}
\Sigma_\omega(\mathbf{k}) = &-\frac{T}{N} \sum_{\omega' \mathbf{k}'} \sum_{\sigma'} \gamma^{\sigma\sigma\sigma'\sigma'}_{\omega'\,\omega\,\Omega=0} G_{\omega'}(\mathbf{k}') \\
&- \frac{1}{2}\frac{T^2}{N^2} \sum_{\Omega\mathbf{q}} \sum_{\omega'\mathbf{k}'} \sum_{\sigma'} \gamma^{\sigma\sigma'\sigma'\sigma}_{\omega'\,\omega\,\Omega} G_{\omega+\Omega}(\mathbf{k}+\mathbf{q}) G_{\omega'}(\mathbf{k}') G_{\omega'+\Omega}(\mathbf{k}'+\mathbf{q}) \, \Gamma^{\sigma\sigma\sigma'\sigma'}_{\omega\omega'\Omega}(\mathbf{k},\mathbf{k}',\mathbf{q}) \\
&- \frac{1}{2}\frac{T^2}{N^2} \sum_{\Omega\mathbf{q}} \sum_{\omega'\mathbf{k}'} \gamma^{\sigma\sigma\bar\sigma\bar\sigma}_{\omega'\,\omega\,\Omega} G_{\omega+\Omega}(\mathbf{k}+\mathbf{q}) G_{\omega'}(\mathbf{k}') G_{\omega'+\Omega}(\mathbf{k}'+\mathbf{q}) \Gamma^{\sigma\bar\sigma\bar\sigma\sigma}_{\omega\omega'\Omega}(\mathbf{k},\mathbf{k}',\mathbf{q}) \, .
\end{aligned}
$$

$$(11.31)$$

It is depicted in Fig. 11.6. Inserting the approximation to the full vertex (11.29), one obtains

$$
\begin{aligned}
\Sigma_\omega(\mathbf{k}) = &-\frac{T}{N} \sum_{\omega' \mathbf{k}'} \sum_{\sigma'} \gamma^{\sigma\sigma\sigma'\sigma'}_{\omega\omega'\,\Omega=0} G_{\omega'}(\mathbf{k}') \\
&- \frac{1}{2}\frac{T^2}{N^2} \sum_{\Omega\mathbf{q}} \sum_{\omega'\mathbf{k}'} \sum_{\sigma'} \gamma^{\sigma\sigma'\sigma'\sigma}_{\omega'\,\omega\,\Omega} G_{\omega+\Omega}(\mathbf{k}+\mathbf{q}) G_{\omega'}(\mathbf{k}') G_{\omega'+\Omega}(\mathbf{k}'+\mathbf{q}) \, \Gamma^{\mathrm{eh}\,\sigma\sigma\sigma'\sigma'}_{\omega\omega'\Omega}(\mathbf{q}) \\
&- \frac{1}{2}\frac{T^2}{N^2} \sum_{\Omega\mathbf{q}} \sum_{\omega'\mathbf{k}'} \gamma^{\sigma\sigma\bar\sigma\bar\sigma}_{\omega'\,\omega\,\Omega} G_{\omega+\Omega}(\mathbf{k}+\mathbf{q}) G_{\omega'}(\mathbf{k}') G_{\omega'+\Omega}(\mathbf{k}'+\mathbf{q}) \Gamma^{\mathrm{eh}\,\sigma\bar\sigma\bar\sigma\sigma}_{\omega\omega'\Omega}(\mathbf{q}) \\
&+ \frac{1}{2}\frac{T^2}{N^2} \sum_{\Omega\mathbf{q}} \sum_{\omega'\mathbf{k}'} \sum_{\sigma'} \gamma^{\sigma\sigma'\sigma'\sigma}_{\omega'\,\omega\,\Omega} G_{\omega+\Omega}(\mathbf{k}+\mathbf{q}) G_{\omega'}(\mathbf{k}') G_{\omega'+\Omega}(\mathbf{k}'+\mathbf{q}) \, \Gamma^{\mathrm{eh}\,\sigma\sigma'\sigma'\sigma}_{\omega\,\omega+\Omega\,\omega'-\omega}(\mathbf{k}'-\mathbf{k}) \\
&+ \frac{1}{2}\frac{T^2}{N^2} \sum_{\Omega\mathbf{q}} \sum_{\omega'\mathbf{k}'} \gamma^{\sigma\sigma\bar\sigma\bar\sigma}_{\omega'\,\omega\,\Omega} G_{\omega+\Omega}(\mathbf{k}+\mathbf{q}) G_{\omega'}(\mathbf{k}') G_{\omega'+\Omega}(\mathbf{k}'+\mathbf{q}) \Gamma^{\mathrm{eh}\,\sigma\sigma\bar\sigma\bar\sigma}_{\omega\,\omega+\Omega\,\omega'-\omega}(\mathbf{k}'-\mathbf{k}) \\
&+ \frac{1}{2}\frac{T^2}{N^2} \sum_{\Omega\mathbf{q}} \sum_{\omega'\mathbf{k}'} \sum_{\sigma'} \gamma^{\sigma\sigma'\sigma'\sigma}_{\omega'\,\omega\,\Omega} G_{\omega+\Omega}(\mathbf{k}+\mathbf{q}) G_{\omega'}(\mathbf{k}') G_{\omega'+\Omega}(\mathbf{k}'+\mathbf{q}) \, \gamma^{\sigma\sigma\sigma'\sigma'}_{\omega\omega'\Omega} \\
&+ \frac{1}{2}\frac{T^2}{N^2} \sum_{\Omega\mathbf{q}} \sum_{\omega'\mathbf{k}'} \gamma^{\sigma\sigma\bar\sigma\bar\sigma}_{\omega'\,\omega\,\Omega} G_{\omega+\Omega}(\mathbf{k}+\mathbf{q}) G_{\omega'}(\mathbf{k}') G_{\omega'+\Omega}(\mathbf{k}'+\mathbf{q}) \gamma^{\sigma\bar\sigma\bar\sigma\sigma}_{\omega\omega'\Omega} \, .
\end{aligned}
$$

$$(11.32)$$

This expression can be further simplified by noting that lines two and three of the right-hand side give exactly the same contribution as lines four and five: Consider line four for the particular spin configuration $\sigma' = \bar\sigma$. Shifting frequencies and corresponding momenta according to $\tilde\omega = \omega$, $\tilde\omega' = \omega + \Omega$ and $\tilde\Omega = \omega' - \omega$ and using the antisymmetry

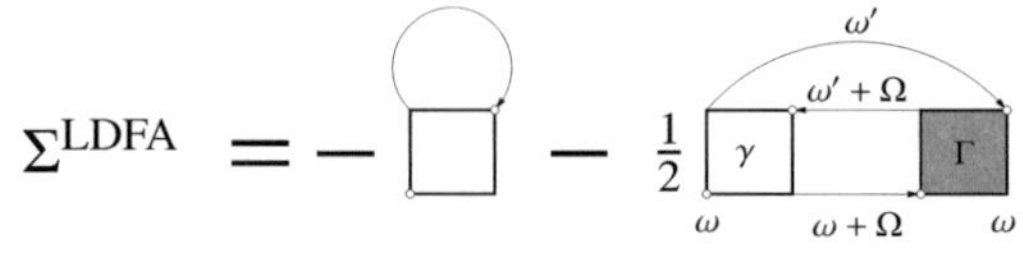

$$\Sigma^{\text{LDFA}} \;=\; -\;\square\;-\;\frac{1}{2}\;\boxed{\gamma}\;\boxed{\;\;}\;\boxed{\Gamma}$$

*Figure 11.6:* Diagrammatic representation of the Schwinger-Dyson equation for the LDFA self-energy. The frequency labels are indicated.

of the local vertex, $\gamma^{\sigma\bar\sigma\bar\sigma\sigma}_{\tilde\omega+\tilde\Omega\,\tilde\omega\,\tilde\omega'-\tilde\omega} = -\gamma^{\sigma\sigma\bar\sigma\bar\sigma}_{\tilde\omega'\,\tilde\omega\,\tilde\Omega}$, this line transforms to

$$-\frac{1}{2}\frac{T^2}{N^2}\sum_{\tilde\omega'\,\tilde\Omega}\sum_{\tilde{\mathbf{k}}'\,\tilde{\mathbf{q}}}\gamma^{\sigma\sigma\bar\sigma\bar\sigma}_{\tilde\omega'\,\tilde\omega\,\tilde\Omega}G_{\tilde\omega'}(\tilde{\mathbf{k}}')G_{\tilde\omega+\tilde\Omega}(\tilde{\mathbf{k}}+\tilde{\mathbf{q}})G_{\tilde\omega'+\tilde\Omega}(\tilde{\mathbf{k}}'+\tilde{\mathbf{q}})\,\Gamma^{\text{eh}\,\sigma\bar\sigma\bar\sigma\sigma}_{\tilde\omega\,\tilde\omega'\,\tilde\Omega}(\tilde{\mathbf{q}})\,, \tag{11.33}$$

which is the same as the third line of Eq. 11.31. In the same way the fifth line of this equation is straightforwardly verified to be the same as the second with $\sigma' = \bar\sigma$, and the fourth line the same as the second for $\sigma' = \sigma$. Here it was implicitly used that the vertex has crossing symmetry with respect to the incoming endpoints; when inserted into the Schwinger-Dyson equation, the frequency and momentum of one of the outgoing endpoints is fixed to the external values (cf. Fig. 11.6). Introducing $\chi^0_{\omega\Omega}(\mathbf{q}) := -(T/N)\sum_{\mathbf{k}}G_\omega(\mathbf{k})G_{\omega+\Omega}(\mathbf{k}+\mathbf{q})$, the Schwinger-Dyson equation takes the compact form

$$\Sigma^{\text{LDFA}}_\sigma(\omega,\mathbf{k}) = -\frac{T}{N}\sum_{\omega'\mathbf{k}'}\sum_{\sigma'}\gamma^{\sigma\sigma\sigma'\sigma'}_{\omega'\,\omega\,\Omega=0}G_{\omega'}(\mathbf{k}')$$
$$+\frac{1}{2}\frac{T}{N}\sum_{\Omega\mathbf{q}}\sum_{\omega'}\sum_{\sigma'}\gamma^{\sigma\sigma'\sigma'\sigma}_{\omega'\,\omega\,\Omega}G_{\omega+\Omega}(\mathbf{k}+\mathbf{q})\chi^0_{\omega'\Omega}(\mathbf{q})\left[2\Gamma^{\text{eh}\,\sigma\sigma'\sigma'\sigma'}_{\omega\omega'\Omega}(\mathbf{q})-\gamma^{\sigma\sigma'\sigma'\sigma'}_{\omega\omega'\Omega}\right]$$
$$+\frac{1}{2}\frac{T}{N}\sum_{\Omega\mathbf{q}}\sum_{\omega'}\gamma^{\sigma\sigma\bar\sigma\bar\sigma}_{\omega'\,\omega\,\Omega}G_{\omega+\Omega}(\mathbf{k}+\mathbf{q})\chi^0_{\omega'\Omega}(\mathbf{q})\left[2\Gamma^{\text{eh}\,\sigma\bar\sigma\bar\sigma\sigma}_{\omega\omega'\Omega}(\mathbf{q})-\gamma^{\sigma\bar\sigma\bar\sigma\sigma}_{\omega\omega'\Omega}\right]\,. \tag{11.34}$$

This equation does not simplify considerably when expressed in terms of the spin diagonalized vertices (10.10). Some notes are in place: With the electron-hole vertex obtained by matrix inversion according to (10.14), this equation is equivalent to the infinite order ladder approximation to the self-energy (see appendix A.8.4). This is also true order by order: inserting the electron-hole vertex obtained from $n$ iterations of the BSE (starting from $\Gamma^{(0)} = \gamma$), yields the ladder diagram expansion of the self-energy up to order $n + 2$ in $\gamma$.

The solution to the BSE in either of the two equivalent electron-hole channels (horizontal or vertical) contains the full ladder sum. Diagrammatically, one may therefore attempt to obtain the ladder approximation to the self energy by simply

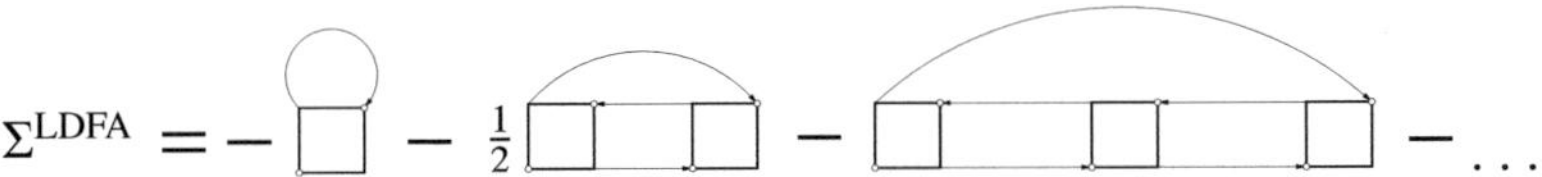

*Figure 11.7:* Diagrammatic representation of the dual self-energy in the ladder approximation. The diagrams are shown with their corresponding signs and prefactors. All higher-order terms in the expansion of $\Sigma^d$ have the same prefactor.

attaching a line either to the vertex in the electron-hole or vertical channel (without adding the Hartree-Fock contribution). The two possibilities are related through (11.28). However, this leads to overcounting of the second-order contribution (6.21). Unlike all other diagrams which are associated with a prefactor of $-1$, the second-order contribution appears with a prefactor of $-1/2$. It is therefore explicitly subtracted in (11.34). This overcounting is also encountered in the case of Hugenholtz diagrams [177, 53]. Although the horizontal and vertical channels are equivalent in the antisymmetrized technique, they have to be added here in order to obtain the correct prefactor of the diagrams (1 instead of 1/2). Note further that the LDFA equations are related to FLEX. In particular, the FLEX equations for the Hubbard model can be recovered by a suitable replacement of the interaction and Green functions. Replacing the bare dual Green functions by $G_\omega^{-1}(\mathbf{k}) = i\omega + \mu - h_\mathbf{k}$ and the vertex such that $-\frac{1}{4}\gamma_{1234} = +\frac{1}{4}U(\delta_{12}\delta_{34} - \delta_{14}\delta_{32})$ (a factor $1/2$ comes from the antisymmetrization) or

$$\gamma_{1234} = -U(\delta_{12}\delta_{34} - \delta_{14}\delta_{32}),\qquad (11.35)$$

generates the FLEX equations for the Hubbard model [174, 175], which here are formulated in terms of an antisymmetrized interaction[2]. As is obvious from the construction, the ladder dual fermion approach is actually the fluctuation-exchange approximation to dual fermions. This terminology is misleading however, as the LDFA goes far beyond the conventional FLEX. Most notably, as will be shown below, the LDFA is also applicable for strong coupling.

The ladder diagrams describe multiple scattering of particle-hole pairs. In the magnetic channel, the collective excitations are magnons. The approximation to the full vertex includes both longitudinal and transverse modes. Note that $\Gamma^{\sigma\sigma'}$ contributes to the longitudinal spin susceptibility $\chi^{zz}$ ($z$ is the quantization axis), while $\Gamma^{\sigma\bar{\sigma}}$ contributes to $\chi^{\pm} := \langle S^{\pm}S^{\mp}\rangle$ (recall that $S^{\pm} = S_x + iS_y$) and hence to the transverse response. Due to the equivalence between two-particle excitations of dual and lattice fermions, one may

---

[2]The negative sign on $U$ in (11.35) is due to the fact that the lowest order term in the dual potential corresponds to a formally attractive interaction

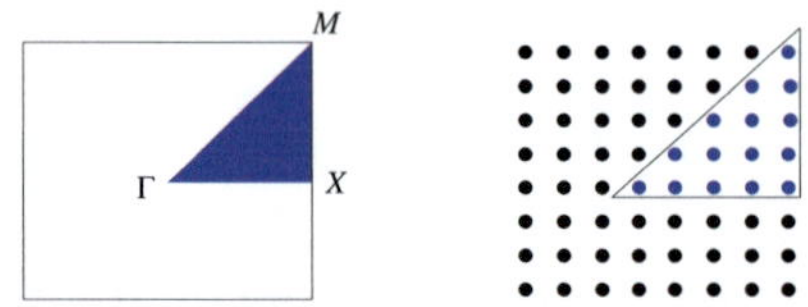


*Figure 11.8:* Irreducible wedge of the Brillouin zone for the two-dimensional square lattice in the continuum case (left) and for a discrete $8 \times 8$ lattice.

expect that the LDFA self-energy constructed from the vertex describes the effects of magnon exchange and RPA screening.

## 11.2.6  Implementation Notes

With an application to multiorbital systems in mind, a computationally efficient implementation for the evaluation of the LDFA self-energy is crucial. The shift of momenta, which led to (11.33), leaves a dependence of the vertex on transferred momentum $\mathbf{q}$ only (instead of $\mathbf{k}' - \mathbf{k}$). Storing the full $\mathbf{q}$-dependent vertex in memory is nevertheless prohibitive. For multiorbital systems, already the size of the local vertex can easily exceed the available memory and symmetries have to be exploited where possible. Fortunately, the vertex $\Gamma^{\mathrm{eh}}$ does not need to be stored. In the current implementation, the LDFA self-energy is calculated as follows: The momentum sum in the particle-hole bubble $\chi^0_{\omega\Omega}(\mathbf{q})$ is evaluated using FFT (see Sec. 6.6). The vertex $\Gamma^{\mathrm{eh}}$ is obtained for a single reciprocal space vector $\mathbf{q}$ by solving the BSE using matrix inversion (or in some cases iteratively, see below). From this, the following intermediate quantity is calculated:

$$\Lambda_{\omega\Omega}(\mathbf{q}) = \frac{1}{2}\frac{T}{N}\sum_{\omega'}\sum_{\sigma'} \gamma^{\sigma\sigma'\sigma'\sigma}_{\omega'\,\omega\Omega}\chi^0_{\omega\Omega}(\mathbf{q})\left[2\Gamma^{\mathrm{eh}\,\sigma\sigma'\sigma'}_{\omega\omega'\Omega}(\mathbf{q}) - \gamma^{\sigma\sigma'\sigma'}_{\omega\omega'\Omega}\right]$$
$$+ \frac{1}{2}\frac{T}{N}\sum_{\omega'} \gamma^{\sigma\sigma\bar\sigma\bar\sigma}_{\omega'\,\omega\Omega}\chi^0_{\omega\Omega}(\mathbf{q})\left[2\Gamma^{\mathrm{eh}\,\sigma\bar\sigma\bar\sigma\sigma}_{\omega\omega'\Omega}(\mathbf{q}) - \gamma^{\sigma\bar\sigma\bar\sigma\sigma}_{\omega\omega'\Omega}\right]. \tag{11.36}$$

In the next step, this quantity is transformed to real space using FFT. The self-energy for a given frequency $\omega$ is then obtained by fast convolution, i.e. as the FFT of

$$\Sigma^{\mathrm{LDFA}}_{\omega}(\mathbf{r}) = -T\sum_{\omega'}\sum_{\sigma'} \gamma^{\sigma\sigma\sigma'\sigma'}_{\omega'\,\omega\,\Omega=0}G_{\omega'}(\mathbf{r})\delta_{\mathbf{r},0} + T\sum_{\Omega}G_{\omega+\Omega}(-\mathbf{r})\Lambda_{\omega\Omega}(\mathbf{r}). \tag{11.37}$$

Instead of storing the entire vertex $\Gamma^{\mathrm{eh}}(\mathbf{q})$, it is therefore sufficient to store $\chi^0_{\omega\Omega}(\mathbf{q})$ and $\Lambda_{\omega\Omega}(\mathbf{q})$, which take much less memory. These must be stored for all k-points, because of the FFT. The self-energy is evaluated most efficiently by updating it for all frequencies $\omega$

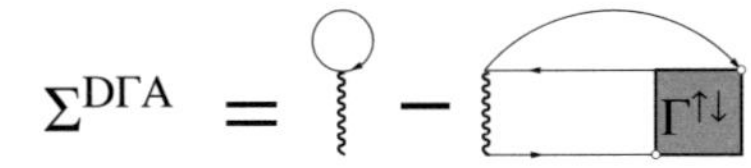

$$\Sigma^{\mathrm{D\Gamma A}} = \quad - \quad \boxed{\Gamma^{\uparrow\downarrow}}$$

*Figure 11.9:* Diagrammatic representation of the Schwinger-Dyson equation for the DΓA self-energy.

simultaneously and instead evaluating and storing $\chi^0_{\omega\Omega}(\mathbf{q})$ and $\Lambda_{\omega\Omega}(\mathbf{q})$ for a single $\Omega$ at a time. Because the BSE can be solved independently for each $\Omega$, using matrix inversion only requires an additional array of size $O(N_\omega^2)$ (and two for iterative solution). This is negligible compared to the memory requirements for the local vertex, which has to be stored in memory for three independent frequencies. Therefore, the computational feasibility is only limited through the memory requirements of the local vertex, which are $O(N_\omega^3)$ times a factor that grows as $n^4$ where $n$ is the number of orbitals[3]. The computational effort can be further reduced by exploiting the symmetry of the lattice, in this case the $C_{4v}$ point group symmetry of the 2D square lattice. The BSE only needs to be evaluated for wave vectors in the irreducible wedge of the Brillouin zone depicted in Fig. 11.8. For the discrete case, the number of k-points in the irreducible wedge for a lattice of $N = L^2$ sites is given by $(L/2 + 1)(L/2 + 2)/2$, so that the computational effort is essentially decreased by a factor of 8 for large lattices.

## 11.2.7  DΓA Self-energy

For comparison with the LDFA, the derivation of the DΓA self-energy is briefly repeated here. In DΓA, the bare interaction of the fermions is simply given by the Hubbard $U$. The Schwinger-Dyson equation for this case hence reads [39, 176]

$$\Sigma^{\mathrm{D\Gamma A}}_\omega(\mathbf{k}) = U\frac{n}{2} - \frac{T^2}{N^2}U\sum_{\omega'\Omega}\sum_{\mathbf{k'q}} G_{\omega+\Omega}(\mathbf{k}+\mathbf{q})G_{\omega'}(\mathbf{k'})G_{\omega'+\Omega}(\mathbf{k'}+\mathbf{q})\Gamma^{\uparrow\uparrow\downarrow\downarrow}_{\omega\omega'\Omega}(\mathbf{k},\mathbf{k'},\mathbf{q}) \,. \tag{11.38}$$

Here only a single spin component enters as the Hubbard interaction only couples electrons with opposite spins. Neglecting the contribution from the electron-electron channel in (11.21), the approximation to the vertex reads $\Gamma = \Gamma^{\mathrm{eh}} + \Gamma^{\mathrm{v}} - \gamma$ (recall that $\Gamma_{\mathrm{loc}} = \gamma$).

---

[3]Or as $n^2$ if $\gamma$ is diagonal, as would be the case for the segment representation of the strong-coupling continuous-time QMC algorithm (Sec. 3.4).

Using (11.29) and expressing the vertices in terms of the spin diagonalized quantities according to (10.10), the vertex $\Gamma$ can be expressed as

$$
\begin{aligned}
\Gamma^{\uparrow\uparrow\downarrow\downarrow}_{\omega\omega'\Omega}(\mathbf{k},\mathbf{k}',\mathbf{q}) &= \Gamma^{\text{eh}\,\uparrow\uparrow\downarrow\downarrow}_{\omega\omega'\Omega}(\mathbf{q}) - \Gamma^{\text{eh}\,\downarrow\uparrow\uparrow\downarrow}_{\omega\,\omega+\Omega\,\omega'-\omega}(\mathbf{k}'-\mathbf{k}) - \gamma^{\uparrow\uparrow\downarrow\downarrow}_{\omega\omega'\Omega} \\
&= \frac{1}{2}\left(\Gamma^{\text{d}}_{\omega\omega'\Omega}(\mathbf{q}) - \Gamma^{\text{m}}_{\omega\omega'\Omega}(\mathbf{q})\right) - \Gamma^{\text{m}}_{\omega\,\omega+\Omega\,\omega'-\omega}(\mathbf{k}'-\mathbf{k}) - \frac{1}{2}\left(\gamma^{\text{d}}_{\omega\omega'\Omega} - \gamma^{\text{m}}_{\omega\omega'\Omega}\right) .
\end{aligned}
\tag{11.39}
$$

Inserting this into (11.38) and shifting frequency and momenta in analogy to the foregoing, the D$\Gamma$A self-energy reads

$$
\begin{aligned}
\Sigma^{\text{D}\Gamma\text{A}}_{\omega}(\mathbf{k}) = U\frac{n}{2} + \frac{1}{2}\frac{T}{N}\sum_{\omega'\Omega}\sum_{\mathbf{q}} U\, G_{\omega+\Omega}(\mathbf{k}+\mathbf{q})\chi^{\Omega}_{0\omega}(\mathbf{q})\Big[&\Gamma^{\text{d}}_{\omega\omega'\Omega}(\mathbf{q}) - 3\Gamma^{\text{m}}_{\omega\omega'\Omega}(\mathbf{q}) \\
&- \gamma^{\text{d}}_{\omega\omega'\Omega}(\mathbf{q}) + \gamma^{\text{m}}_{\omega\omega'\Omega}(\mathbf{q})\Big] ,
\end{aligned}
\tag{11.40}
$$

with the usual definition $\chi^{0}_{\omega\Omega}(\mathbf{q}) = -T/N\sum_{\mathbf{k}'} G_{\omega'}(\mathbf{k}')\,G_{\omega'+\Omega}(\mathbf{k}'+\mathbf{q})$. The Hartree term is absorbed into the chemical potential. Note that the second term differs from the one of Ref. [39] by a sign, because the definition of the vertices differs by a sign. In the D$\Gamma$A, the *irreducible* vertex of the impurity problem and the DMFT Green function enter the BSE.

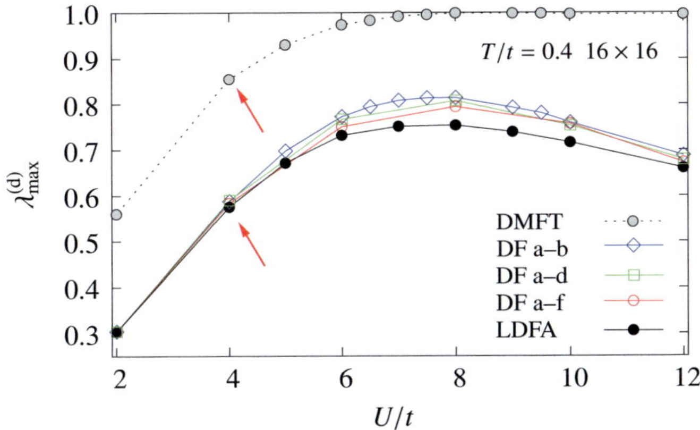


*Figure 11.10:* Leading eigenvalue of the Bethe-Salpeter equation obtained within various approximations in the $\mathbf{q} = (\pi, \pi)$ spin channel as a function of the interacton $U$. $\lambda$ ($\lambda^d$) denotes lattice (dual) fermion eigenvalues. The diagrams included are indicated in the legend (labels are the same as in Fig. 6.3). The dual perturbation theory converges fast in particular for weak and strong coupling. A straightforward diagrammatic expansion around DMFT breaks down for large $U$.

## 11.3  Convergence Properties

For a perturbative approach, the convergence properties are of paramount importance. These are examined in the vicinity of the antiferromagnetic instability (AFI) in the 2D Hubbard model. For small $U$, the theory is expected to converge fast, since in this limit $\gamma \sim U$. For large $U$, the reducible vertex also becomes large, and no analytical arguments are available to prove convergence apart from the case of an expansion around the atomic limit (see Sec. 6.1). The convergence properties can be characterized using the eigenvalue problem derived from the BSE (10.11)

$$-\frac{T}{N} \sum_{\omega' \mathbf{k}'} \Gamma^{\text{irr, m}}_{\omega\omega'\Omega} G_{\omega'}(\mathbf{k}) G_{\omega'+\Omega}(\mathbf{k+q})\, \phi_{\omega'} = \lambda \phi_{\omega'} \, . \quad (11.41)$$

The matrix is the building block of the particle-hole ladder and may be thought of as the effective two-fermion interaction. For dual fermions, the irreducible vertex is given by the bare dual interaction $\Gamma^{\text{irr, m}}_{\omega\omega'\Omega} = \gamma^{m}_{\omega\omega'\Omega} = \gamma^{\uparrow\uparrow}_{\omega\omega'\Omega} - \gamma^{\uparrow\downarrow}_{\omega\omega'\Omega}$ in the magnetic spin channel and $G$ stands for the full dual Green function. Here the focus is on the leading eigenvalues in the vicinity of the AFI and hence $\mathbf{q} = (\pi, \pi)$ and $\Omega = 0$. An eigenvalue of $\lambda_{\max} = 1$ implies a divergence of the ladder sum and hence a breakdown of the perturbation theory.

The results are displayed in Fig. 11.10. For weak coupling, the leading eigenvalue is small and implies a fast convergence of the diagrams in the electron-hole ladder. More significantly, the eigenvalues decrease and converge to the same intercept in the large $U$ limit. This nicely illustrates that the dual perturbation theory smoothly interpolates between a standard perturbation expansion at small, and the cumulant expansion at large $U$, ensuring fast convergence in both regimes. From the figure it is clear that this also improves the convergence properties for intermediate coupling ($U \sim W$). Even here corrections from approximations involving higher-order diagrams remain small, including those from the LDFA. Diagrams involving the three-particle vertex give a negligible contribution.

For a straightforward diagrammatic expansion around DMFT, the building block of the particle-hole ladder is constructed from the irreducible impurity vertex $\gamma^{\mathrm{irr,\,m}}_{\omega\omega'\Omega}$ and DMFT Green functions. As seen in Fig. 11.10, the corresponding leading eigenvalue (and the effective interaction) is much larger than for dual fermions over the whole parameter range (e.g. at red arrows). When transforming the leading eigenvalue back to lattice fermions (red arrow in Fig. 11.12), it is close to the DMFT value for these parameters (the data labeled DMFT at the red arrows is essentially the same in both plots). Hence convergence is enhanced for a perturbation theory in terms of dual fermions. Remarkably, for the intermediate to strong coupling region, standard perturbation theory has to break down (since the eigenvalue approaches one), while for a theory in terms of dual fermions, this is not the case. The fact that the leading eigenvalue for dual fermions is smaller appears to be a generic feature. It is also observed away from half-filling and for the electron-electron channel (not shown). These results further confirm the expectation of Sec. 6.1, that the perturbation theory converges in the opposite limits of weak and strong coupling.


## 11.4 Comparison to QMC

As a first test, the LDFA is compared to results obtained by CTQMC, which is employed as a lattice solver as in Sec. 6.9, albeit for the $4 \times 4$ lattice (with periodic boundary conditions). The LDFA calculation procedure is essentially the same as for regular DF calculations, i.e. the procedures to adjust the hybridization function in the outer loop and the self-consistent renormalization of the Green function in the inner loop remain intact. The only difference is the calculation of the self-energy, which is obtained from the vertex $\Gamma^{\mathrm{eh}}$ according to (11.34). The BSE has to be solved in each iteration of the inner loop, which however remains numerically tractable. One can see that the LDFA generally improves the solution compared to the second-order dual perturbation theory. In particular, the nearest- and next-nearest-neighbor Green functions are better approxi-

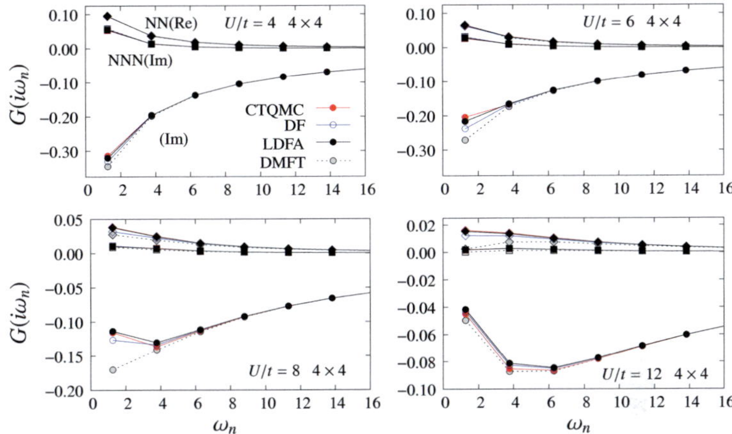

*Figure 11.11:* Comparison of the dual fermion approach (DF) and the LDFA to a CTQMC lattice calculation for a $4 \times 4$ lattice with periodic boundary conditions at $T/t = 0.4$. (N)NN denotes the (next-)nearest-neighbor Green functions. In LDFA, in particular the (N)NN Green functions are better represented.

mated. These results where obtained at relatively high temperature. The corrections are expected to be much more significant at lower temperatures, where spatial fluctuations gain importance.

The reliability of the method is therefore assessed in the vicinity of the antiferromagnetic instability (AFI) in the 2D Hubbard model. In this regime, antiferromagnetic long-range fluctuations are expected to be of vital importance. As seen in Sec. 10.3.1, the dual fermion approximation appeared insufficient for two-particle quantities. In order to test whether this issue is resolved within the LDFA, the leading eigenvalue of the BSE is compared to results from Hirsch-Fye QMC calculations [164].

Being interested in the antiferromagnetic instability, the vertex is considered in the spin channel and the channel index is henceforth omitted. The equivalence of two-particle excitations only implies that the leading eigenvalues in terms of dual and lattice fermions must be identical at the instability, i.e. $\lambda_{max} = \lambda_{max}^{d} = 1$. In general they are different. For a direct comparison of the eigenvalues, the irreducible vertex therefore has to be transformed to the corresponding quantity in terms of lattice fermions. To this end, the dual vertex $\Gamma^{d}$ is first obtained by inverting a BSE according to Eq. 10.14. It has full frequency dependence, but depends on transferred momentum $\mathbf{q}$ only. The result is transformed to lattice fermions using the relation between the dual and lattice fermion vertices (6.37). For the present case, it reads

$$\Gamma_{\omega\omega'\Omega}(\mathbf{k},\mathbf{k}',\mathbf{q}) = L'_{\omega+\Omega}(\mathbf{k}+\mathbf{q})\, L'_{\omega}(\mathbf{k})\, \Gamma^{d}_{\omega\omega'\Omega}(\mathbf{q})\, R'_{\omega'}(\mathbf{k}')\, R'_{\omega'+\Omega}(\mathbf{k}'+\mathbf{q})\,. \qquad (11.42)$$

Through the functions $L'$ and $R'$, it explicitly depends on momenta $\mathbf{k}$ and $\mathbf{k}'$. This dependence enters only via single-particle quantities and does not contain information about two-particle excitations. A closer analysis shows that the number of numerically non-zero eigenvalues (i.e. eigenvalues with modulus larger than the effective machine accuracy) for a given $\mathbf{q}$ exactly equals the number of frequencies. Hence the $\mathbf{k}$, $\mathbf{k}'$-dependence is indeed trivial and the vertex is not invertible. In order to obtain the irreducible vertex, the dependence on these wave vectors must hence be integrated out:

$$\Gamma_{\omega\omega'\Omega}(\mathbf{q}) := \frac{1}{N^2} \sum_{\mathbf{k},\mathbf{k}'} \Gamma_{\omega\omega'\Omega}(\mathbf{k},\mathbf{k}'\mathbf{q})\,. \qquad (11.43)$$

Note that in the calculation of the susceptibility from $\Gamma$ according to Eqs. 10.18, 10.19, the $\mathbf{k}$- and $\mathbf{k}'$-dependence is also integrated out. The resulting frequency dependent vertex is invertible for each $\mathbf{q}$ and allows to obtain the irreducible vertex by inverting a BSE similar to (10.14). This yields the irreducible lattice vertex in the dual fermion approximation. The leading eigenvalues are obtained by solving an eigenvalue problem of the form (11.41). The Green functions which enter here are obtained from dual Green functions in the respective approximation using the identity (6.26).

Results for the leading eigenvalues are shown in Fig. 11.12. While at higher temperatures all approximations give similar results, DMFT fails in the vicinity of the AFI. The instability at a Néel temperature of $T_{\mathrm{N}}^{\mathrm{DMFT}}/t = 0.233$ is an artefact of the mean-field approximation, which tends to stabilize the AF order. Including short-range spatial correlations beyond DMFT, through the leading nonlocal diagram b) of Fig. 6.3, only slightly improves the solution and reduces the critical temperature down to $T_{\mathrm{N}}^{\mathrm{DF}}/t = 0.215$. Note that the critical temperatures can also be read off from the left panel of Fig. 10.6. The small reduction of $T_{\mathrm{N}}$ is in accordance with the leading eigenvalue being close to unity, indicating a decelerated convergence and pointing to the importance of long-wavelength fluctuations in the vicinity of the AFI. These corrections are obviously included through the LDFA, which complies with QMC close to the AFI even on a two-particle level. This is quite remarkable because the results have been obtained perturbatively, starting from DMFT as a local approximation (at these parameters the DMFT hybridization function remains essentially unchanged).

While the self-consistent renormalization of the LDFA self-energy is found to have a small effect for intermediate temperatures, which is in accordance to the conjecture of Ref. [39], it becomes more significant as the temperature is lowered towards the instability. Here the non-selfconsistent calculation significantly overestimates the eigenvalues. In this regime it is necessary to account for multiple exchange of spin fluctuations through self-consistent renormalization. It becomes essential below $T_{\mathrm{N}}^{\mathrm{DF}}$ for which the

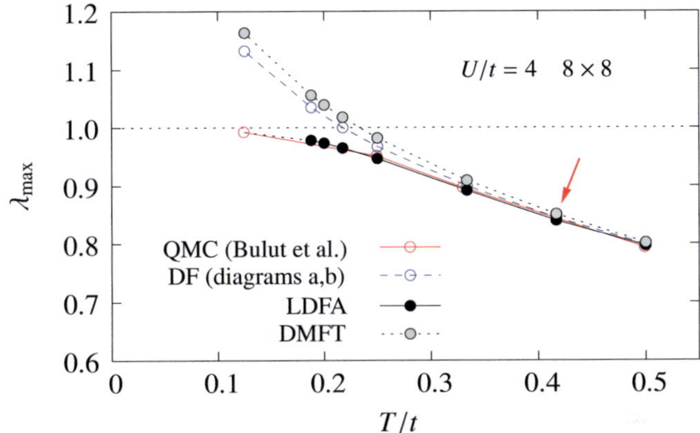


*Figure 11.12:* Leading eigenvalue in the $\mathbf{q} = (\pi, \pi)$ spin channel for different approximations. The QMC results have been obtained from Ref. [164] using an image digitizer tool [100]. Close to the instability, it is essential to include the exchange of spin fluctuations through the ladder approximation.

leading eigenvalue of (11.41) exceeds one. The ladder sum is hence divergent and forbids a straightforward summation of the infinite electron-hole ladder. Formally, the BSE can still be inverted for $\lambda > 1$ to construct the LDFA self-energy, but this leads to wrong results. In the DF calculation, the Green functions in the matrix (11.41) have been renormalized self-consistently including only diagrams up to second order. In accordance with the fact that long-range fluctuations suppress $T_\mathrm{N}$, the eigenvalue decreases when higher-order diagrams are included. This is facilitated by solving the BSE iteratively up to some finite order $n$. For not too low temperatures, it allows to push the eigenvalue below the threshold $\lambda_{\mathrm{max}} = 1$, so that the ladder sum becomes convergent. In this case, the self-consistent renormalization is needed to obtain a well-defined result by removing the arbitrariness to include diagrams up to some large but finite order. In the calculation the actual threshold is taken to be somewhat smaller, e.g. $\lambda_{\mathrm{max}} \lesssim 0.99$, in order to avoid the inversion of numerically ill-conditioned matrices. The eigenvalues in all possible channels (in the present case the $\mathbf{q} = (\pi, \pi)$ spin channel is sufficient) are verified to fulfill this condition before summing up the infinite ladder.

The results below $T_\mathrm{N}^{\mathrm{DF}}$ obtained this way are in very good agreement with the QMC results. This also illustrates that contributions from the particle-particle fluctuation channel are indeed negligible, as anticipated. The renormalization procedure allows to access a region with very strong antiferromagnetic fluctuations. Approaching the AFI

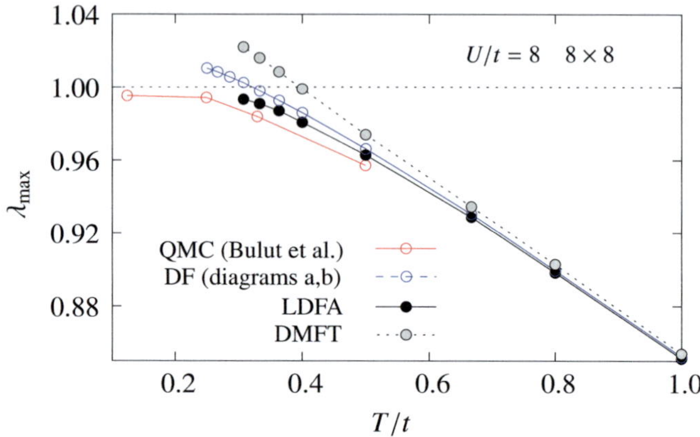


*Figure 11.13:* Same as Fig. 11.12, albeit for $U/t = 8$ (note the different scales).

closer however becomes computationally demanding because an increasingly large but finite number of terms has to be incorporated into the ladder in order to push the eigenvalue below the threshold. It is noteworthy that for the lowest temperatures for which this can be achieved, the self-consistent renormalization does not converge. This is an indication that the true solution cannot be obtained perturbatively starting from the paramagnetic state. Note that it is only possible to approach the transition from above and no converged solution which erroneously predicts a transition to the ordered state can be obtained. Lower temperatures can be accessed by starting in the symmetry-broken state, where the ladder diagram series is found to be convergent (see Sec. 6.10.1). Away from half-filling lower temperatures can be accessed due to the released nesting condition (see Sec. 10.4).

Fig. 11.13 shows the corresponding results for an interaction $U/t = 8$, which is equal to the bandwidth. This is the most unfavorable case for the perturbation theory in the sense that there is no small parameter in the problem (neither $\gamma$ nor $G^d$ are small). Indeed, the maximum of the eigenvalues of the effective two-fermion interaction in Fig. 11.10 was found at $U = W$. Even here the results are in good agreement with QMC (note the different scale with respect to Fig. 11.12). The corrections to the DF calculation become more significant for lower temperatures. It can be clearly seen that the LDFA curve turns towards the QMC curve and there is no indication of a mean-field like transition. Consistent with the results of Sec. 6.8, the system is in the insulating state. The results have been obtained after a self-consistent determination of the hybridization function. This, however, has only a small effect on the eigenvalues in the spin channel.

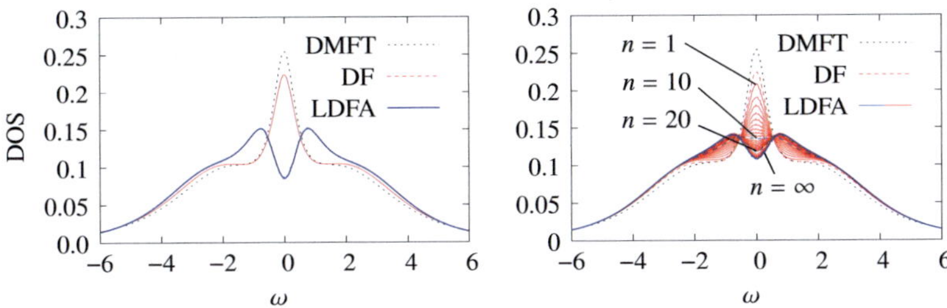

*Figure 11.14:* Left: Local density of states (DOS) obtained within DMFT, a dual fermion calculation with diagrams a) and b) and the LDFA with selfconsistent renormalization of the self-energy for $T/t = 0.188$. The LDFA DOS exhibits the antiferromagnetic pseudogap. Right: Local LDFA DOS for different approximations to the LDFA self-energy obtained by $n$ iterations of the BSE and selfconsistent renormalization for $T/t = 0.2$. Solutions are shown for $n = 1, \ldots, 20$ in steps of 1 and from $n = 25, \ldots, 100$ in steps of 5. The self-energy includes all ladder diagrams up to order $n + 2$ in $\gamma$. The results are compared to DMFT and DF. DMFT includes none of the diagrams and exhibits the quasiparticle peak. DF is equivalent to the LDFA with $n = 0$. This figure demonstrates that it is necessary to sum up the infinite ladder as diagrams at virtually every order contribute to the pseudogap. The solution obtained by inverting the BSE appears to give the correct limiting behavior for $n \to \infty$.

Susceptibilities have been compared for the dual fermion approach, DMFT and the dynamical vertex approximation (DΓA) in Ref. [137]. For $\mathbf{q} = (0, 0)$, the DΓA solution exhibits a maximum, whereas the DF and DMFT solution increases monotonically as the temperature is lowered. The downturn at low temperature has been attributed to an artefact of the non-selfconsistent calculation of the DΓA self-energy. The LDFA susceptibility from a fully selfconsistent calculation also exhibits a maximum at a temperature which is in agreement with finite-size lattice QMC calculations [178] (not shown). The maximum actually reflects Heisenberg-like behavior. For sufficiently large $U$, a maximum is observed at a temperature $T \approx J$ where $J = 4t^2/U$ is the effective exchange coupling [178]. At smaller $U/t = 4$ this maximum is still present, but does not occur at $4t^2/U$ because $t/U$ is not small. A correct description of this feature requires to account for nonlocal spin fluctuations which are not present in DMFT or the DF approximation.

## 11.5 Illustrative Results

Fig. 11.14 shows the local maximum entropy density of states for $U/t = 4$ and $T/t = 0.19$, i.e. in the temperature regime where the fluctuations are particularly strong. The DMFT solution exhibits the quasiparticle peak at zero energy. The height is only slightly reduced when nonlocal corrections are taken into account in the DFA. The pseudogap opens as the strong spin fluctuations are taken into account through the LDFA. As shown in Ref. [179], a concomitant feature in the self-energy signals breakdown of Fermi-liquid behavior. The pseudogap has been obtained within large scale cluster calculations [179], or semiclassically [28]. It is also present in the symmetry-broken state [139]. Here it is obtained within a translationally invariant solution and in the paramagnetic state, for $U$ well below the Mott transition point. Pseudogap behavior was also reported in Ref. [40] for $U/t = 8$. In the dual fermion approach a pseudogap is obtained already in second-order perturbation theory at this value of $U$ (for $\Delta$ fixed to its DMFT value). In view of the results of Sec. 6.8, this is actually the precursor of the short-range AF correlation assisted Mott transition.

The right panel of Fig. 11.14 shows the DOS for different approximations to the LDFA self-energy. The Green function has been obtained by self-consistent renormalization. One can see that it is necessary to include the full ladder sum because diagrams at essentially all orders contribute. The curves converge to the result obtained by inverting the BSE, which corresponds to the limit $n \to \infty$. The fact that the eigenvalue is close to 1 indicates that multiple quasiparticle scattering is important in this regime. It can also be seen that the pseudogap is less pronounced compared to the left panel. The temperature $T/t = 0.2$ is somewhat higher and the fluctuations are weaker. The depth of the pseudogap also appears to be somewhat underestimated compared to the lattice CTQMC solution. Many classes of diagrams are still neglected, e.g. those generated by a full parquet solution. However, the ladder approximation captures the relevant physics. No pseudogap is found in the DOS for temperatures above $T_N^{DF}$, where the fluctuations are less strong. It should however be visible in the k-resolved spectral function [40]. The pseudogap develops on the noninteracting Fermi surface in reciprocal space, which coincides with the magnetic Brillouin zone. Hot spots are hence located everywhere on the Fermi surface. At these points the quasiparticles couple strongly to the magnetic excitations.

The 3D Hubbard model provides a possibility to validate the applicability of the LDFA in the strong-coupling region. The $U$-$T$-phasediagram obtained from various approximations is shown in Fig. 11.15. For small values of $U$, similar results for the phase boundary to the antiferromagnetic state are obtained irrespective of the method. The curves diverge beyond $U > W/2$, (the bandwidth is $W = 12t$). In this region, nonlocal fluctuations gain importance. For large $U$, second-order (iterated) perturbation theory (IPT) breaks down. Charge fluctuations are suppressed in this regime and DMFT approaches the result from Weiss mean-field theory, as expected. The Néel temperature for the dual fermion results is defined as the temperature for which the leading eigenvalue in the $\mathbf{q} = (\pi, \pi, \pi)$ spin channel equals 1. For intermediate coupling, already the second-order dual perturbation theory (DFA) gives a significant correction to DMFT. This is mainly the effect of antiferromagnetic short-range correlations not present in DMFT. For large $U$, the dual perturbation theory does not break down, but seems to

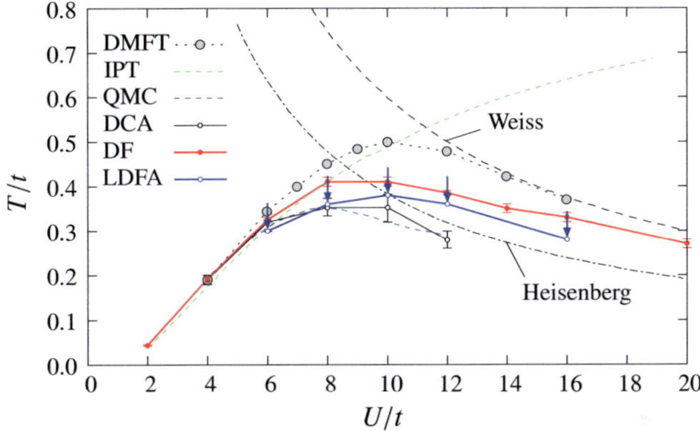



*Figure 11.15:* $U$-$T$ phasediagram for the 3D Hubbard model at half-filling with the antiferromagnetic phase boundary obtained within various approximations: dynamical mean-field theory (DMFT), second-order (iterated) perturbation theory (IPT), finite size QMC, dynamical cluster approximation (DCA), dual fermion (DF) and ladder dual fermion approximation. Data for the Heisenberg model and from Weiss mean-field theory for the Heisenberg model are shown for comparison. All data except DF and LDFA are taken from Ref. [180]. The DF and LDFA results are obtained for an $8 \times 8 \times 8$ lattice. In contrast to conventional second-order perturbation theory, the dual perturbation theory does not break down for large $U$. The LDFA results should be interpreted as upper bounds as indicated by arrows. For large $U$, it approaches Heisenberg rather than Weiss mean-field-like behavior.

exhibit mean-field-like rather than Heisenberg behavior. For the interpretation of the LDFA results, note that for a consistent calculation of $T_N$ within LDFA, the Green functions should be renormalized self-consistently with the full ladder sum before calculating eigenvalues. This in turn is only possible if the leading eigenvalue is smaller than 1. As $T_N$ is approached, the number of diagrams needed to push the leading eigenvalue below the threshold diverges, so that the critical temperature can only be approached from above. In this sense the LDFA results constitute an upper bound on $T_N$. Note also that the LDFA and DF results should not strictly be compared to the DCA and QMC results, which have been obtained by finite-size scaling. However, it can be seen that including the full ladder still gives a significant correction to the DF approximation at strong coupling and appears to favor Heisenberg-like behavior. Hence the LDFA gives consistent results even for strong coupling. This is in strong contrast to FLEX, which is applicable only for weak coupling [175].

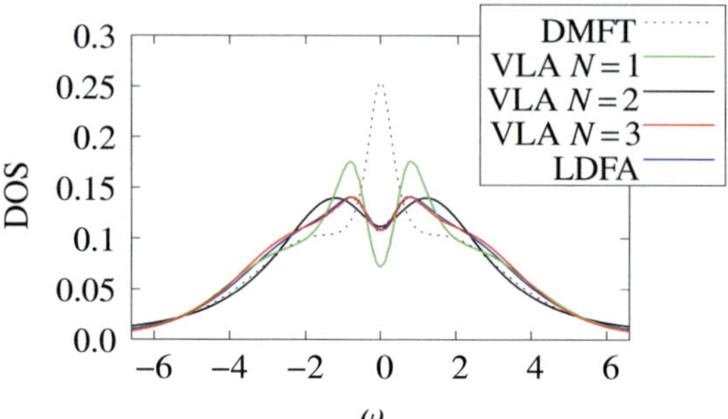

*Figure 11.16:* Local density of states obtained within the variational lattice approach (VLA), which combines the LDFA with exact diagonalization, for $U/t = 4$ and $T/t = 0.2$. The number $N$ of bath parameters $V_i$ ($\epsilon_i$) is indicated.

## 11.6   Variational Approach

Before concluding the chapter, we propose an efficient method for the description of quantum lattice models. It is based on the observation that the dynamical field $\Delta_\omega$ in the dual fermion approach is an arbitrary quantity. The freedom to choose a particular reference problem for the perturbation theory can be exploited to combine the LDFA with exact diagonalization. Taking the hybridization to be the discrete expression (9.3) is an application of the superperturbation to the lattice problem. For the local impurity problem, the perturbation is essentially $(\Delta_\omega^{(N)} - \Delta_\omega)$, while here it is $(\Delta_\omega^{(N)} - h_{\mathbf{k}})$.

The determination of the bath parameters proceeds in the same way as described in chapter 9. Here their values have been fixed by minimizing the distance function (9.4) with respect to the hybridization obtained from the noninteracting problem. The parameterization may be optimized by imposing a self-consistency condition. In this sense, the approach is variational, but rather different in spirit from the VCA. One possibility is to minimize the distance function with respect to the hybridization $\Delta_{\text{new}}$ obtained from the general self-consistency condition (6.46). This facilitates the description of the Mott transition. Note that in this approach the lowest order diagram and higher orders with local propagators always contribute because the condition (6.44), i.e. $\sum_{\mathbf{k}} G_{0\,\omega}^{\text{d}}(\mathbf{k}) = 0$ in general cannot be fulfilled using a finite number of bath parameters.

As an illustration, Fig. 11.16 demonstrates that this approximation also captures the physics of the pseudogap formation. As the number $N$ of bath sites is increased, the result converges rapidly to the LDFA result, which corresponds to $N \to \infty$. The results differ in the depth of the pseudogap and the appearance of the shoulders at high energies. The dependence on the number of bath parameters indicates an even-odd effect. Similar behavior is observed for the superperturbation. Note that for $N$ even, no bath level is located at the Fermi level, which is unfavorable for the description of the metallic state. For the special case $\Delta^{(N)} \equiv 0$, this approach reduces to the strong coupling expansion of Refs. [29, 30]. This approach is more stable than an expansion in which the atomic limit vertex appears as the bare interaction. The causality problem that arises for an expansion around the atomic limit [142] is alleviated or even absent for a finite number of bath sites, and has not been observed in LDFA calculations (i.e. for a continuous bath).

## 11.7 Conclusions and Outlook

The ladder dual fermion approach has been motivated by the need to include the feedback of the two-particle (here magnetic) excitations onto the self-energy. The results confirm the importance of including diagrams from the particle-hole fluctuation channels in the vicinity of the magnetic instability. The leading eigenvalue of the particle-particle channel is considerably smaller in this regime, indicating that these contributions can be neglected. The critical temperature is suppressed and good agreement is found compared to quantum Monte Carlo data. In accordance with DΓA [181] and related approaches [40], the density of states displays the pseudogap.

These results are promising and illustrate that, despite of the abstract nature of the transformation to dual fermions, the construction of approximations to the dual self-energy may be guided by physical considerations. Indeed, sensible results have been obtained in all cases where this has been done. By a phase-space argument, the contribution of three-particle and higher-order scattering processes is expected to be small. Because of the equivalence of $n$-particle excitations for dual and lattice fermions, it provides justification for the (necessary) truncation of the dual potential after the leading-order term. This is in fact similar to the DΓA, where three-particle and higher-order fully irreducible vertices are neglected. The smallness of diagrams including the three-particle vertex has been confirmed in the calculations. This is important for practical calculations, because the computation of the three-particle vertex is computationally highly demanding and currently unfeasible for low temperatures.

Diagrammatic extensions of DMFT share the fact that the expansion is a series in powers of a time-dependent interaction. A time-dependent vertex is essential to account for the strong correlations, and the fact that the Coulomb interaction acts on short time scales. The use of the static bare $U$ is responsible for the breakdown of FLEX. Nevertheless, it has been shown that a straightforward expansion around DMFT in terms of the irreducible vertex still breaks down for large interaction. This is not the case for an expansion in terms of dual fermions. The advantage can be attributed to the fact that it efficiently separates the strong local correlations from weaker spatial correlations. As a result, the bare interaction of dual fermions is the *reducible* (screened) vertex of the impurity. In addition, the dual Green function near the atomic limit is small (see Sec. 6.1).

The LDFA is applicable for the full interaction range, from weak to strong coupling and includes long-range correlations. These properties suggest various applications. It is certainly important to apply it to real materials in conjunction with bandstructure methods. It is further interesting to study the effect of the exchange of spin fluctuations in the vicinity of the pairing instability and the corresponding modification of the phasediagram (see Sec. 10.4). Another possibility is the case of half-metallic antiferromagnets, where non-quasi-particle states at the Fermi level are related to magnons [182].

The variational approach may be used to address the case of Nagaoka ferromagnetism [183] for large but finite interaction. The ferromagnetism in the nearly half-filled band is a consequence of the absence of a bound state of the hole to magnetic excitations [184]. Because of the efficiency of the variational approach, it could be applied to cold atoms in optical lattices, even when translational invariance is broken.

# Chapter 12

# General Conclusions and Outlook

Numerical approaches represent an invaluable tool for the understanding of strongly correlated systems. They provide important insights where analytical techniques are of limited use due to the high complexity and possibly nonperturbative character of the problem.

Dynamical mean-field theory (DMFT) has provided important insights, in particular into the physics of the Mott transition. While it accounts for local temporal quantum fluctuations, spatial correlations are neglected. The effect of nonlocal correlations may however be profound. This has been demonstrated through various examples in this thesis: By calculating the nearest-neighbor spin-correlations for the two-plane Hubbard model on the Bethe lattice, it has been shown that the model undergoes a metal-insulator transition driven by short-range singlet correlations between the planes. The calculation of the dynamical susceptibility revealed that this transition is accompanied by a simultaneous opening of a spin-gap. For an Anderson impurity problem in the Kondo regime, the spatial dependence of the spin and charge correlations of the conduction electrons have been found to display a signature of the Kondo screening cloud. The influence of spatial correlations is expected to be particularly strong in low dimensions. For the two-dimensional Hubbard model it was demonstrated that including nonlocal fluctuations in the ladder dual fermion approach leads to a qualitative modification of the mean-field result, i.e. to the formation of a pseudogap in the single-particle density of states and non-Fermi-liquid behavior.

Cluster extensions to dynamical mean-field theory capture short-range physics and allow a systematic improvement by increasing the cluster size. Long-range correlations, however, require large scale calculations at the limit of todays computational feasibility and their study is therefore limited to simple (one-band) model Hamiltonians. Even if computational resources continue to increase as in the past few years, it is not hard to predict that the accessible length scales will only be marginally affected due to the exponential growth of the computational complexity with the system size.

In order to facilitate a realistic description of strongly correlated materials, it is hence vital to develop novel theoretical concepts and approaches. The symbiosis of numerical and analytical approaches is a very promising perspective for the solution of quantum lattice models, which is just beginning to emerge. The combination of numerical methods with diagrammatic techniques has previously been used to gain insight into the pairing mechanism in the Hubbard model, and to include spatial correlations beyond dynamical mean-field theory, for which different schemes have been proposed. Such an approach has the particular advantage that certain (e.g. long-range) contributions can be selectively included or discluded from the approximation. It allows to separate and unravel the responsible physical mechanisms and potentially can provide more insight than a full (numerical) solution of the problem.

The dual fermion approach is a promising alternative along these lines. Starting from a reference system, i.e. an optimal quantum impurity problem, the solution to the lattice model is approximated diagrammatically. The approach is related to DMFT and facilitates a systematic study of the effects of spatial correlations beyond. An important prerequisite is the calculation of two-particle quantities.

In this thesis, it has been shown that using the recently developed continuous-time quantum Monte Carlo algorithms, imaginary-time correlation functions and, more importantly, the complete two-particle impurity vertex function can be reliably computed. The calculation of two-particle quantities for multiband problems is technically challenging and corresponding work is in progress. It is to be expected that their importance – and that of novel approaches based on them – will grow in the near future.

With an application to real materials in mind, the dual fermion approach has been generalized to the multiorbital case. The diagrammatic rules have been derived and it has been clarified how the approach can be applied to broken symmetry phases and to calculate generalized susceptibilities. The equivalence between $n$-particle excitations of lattice and dual fermions and the relation to DMFT have been established and it has been proven that in the DMFT limit, the calculation of susceptibilities is equivalent to the procedure in DMFT. This allows a systematic study of the effect of spatial correlations beyond dynamical mean-field theory. A general and versatile framework for an object-oriented implementation of diagrammatic techniques has been developed and tested.

The approach has further been implemented in a way which allows a self-consistent optimization of the underlying reference problem. It has been demonstrated that this is essential in particular for low-dimensional systems. In the one- and two-dimensional Hubbard model, the insulator cannot be recovered perturbatively when DMFT erroneously predicts a metallic state, unless the DMFT hybridization function (or Weiss field) is adjusted self-consistently. In agreement with cluster DMFT calculations, it was found that short-range antiferromagnetic correlations strongly reduce the critical interaction of the first-order metal-insulator transition and qualitatively alter the shape of the transition lines compared to DMFT. This cannot be achieved by straightforward diagrammatic extensions to DMFT.

It was further demonstrated that despite of the abstract nature of the transformation to dual fermions and the necessary approximations, in particular the truncation of the dual potential, the approach yields sensible results. The construction of diagrammatic approximations may be guided by physical considerations. An example is the paramagnon contribution to the self-energy which was included through a ladder approximation for the dual self-energy. A very important result for the general applicability of the approach is that a perturbation expansion in terms of dual fermions does not break down for intermediate to strong coupling, in contrast to straightforward diagrammatic expansions around DMFT. The convergence is considerably enhanced by the transformation to dual fermions and the approach is applicable in particular for strong coupling.

The method has been improved along two lines: The first possibility regards the reference system and results in the cluster dual fermion approach. It has been established that cluster DMFT appears as the lowest order approximation in this approach and the rules for the construction of translationally invariant quantities has been given. The second possibility is to enhance the approximation to the self-energy. The dual fermion approach has been supplemented by an infinite-order ladder approximation. It was found to give highly nontrivial results, such as the formation of the pseudogap in the density of states. Remarkably, it is applicable over the whole interaction range, from weak to strong coupling. These results are of broad interest, because they demonstrate that it is possible to construct a well-behaved diagrammatic expansion beyond dynamical mean-field theory even for low-dimensional systems.

The approach is complementary to cluster approaches because it provides a translationally invariant solution and high momentum resolution. It treats correlations on all length scales on equal footing and allows to take long-range correlations into account. The ladder dual fermion approach can be implemented in a computationally feasible manner even for multiband models. Since it is based on the solution of an impurity problem, it is not severely hampered by the fermionic sign-problem and can account for general local interactions. In conjunction with band-structure methods, it will for the first time allow a realistic description of effects due to long-range correlations within calculations of correlated materials. For these reasons, it is well suited for the application to systems of present-day interest, such as high-temperature superconducting materials, quasi-one-dimensional organic conductors or cold atoms in optical lattices. Corresponding work is currently in progress.

It has further been clarified that the transformation to dual fermions, independent of its application to lattice models, is a concept of much broader scope: The full problem is replaced by a nontrivial, numerically solvable reference problem which shares the same interaction part. Invoking the Hubbard-Stratonovich transformation allows to formulate a perturbation theory in terms of auxiliary, dual fermions although the reference problem is non-Gaussian. This gives rise to the notion of a superperturbation – a concept that is fairly general.

Based on this observation, the reference problem is chosen such that it can be solved with small computational effort using exact diagonalization. This yields an efficient perturbative method for quantum lattice models which inherits the advantageous properties of the dual fermion approach. The idea has further been applied to the local quantum impurity problem to devise a very efficient and remarkably accurate solver for the Anderson impurity model. The solver is free of statistical errors and allows a stable analytical continuation of dynamical quantities to the real axis. It can be used to study multiplet effects in solids and in the context of DMFT. It has been explicitly shown that the perturbation theory becomes equivalent to standard perturbation theory in the strongly hybridized weak-coupling limit and to the hybridization expansion in the opposite, weakly hybridized strong-coupling limit. The role of the reference system has been elucidated. For example, it has been shown that the Kondo resonance cannot be reproduced through an expansion around the atomic limit, but is already obtained by choosing a minimal reference system which contains the essential physics, in this case the singlet formation of the impurity site with a single bath site.

The results presented in this thesis illustrate the versatility and strength of the proposed numerical approaches for future studies of strongly correlated systems.

# Appendix A

# Dual Fermion Formalism: General Formulation

## A.1 Hubbard-Stratonovich Transformation

The dual fermions are introduced via a continuous Hubbard-Stratonovich transformation for Grassmann variables. Consider the Gaussian integral identity

$$\frac{1}{\det a} \int \exp\left(-\eta_\alpha^* a_{\alpha\beta} \eta_\beta + \eta_\alpha^* \xi_\alpha + \xi_\alpha^* \eta_\alpha\right) \prod_\gamma d\eta_\gamma^* d\eta_\gamma = \exp\left(\xi_\alpha^*(a^{-1})_{\alpha\beta}\xi_\beta\right) . \tag{A.1}$$

Here $\eta^*, \eta, \xi^*, \xi$ are Grassmann numbers and $a$ is a complex matrix with elements $a_{\alpha\beta}$. This identity is readily proven by imposing the linear shift transformation

$$\eta_\alpha^* \to \eta_\alpha^* + \xi_\beta^*(a^{-1})_{\beta\alpha} , \qquad \eta_\alpha \to \eta_\alpha + (a^{-1})_{\alpha\beta}\xi_\beta , \tag{A.2}$$

upon which the exponential on the left-hand side transforms to

$$\exp\left(-\eta_\alpha^* a_{\alpha\beta}\eta_\beta + \eta_\alpha^*\xi_\alpha + \xi_\alpha^*\eta_\alpha\right) \quad \to \quad \exp\left(-\eta_\alpha^* a_{\alpha\beta}\eta_\beta + \xi_\alpha^*(a^{-1})_{\alpha\beta}\xi_\beta\right) . \tag{A.3}$$

Further noting that the Jacobian of this linear transformation is unity and

$$\int \exp\left(-\eta_\alpha^* a_{\alpha\beta}\eta_\beta\right) \prod_\gamma d\eta_\gamma^* d\eta_\gamma = \det a \tag{A.4}$$

proves Eq. A.1. The last identity is seen to hold by rewriting the left-hand side as

$$\int \prod_{\alpha\beta} \exp\left(-\eta_\alpha^* a_{\alpha\beta}\eta_\beta\right) \prod_\gamma d\eta_\gamma^* d\eta_\gamma = \int \prod_{\alpha\beta}\left(1 - \eta_\alpha^* a_{\alpha\beta}\eta_\beta\right) \prod_\gamma d\eta_\gamma^* d\eta_\gamma . \tag{A.5}$$

Note that here no summation over repeated indices is implied. After expanding the product, only those terms which contain each Grassmann number exactly once contribute to the integral. A possible sign arising from expanding the product cancels by reversing the order of integration (the differentials anticommute). Each term can be brought into an order in which $\eta_\beta^*$ appears to the right of $\eta_\alpha^*$ for $\beta > \alpha$ by moving pairs of Grassmann numbers and hence no additional sign arises. Since expanding the product generates all possible orders of the terms $\eta_\alpha^* a_{\alpha\beta}\eta_\beta$, the result can be represented in the form

$$\int \sum_{p\in\mathcal{P}} \prod_\alpha \eta_\alpha^* a_{\alpha p_\alpha} \eta_{p_\alpha} \prod_\gamma d\eta_\gamma d\eta_\gamma^* \,. \tag{A.6}$$

Here $p$ is a permutation in the symmetric group $\mathcal{P}$ and $p_\alpha$ denotes the permutation of the index $\alpha$. Bringing the $\eta$'s into the same order as the differentials picks up the sign of the permutation. Hence

$$\sum_{p\in\mathcal{P}} \mathrm{sgn}(p) \prod_\alpha a_{\alpha p_\alpha} \int \eta_\alpha^* \eta_\alpha \prod_\gamma d\eta_\gamma d\eta_\gamma^* = \sum_{p\in\mathcal{P}} \mathrm{sgn}(p) \prod_\alpha a_{\alpha p_\alpha} = \det a \,. \tag{A.7}$$

Eq. A.1 is generalized using the substitutions

$$\xi_\alpha^* \to c_\beta^* b_{\beta\alpha} \qquad \xi_\alpha \to b_{\alpha\beta} c_\beta \qquad \eta_\alpha^* \to f_\alpha^* \qquad \eta_\alpha \to f_\alpha \,, \tag{A.8}$$

where now $c, c^*, f$ and $f^*$ are used to denote Grassmann fields and $b$ is a complex matrix (note that the generators $\xi$ and $\xi^*$ are independent and one could choose $\xi_\alpha \to \bar{b}_{\alpha\beta} c_\beta$). The desired identity reads

$$\boxed{\frac{1}{\det a} \int \exp\left(-f_\alpha^* a_{\alpha\beta} f_\beta + f_\alpha^* b_{\alpha\beta} c_\alpha + c_\alpha^* b_{\alpha\beta} f_\beta\right) \prod_\gamma df_\gamma^* df_\gamma = \exp\left(c_\alpha^* b_{\alpha\beta}(a^{-1})_{\beta\gamma} b_{\gamma\delta} c_\delta\right) \,.} \tag{A.9}$$

## A.2    Dual Fermion Matrix Formalism

In this section, the dual fermion formalism is derived in its general form. The framework of the coherent state path integral formalism introduced in Sec. 2.3 is used here. For clarity, the dependence on momentum $k$ and position $i$ as well as the time dependence are written explicitly. The momentum sums are to be understood as $\sum_k = 1/N \sum_\mathbf{k}$, where $N$ is the number of k-points. Greek indices label all remaining degrees of freedom.

The imaginary time lattice action can be written in the following general form

$$S[c^*,c] = \int_0^\beta d\tau \sum_{k,\alpha\beta} c_\alpha^*(\tau) \left[(\partial_\tau - \mu)\delta_{\alpha\beta} + h_{k\,\alpha\beta}\right] c_\beta(\tau) + \sum_i S_{\mathrm{int}}^{(i)}[c^*,c] \,. \tag{A.10}$$

The action for an impurity embedded in a time dependent bath is given by

$$S_{\text{imp}}[c^*, c] = -\int_0^\beta d\tau \int_0^\beta d\tau' \sum_{\alpha\beta} c_\alpha^*(\tau) \left[ \mathcal{G}_{\text{th}}^{-1}(\tau - \tau') - \Delta(\tau - \tau') \right]_{\alpha\beta} c_\beta(\tau') + S_{\text{int}}[c^*, c] \, .$$

(A.11)

with the thermal Green function defined by $[\partial_\tau - \mu]\mathcal{G}_{\text{th}}(\tau - \tau') = -\delta_{\alpha\beta}\delta(\tau - \tau')$. By adding and subtracting a local (i.e. $k$-independent) hybridization function at each site $i$, the lattice action is represented as a collection of impurities and a bilinear coupling between them:

$$S[c^*, c] = \sum_i S_{\text{imp}}^{(i)}[c_i^*, c_i] - \int_0^\beta d\tau \int_0^\beta d\tau' \sum_{k,\alpha\beta} c_\alpha^*(\tau) \left[ \Delta(\tau - \tau') - h_k\delta(\tau - \tau') \right]_{\alpha\beta} c_\beta(\tau') \, .$$

(A.12)

The Hubbard-Stratonovich transformation (A.9) is now applied to transform the second term of (A.12) in the coherent state path integral representation of the partition function,

$$\mathcal{Z} = \int \exp\left(-S[c^*, c]\right) \mathcal{D}[c^*, c] \, .$$

(A.13)

More precisely, choose

$$b_{\alpha\beta} = -g_{\alpha\beta}^{-1} \, , \qquad a_{\alpha\beta} = [g^{-1}(\Delta - h)^{-1}g^{-1}]_{\alpha\beta} \quad \Rightarrow \quad ba^{-1}b = \Delta - h \, ,$$

(A.14)

where $g_{\alpha\beta}$ denotes the local impurity Green function. To simplify the notation, the summation over e.g. $\alpha$ from here on comprises time integration (and correspondingly the time index is omitted), or equivalently frequency summation after transformation to the Matsubara representation:

$$c(\tau) = \frac{1}{\sqrt{\beta}} \sum_{\omega_n} c(\omega_n)e^{-i\omega_n\tau} \, , \qquad c^*(\tau) = \frac{1}{\sqrt{\beta}} \sum_{\omega_n} c^*(\omega_n)e^{i\omega_n\tau} \, .$$

(A.15)

Application of (A.9) yields the partition function in the form

$$\mathcal{Z} = \mathcal{Z}_f \int \int \exp(-S[c^*, c; f^*, f])\mathcal{D}[c^*, c; f^*, f] \, ,$$

(A.16)

where

$$\mathcal{Z}_f = \det([g(\Delta - h)g]_{\alpha\beta}) \, .$$

(A.17)

The transformed action is given by

$$S[c^*, c; f^*, f] = \sum_i S_{\text{site}}^{(i)}[c^*, c; f^*, f] + \sum_{k,\alpha\beta} f_\alpha^*[g^{-1}(\Delta - h_k)^{-1}g^{-1}]_{\alpha\beta}f_\beta \, .$$

(A.18)

And the local action $S_{\text{site}}^{(i)}$ reads

$$S_{\text{site}}^{(i)}[c^*, c; f^*, f] = S_{\text{imp}}^{(i)}[c^*, c] + \sum_{\alpha\beta} [f_{i\alpha}^* g_{\alpha\beta}^{-1} c_{i\beta} + c_{i\alpha}^* g_{\alpha\beta}^{-1} f_{i\beta}] \,. \tag{A.19}$$

Note that the summation over all states labeled by $k$ in (A.19) was replaced by the equivalent summation over all sites $i$ by a change of basis.

In the next section, it is shown that the original fermions represented by $c^*$ and $c$ can be integrated out formally exactly from the action $S_{\text{site}}$, which gives rise to the dual potential $V[f^*, f]$. One has

$$\int \exp\Big( -\sum_i S_{\text{site}}^{(i)}[c_i^*, c_i; f_i^*, f_i]\Big)\mathcal{D}[c^*, c] = \prod_i \mathcal{Z}_{\text{imp}}^{(i)} \exp\Big( -\sum_{\alpha\beta} f_{i\alpha}^* g_{\alpha\beta}^{-1} f_{i\beta} - V_i[f_i^*, f_i]\Big) \,. \tag{A.20}$$

Applying this to Eq. A.16 yields

$$\mathcal{Z} = \tilde{\mathcal{Z}}_f \int \exp(-S[f^*, f])\mathcal{D}[f^*, f] \,, \tag{A.21}$$

with $\tilde{\mathcal{Z}}_f = \mathcal{Z}_f \prod_i \mathcal{Z}_{\text{imp}}^{(i)}$. Note that the integral on the right hand side of (A.21) is the partition function of the dual ensemble, $\mathcal{Z}_{\text{d}} = \mathcal{Z}/\tilde{\mathcal{Z}}_f$. The dual action and the bare dual Green function (in matrix form) are given by

$$S[f^*, f] = -\sum_{k,\alpha\beta} f_\alpha^* [-g^{-1} - g^{-1}(\Delta - h_k)^{-1} g^{-1}]_{\alpha\beta} f_\beta + \sum_i V_i[f_i^*, f_i] \tag{A.22}$$

and

$$G_0^{\text{d}} = -g\left[g + (\Delta - h)^{-1}\right]^{-1} g \,. \tag{A.23}$$

Finally, a few useful matrix relations are derived. To this end, define the dual self energy as

$$\Sigma^{\text{d}} = G_0^{\text{d}-1} - G^{\text{d}-1} \,. \tag{A.24}$$

The impurity and lattice self-energies are correspondingly given by ($G_{(0)}$ is the (bare) lattice Green function)

$$\Sigma_{\text{imp}} = g_0^{-1} - g^{-1} \,, \qquad \Sigma = G_0^{-1} - G^{-1} \,. \tag{A.25}$$

The ($k$-dependent) correction to the impurity self-energy can be defined as

$$\begin{aligned}
\Sigma' &= \Sigma - \Sigma_{\text{imp}} \\
&= G_0^{-1} - G^{-1} - \Sigma_{\text{imp}} \\
&= [(i\omega + \mu)\mathbb{1} - h] - G^{-1} - [(i\omega + \mu)\mathbb{1} - \Delta] + g^{-1} \\
&= g^{-1} + (\Delta - h) - G^{-1} \,.
\end{aligned} \tag{A.26}$$

Next, the Green function is expressed in terms of the dual self-energy. Using the relation between the dual and lattice Green functions, Eq. A.115, one finds

$$
\begin{aligned}
G &= \left[(\Delta - h)^{-1} g^{-1} G^{\mathrm{d}} g^{-1} (\Delta - h)^{-1}\right] + (\Delta - h)^{-1} \\
&= \left[(\Delta - h)^{-1} g^{-1} G^{\mathrm{d}} g^{-1} (\Delta - h)^{-1}\right] \left\{ \mathbb{1} + \left[(\Delta - h)^{-1} g^{-1} G^{\mathrm{d}} g^{-1} (\Delta - h)^{-1}\right]^{-1} (\Delta - h)^{-1} \right\} \\
&= \left[(\Delta - h)^{-1} g^{-1} G^{\mathrm{d}} g^{-1} (\Delta - h)^{-1}\right] \left\{ \mathbb{1} + (\Delta - h) g \, G^{\mathrm{d}-1} g \right\} \, .
\end{aligned}
\tag{A.27}
$$

The term in curly brackets is evaluated as follows:

$$
\begin{aligned}
\mathbb{1} + (\Delta - h) g \, G^{\mathrm{d}-1} g &= \mathbb{1} + (\Delta - h) g \, [G_0^{\mathrm{d}-1} - \Sigma^{\mathrm{d}}] g \\
&= \mathbb{1} - (\Delta - h)\,[g + (\Delta - h)^{-1}] - (\Delta - h) g \Sigma^{\mathrm{d}} g \\
&= -(\Delta - h) g \left[\mathbb{1} + \Sigma^{\mathrm{d}} g\right] \, .
\end{aligned}
\tag{A.28}
$$

Putting this back into Eq. A.27 yields

$$
G = -(\Delta - h)^{-1} g^{-1} G^{\mathrm{d}} \left[\mathbb{1} + \Sigma^{\mathrm{d}} g\right] = (\Delta - h)^{-1} g^{-1} \left[\Sigma^{\mathrm{d}} - G_0^{\mathrm{d}-1}\right]^{-1} \left[\mathbb{1} + \Sigma^{\mathrm{d}} g\right] \, .
\tag{A.29}
$$

Inverting $G$ and inserting the expression for $G_0^{\mathrm{d}}$ gives

$$
G^{-1} = \left[\mathbb{1} + \Sigma^{\mathrm{d}} g\right]^{-1} \left[g^{-1} + (\Delta - h) + \Sigma^{\mathrm{d}} g(\Delta - h)\right] \, .
\tag{A.30}
$$

This may be used to establish a relation to $\Sigma'$ by expressing $G$ using (A.26):

$$
\begin{aligned}
\Sigma' &= g^{-1} + (\Delta - h) - \left[\mathbb{1} + \Sigma^{\mathrm{d}} g\right]^{-1} \left[g^{-1} + (\Delta - h) + \Sigma^{\mathrm{d}} g(\Delta - h)\right] \\
&= \left[\mathbb{1} + \Sigma^{\mathrm{d}} g\right]^{-1} \left[(\mathbb{1} + \Sigma^{\mathrm{d}} g)(g^{-1} + (\Delta - h)) - [g^{-1} + (\Delta - h) + \Sigma^{\mathrm{d}} g(\Delta - h)]\right] \\
&= \left[\mathbb{1} + \Sigma^{\mathrm{d}} g\right]^{-1} \Sigma^{\mathrm{d}} \, ,
\end{aligned}
\tag{A.31}
$$

or

$$
\boxed{(\Sigma')^{-1} = \left(\Sigma^{\mathrm{d}}\right)^{-1} + g \, .}
\tag{A.32}
$$

It is straightforward to check that this is equivalent to

$$
g^{-1} - \Sigma' = [g(\mathbb{1} + \Sigma^{\mathrm{d}} g)]^{-1} \, .
\tag{A.33}
$$

Now use Dyson's equation for dual fermions in connection with the explicit expression (A.23) for the bare dual Green function to get

$$
\begin{aligned}
(G^{\mathrm{d}})^{-1} &= (G_0^{\mathrm{d}})^{-1} - \Sigma^{\mathrm{d}} \\
&= -g^{-1} \left[(\Delta - h)^{-1}\right] + g \Big] g^{-1} - \Sigma^{\mathrm{d}} \\
&= -g^{-1} \left[(\Delta - h)^{-1} + g(\mathbb{1} + \Sigma^{\mathrm{d}} g)\right] g^{-1} \, .
\end{aligned}
\tag{A.34}
$$

Using (A.33) yields

$$G^{\mathrm{d}} = -g\left[(\Delta - h)^{-1} + (g^{-1} - \Sigma')^{-1}\right]^{-1} g \, . \tag{A.35}$$

In order to derive the relation between the lattice Green function and the self-energy correction $\Sigma'$, use $g^{-1} = (i\omega + \mu)\mathbb{1} - \Delta - \Sigma_{\mathrm{imp}}$ to write

$$\begin{aligned}
G^{-1} = G_0^{-1} - \Sigma &= G_0^{-1} - \Sigma_{\mathrm{imp}} - \Sigma' \\
&= -h + \Delta + g^{-1} - \Sigma' \, ,
\end{aligned} \tag{A.36}$$

or

$$\boxed{G = \left[g^{-1} - \Sigma' + (\Delta - h)\right]^{-1} \, .} \tag{A.37}$$

as expected from the definition of $\Sigma'$.

An important relation between the DMFT lattice Green function and the bare dual Green function is established by noting that in DMFT $\Sigma' = 0$. From (A.37), one has

$$G^{\mathrm{DMFT}} = \left[g^{-1} + (\Delta - h)\right]^{-1} \, . \tag{A.38}$$

Hence

$$\begin{aligned}
g - G^{\mathrm{DMFT}} &= g - \left[g^{-1} + (\Delta - h)\right]^{-1} \\
&= \left\{g\left[g^{-1} + (\Delta - h)\right] - \mathbb{1}\right\}\left[g^{-1} + (\Delta - h)\right]^{-1} \\
&= g(\Delta - h)\left[g^{-1} + (\Delta - h)\right]^{-1} = \left[g^{-1}(\Delta - h)^{-1}g^{-1} + g^{-1}\right]^{-1} \\
&= g\left[(\Delta - h)^{-1} + g\right]^{-1} g = -G_0^{\mathrm{d}} \, .
\end{aligned} \tag{A.39}$$

Using (A.31), $\Sigma' = 0$ implies $\Sigma^{\mathrm{d}} = 0$ and $G_0^{\mathrm{d}}$ is the same as $G^{\mathrm{d}}$ in DMFT. Summing (A.39) over $k$ yields (recall that $\sum_k = (1/N)\sum_{\mathbf{k}}$ and that $g$ is local)

$$G_{\mathrm{loc}}^{\mathrm{DMFT}} - g = G_{0\,\mathrm{loc}}^{\mathrm{d}} \, . \tag{A.40}$$

The condition of the local part of the bare dual Green function to be zero is equivalent to the DMFT self-consistency condition. The general update for the hybridization function (not restricted to DMFT) can be written in different ways. One only needs to require that self-consistency is reached when $G_{\mathrm{loc}}^{\mathrm{d}} \equiv 0$. Starting from (2.65) with $G_{\mathrm{loc}} = G_{\mathrm{loc}}^{\mathrm{DMFT}}$:

$$\begin{aligned}
\Delta_{\mathrm{new}} &= \Delta_{\mathrm{old}} + \xi\left[g^{-1} - G_{\mathrm{loc}}^{-1}\right] \\
&= \Delta_{\mathrm{old}} + \xi\left[g^{-1}G_{\mathrm{loc}}G_{\mathrm{loc}}^{-1} - g^{-1}gG_{\mathrm{loc}}^{-1}\right] \\
&= \Delta_{\mathrm{old}} + \xi\left[g^{-1}(G_{\mathrm{loc}} - g)G_{\mathrm{loc}}^{-1}\right] \\
&= \Delta_{\mathrm{old}} + \xi\left[g^{-1}G_{\mathrm{loc}}^{\mathrm{d}}G_{\mathrm{loc}}^{-1}\right] .
\end{aligned} \tag{A.41}$$

For regular dual fermion calculations $G_{\mathrm{loc}} - g = G_{\mathrm{loc}}^{\mathrm{d}}$ does not hold, but the last line of (A.41) can nevertheless be used for the hybridization function update since a self-consistent solution is obtained when $G_{\mathrm{loc}}^{\mathrm{d}} \equiv 0$.

## A.3  Dual Potential

In this appendix, the the dual potential is derived. The defining equation for the dual potential as derived in the previous section is given by

$$\frac{1}{\mathcal{Z}_{\text{imp}}} \int \exp\left(-S_{\text{site}}[c_i^*, c_i; f_i^*, f_i]\right) \mathcal{D}[c_i^*, c_i] = \exp\left(-f_{i\alpha}^* g_{\alpha\beta}^{-1} f_{i\beta} - V_i[f_i^*, f_i]\right) . \tag{A.42}$$

As before, the index $i$ labels sites (or clusters). The dual potential is obtained by expanding both sides of Eq. A.42 and equating the expressions by order. This can be done for each site separately. Therefore the index $i$ is omitted in the following. The local action $S_{\text{site}}$ consists of the impurity action and the local coupling to the dual degrees of freedom:

$$S_{\text{site}}[c^*, c; f^*, f] = S_{\text{imp}}[c^*, c] + f_{\alpha}^* g_{\alpha\beta}^{-1} c_\beta + c_{\alpha}^* g_{\alpha\beta}^{-1} f_\beta . \tag{A.43}$$

First consider the left-hand side of Eq. A.42. Keeping in mind that the exponential appears under the path integral, the two contributions may be factorized,

$$\exp\left(-S_{\text{site}}[c^*, c; f^*, f]\right) = \exp\left(-S_{\text{imp}}[c^*, c]\right) \exp\left(-[f_{\alpha}^* g_{\alpha\beta}^{-1} c_\beta + c_{\alpha}^* g_{\alpha\beta}^{-1} f_\beta]\right) . \tag{A.44}$$

The second exponential is formally expanded as a Taylor series:

$$\exp\left(-[f_{\alpha}^* g_{\alpha\beta}^{-1} c_\beta + c_{\alpha}^* g_{\alpha\beta}^{-1} f_\beta]\right) = \sum_{n=0}^{\infty} \frac{(-1)^n}{n!} [f_{\alpha}^* g_{\alpha\beta}^{-1} c_\beta + c_{\alpha}^* g_{\alpha\beta}^{-1} f_\beta]^n . \tag{A.45}$$

Recalling the definition of the impurity average,

$$\langle \ldots \rangle_{\text{imp}} := \frac{1}{\mathcal{Z}_{\text{imp}}} \int \exp\left(-S_{\text{imp}}[c^*, c]\right) \ldots \mathcal{D}[c^*, c] , \tag{A.46}$$

one sees that integrating out the original fermionic degrees of freedom represented by $c_{\alpha}^*, c_{\alpha}$ in Eq. A.44 corresponds to averaging over the impurity degrees of freedom. Averages containing an odd number of Grassmann fields must vanish: $\langle c \rangle_{\text{imp}} = \langle c^* \rangle_{\text{imp}} = \langle ccc^* \rangle_{\text{imp}} = 0$, etc. As a consequence, odd powers drop out of the expansion upon averaging and the left-hand side of Eq. A.42 becomes

$$\sum_{\substack{n=0 \\ n \text{ even}}}^{\infty} \frac{1}{n!} \langle [f_{\alpha}^* g_{\alpha\beta}^{-1} c_\beta + c_{\alpha}^* g_{\alpha\beta}^{-1} f_\beta]^n \rangle_{\text{imp}} . \tag{A.47}$$

This expression is evaluated by expanding the product and performing the impurity average. To this end, recall the definition of the impurity correlation functions

$$g_{\alpha\beta} = -\langle c_\alpha c_\beta^* \rangle_{\text{imp}} , \tag{A.48}$$

$$\chi_{\alpha\beta\gamma\delta}^{(4)} = \langle c_\alpha c_\beta^* c_\gamma c_\delta^* \rangle_{\text{imp}} , \tag{A.49}$$

$$\chi_{\alpha\beta\gamma\delta\epsilon\zeta}^{(6)} = -\langle c_\alpha c_\beta^* c_\gamma c_\delta^* c_\epsilon c_\zeta^* \rangle_{\text{imp}} . \tag{A.50}$$

To exemplify the procedure, consider the lowest nontrivial order in detail. For $n = 2$ one has

$$\frac{1}{2}\langle [f_\alpha^* g_{\alpha\beta}^{-1} c_\beta + c_\alpha^* g_{\alpha\beta}^{-1} f_\beta]^2 \rangle_{\text{imp}} = \frac{1}{2}\langle f_\alpha^* g_{\alpha\beta}^{-1} c_\beta f_\gamma^* g_{\gamma\delta}^{-1} c_\delta + f_\alpha^* g_{\alpha\beta}^{-1} c_\beta c_\gamma^* g_{\gamma\delta}^{-1} f_\delta$$
$$+ c_\alpha^* g_{\alpha\beta}^{-1} f_\beta f_\gamma^* g_{\gamma\delta}^{-1} c_\delta + c_\alpha^* g_{\alpha\beta}^{-1} f_\beta c_\gamma^* g_{\gamma\delta}^{-1} f_\delta \rangle_{\text{imp}} . \tag{A.51}$$

Only the second and third term on the right-hand side survive averaging[1]:

$$\frac{1}{2}\langle [f_\alpha^* g_{\alpha\beta}^{-1} c_\beta + c_\alpha^* g_{\alpha\beta}^{-1} f_\beta]^2 \rangle_{\text{imp}} = \frac{1}{2} g_{\alpha\beta}^{-1} g_{\gamma\delta}^{-1} \langle c_\beta c_\gamma^* \rangle_{\text{imp}} f_\alpha^* f_\delta + \frac{1}{2} g_{\alpha\beta}^{-1} g_{\gamma\delta}^{-1} \langle c_\delta c_\alpha^* \rangle_{\text{imp}} f_\gamma^* f_\beta . \tag{A.52}$$

Using the definition of the Green function, Eq. A.48, this becomes

$$-\frac{1}{2} g_{\alpha\beta}^{-1} g_{\gamma\delta}^{-1} g_{\beta\gamma} f_\alpha^* f_\delta - \frac{1}{2} g_{\alpha\beta}^{-1} g_{\gamma\delta}^{-1} g_{\delta\alpha} f_\gamma^* f_\beta = -g_{\alpha\beta}^{-1} f_\alpha^* f_\beta . \tag{A.53}$$

The fourth-order term is given by

$$\frac{1}{4!}\langle [f_\alpha^* g_{\alpha\beta}^{-1} c_\alpha + c_\alpha^* g_{\alpha\beta}^{-1} f_\beta]^4 \rangle_{\text{imp}} . \tag{A.54}$$

Multiplying out the product yields $2^4 = 16$ terms, whereof only 6 contain the same number of $c$'s and $c^*$'s and hence survive averaging. All of the remaining terms can be brought into the form

$$g_{\alpha\beta}^{-1} g_{\gamma\delta}^{-1} g_{\epsilon\zeta}^{-1} g_{\eta\theta}^{-1} \langle c_\beta c_\gamma^* c_\zeta c_\eta^* \rangle_{\text{imp}} f_\alpha^* f_\delta f_\epsilon^* f_\theta , \tag{A.55}$$

by an even number of permutations of Grassmann fields and renaming indices. Hence all terms are equivalent. The fourth order contribution is thus ($\frac{1}{4!} \cdot 6 = \frac{1}{4}$):

$$\frac{1}{4} g_{\alpha\beta}^{-1} g_{\gamma\delta}^{-1} g_{\epsilon\zeta}^{-1} g_{\eta\theta}^{-1} \chi_{\beta\gamma\zeta\eta}^{(4)} f_\alpha^* f_\delta f_\epsilon^* f_\theta . \tag{A.56}$$

For the sixth-order contribution, 20 out of $2^6 = 64$ terms survive averaging. This gives a prefactor of the corresponding contribution of $\frac{1}{6!} \cdot 20 = \frac{1}{36}$. In general, the prefactors can be deduced from the binomial theorem, which states that $(x + y)^n = \sum_{k=0}^{n} \binom{n}{k} x^{n-k} y^k$, where $\binom{n}{k} = n!/[k!(n-k)!]$. The number of terms with the same number of $c$'s and $c^*$'s is hence given by the binomial coefficient for which the powers of $x$ and $y$ are equal, i.e. $n - k = k$, or $k = n/2$. Hence the coefficient is $\binom{n}{n/2} = n!/[(n/2)!]^2$. The factor $n!$ cancels and the prefactor is given by $1/[(n/2)!]^2$. This is exactly the number of possibilities to permute the $f$'s and $f^*$'s among themselves, which plays a role for the determination of

---

[1]This is true given the impurity is in the normal (i.e. non-superconducting) state and hence the anomalous averages $\langle c_\alpha c_\beta \rangle_{\text{imp}}$ and $\langle c_\alpha^* c_\beta^* \rangle_{\text{imp}}$ are zero.

combinatorial factors of the diagrams in Sec. A.4.3. For the present order, $n = 6$ and $1/(3!)^2 = 1/36$. It is further easy to see that at every order, all terms can be brought into the standard sequence with an even number of permutations and hence have the same sign. For the sixth order term on the left-hand side one finds

$$\frac{1}{6!}\langle [f_\alpha^* g_{\alpha\beta}^{-1} c_\alpha + c_\alpha^* g_{\alpha\beta}^{-1} f_\beta]^6 \rangle_{\mathrm{imp}} f_\alpha^* f_\delta f_\epsilon^* f_\theta f_\kappa^* f_\nu = -\frac{1}{36} g_{\alpha\beta}^{-1} g_{\gamma\delta}^{-1} g_{\epsilon\zeta}^{-1} g_{\eta\theta}^{-1} g_{\kappa\lambda}^{-1} g_{\mu\nu}^{-1} \chi^{(6)}_{\beta\gamma\zeta\eta\lambda\mu} f_\alpha^* f_\delta f_\epsilon^* f_\theta f_\kappa^* f_\nu . \tag{A.57}$$

The minus sign on the right hand side enters through the definition of the impurity correlation function, Eq. A.50. Now consider the right-hand side of Eq. A.42. Since only even powers of $f^* f$ appear on the left-hand side of this equation, the dual potential must have the following form:

$$V[f^*, f] = a^{(2)}_{\alpha\beta} f_\alpha^* f_\beta + a^{(4)}_{\alpha\beta\gamma\delta} f_\alpha^* f_\beta f_\gamma^* f_\delta + a^{(6)}_{\alpha\beta\gamma\delta\epsilon\zeta} f_\alpha^* f_\beta f_\gamma^* f_\delta f_\epsilon^* f_\zeta + \dots \tag{A.58}$$

Expanding the right-hand side of Eq. A.42 and retaining only terms up to third order in $f^* f$ yields:

$$1 - \left[ f_\alpha^* g_{\alpha\beta}^{-1} f_\beta + a^{(4)}_{\alpha\beta\gamma\delta} f_\alpha^* f_\beta f_\gamma^* f_\delta + a^{(6)}_{\alpha\beta\gamma\delta\epsilon\zeta} f_\alpha^* f_\beta f_\gamma^* f_\delta f_\epsilon^* f_\zeta \right] + \frac{1}{2} \left[ f_\alpha^* g_{\alpha\beta}^{-1} f_\beta f_\gamma^* g_{\gamma\delta}^{-1} f_\delta + \right.$$

$$\left. g_{\alpha\beta}^{-1} a^{(4)}_{\gamma\delta\epsilon\zeta} f_\alpha^* f_\beta f_\gamma^* f_\delta f_\epsilon^* f_\zeta + a^{(4)}_{\alpha\beta\gamma\delta} g_{\epsilon\zeta}^{-1} f_\alpha^* f_\beta f_\gamma^* f_\delta f_\epsilon^* f_\zeta \right] - \frac{1}{3!} \left[ g_{\alpha\beta}^{-1} g_{\gamma\delta}^{-1} g_{\epsilon\zeta}^{-1} f_\alpha^* f_\beta f_\gamma^* f_\delta f_\epsilon^* f_\zeta \right] . \tag{A.59}$$

The coefficients of the dual potential are obtained by equating the left-hand side of Eq. A.42 with Eq. A.59 by order. The bilinear term appears identically on both sides of the Eq. A.42 so that the equation is identically fulfilled up to first order in $f^* f$. Hence it follows that $a^{(2)}_{\alpha\beta} = 0$. For the next order one finds that:

$$a^{(4)}_{\alpha\delta\epsilon\theta} f_\alpha^* f_\delta f_\epsilon^* f_\theta = -\frac{1}{4} g_{\alpha\beta}^{-1} g_{\gamma\delta}^{-1} g_{\epsilon\zeta}^{-1} g_{\eta\theta}^{-1} \chi^{(4)}_{\beta\gamma\zeta\eta} f_\alpha^* f_\delta f_\epsilon^* f_\theta + \frac{1}{2} g_{\alpha\delta}^{-1} g_{\epsilon\theta}^{-1} f_\alpha^* f_\delta f_\epsilon^* f_\theta . \tag{A.60}$$

$\chi^{(4)}$ is the average of four anticommuting Grassmann fields and his hence fully antisymmetric w.r.t. permutation of its indices. The second term on the right is antisymmetrized as follows:

$$\frac{1}{2} g_{\alpha\delta}^{-1} g_{\epsilon\theta}^{-1} f_\alpha^* f_\delta f_\epsilon^* f_\theta = \frac{1}{4} g_{\alpha\delta}^{-1} g_{\epsilon\theta}^{-1} \left( 2 f_\alpha^* f_\delta f_\epsilon^* f_\theta \right) = \frac{1}{4} g_{\alpha\delta}^{-1} g_{\epsilon\theta}^{-1} \left( f_\alpha^* f_\delta f_\epsilon^* f_\theta - f_\alpha^* f_\theta f_\epsilon^* f_\delta \right)$$

$$= \frac{1}{4} \left[ g_{\alpha\delta}^{-1} g_{\epsilon\theta}^{-1} - g_{\alpha\theta}^{-1} g_{\epsilon\delta}^{-1} \right] f_\alpha^* f_\delta f_\epsilon^* f_\theta$$

$$= \frac{1}{4} g_{\alpha\beta}^{-1} g_{\epsilon\zeta}^{-1} \left( g_{\beta\gamma} g_{\zeta\eta} - g_{\beta\eta} g_{\zeta\gamma} \right) g_{\gamma\delta}^{-1} g_{\eta\theta}^{-1} f_\alpha^* f_\delta f_\epsilon^* f_\theta . \tag{A.61}$$

This finally gives

$$a^{(4)}_{\alpha\delta\epsilon\theta} f_\alpha^* f_\delta f_\epsilon^* f_\theta = -\frac{1}{4} \gamma^{(4)}_{\alpha\delta\epsilon\theta} f_\alpha^* f_\delta f_\epsilon^* f_\theta , \tag{A.62}$$

where $\gamma^{(4)}$ is the fully antisymmetric two-particle impurity vertex:

$$\boxed{\gamma^{(4)}_{\alpha\delta\epsilon\theta} = g^{-1}_{\alpha\beta} g^{-1}_{\epsilon\zeta} \tilde{\chi}^{(4)}_{\beta\gamma\zeta\eta} g^{-1}_{\gamma\delta} g^{-1}_{\eta\theta}} \,, \tag{A.63}$$

with

$$\boxed{\tilde{\chi}^{(4)}_{\beta\gamma\zeta\eta} = \chi^{(4)}_{\beta\gamma\zeta\eta} - g_{\beta\gamma} g_{\zeta\eta} + g_{\beta\eta} g_{\zeta\gamma}} \,. \tag{A.64}$$

It is readily seen that the above expression is antisymmetric w.r.t. a permutation of the indices coupled to either $f$ or $f^*$, e.g. by permuting $\beta \leftrightarrow \zeta$: $\tilde{\chi}^{(4)}_{\zeta\gamma\beta\eta} = -\tilde{\chi}^{(4)}_{\beta\gamma\zeta\eta}$, or $\gamma \leftrightarrow \eta$.

For the third order one has (note that the first two terms in the second line of Eq. A.59 are equivalent):

$$-\frac{1}{36} g^{-1}_{\alpha\beta} g^{-1}_{\gamma\delta} g^{-1}_{\epsilon\zeta} g^{-1}_{\eta\theta} g^{-1}_{\kappa\lambda} g^{-1}_{\mu\nu} \chi^{(6)}_{\beta\gamma\zeta\eta\lambda\mu} f^*_\alpha f_\delta f^*_\epsilon f_\theta f^*_\kappa f_\nu =$$

$$- a^{(6)}_{\alpha\delta\epsilon\theta\kappa\nu} f^*_\alpha f_\delta f^*_\epsilon f_\theta f^*_\kappa f_\nu + g^{-1}_{\alpha\delta} a^{(4)}_{\epsilon\theta\kappa\nu} f^*_\alpha f_\delta f^*_\epsilon f_\theta f^*_\kappa f_\nu - \frac{1}{3!} g^{-1}_{\alpha\delta} g^{-1}_{\epsilon\theta} g^{-1}_{\kappa\nu} f^*_\alpha f_\delta f^*_\epsilon f_\theta f^*_\kappa f_\nu \,. \tag{A.65}$$

Solving for $a^{(6)}$ and inserting the result for $a^{(4)}$ yields

$$a^{(6)}_{\alpha\delta\epsilon\theta\kappa\nu} f^*_\alpha f_\delta f^*_\epsilon f_\theta f^*_\kappa f_\nu = \frac{1}{36} g^{-1}_{\alpha\beta} g^{-1}_{\gamma\delta} g^{-1}_{\epsilon\zeta} g^{-1}_{\eta\theta} g^{-1}_{\kappa\lambda} g^{-1}_{\mu\nu} \chi^{(6)}_{\beta\gamma\zeta\eta\lambda\mu} f^*_\alpha f_\delta f^*_\epsilon f_\theta f^*_\kappa f_\nu$$

$$- \frac{1}{4} g^{-1}_{\alpha\delta} \gamma^{(4)}_{\epsilon\theta\kappa\nu} f^*_\alpha f_\delta f^*_\epsilon f_\theta f^*_\kappa f_\nu - \frac{1}{6} g^{-1}_{\alpha\delta} g^{-1}_{\epsilon\theta} g^{-1}_{\kappa\nu} f^*_\alpha f_\delta f^*_\epsilon f_\theta f^*_\kappa f_\nu \,. \tag{A.66}$$

In order to combine the terms, the second term on the right-hand side is antisymmetrized as follows (note that there are $3^2 = 9$ possibilities to permute the first two indices):

$$-\frac{1}{36} g^{-1}_{\alpha\delta} \gamma^{(4)}_{\epsilon\theta\kappa\nu} ( + f^*_\alpha f_\delta f^*_\epsilon f_\theta f^*_\kappa f_\nu - f^*_\alpha f_\theta f^*_\epsilon f_\delta f^*_\kappa f_\nu - f^*_\alpha f_\nu f^*_\epsilon f_\theta f^*_\kappa f_\delta$$

$$+ f^*_\epsilon f_\theta f^*_\alpha f_\delta f^*_\kappa f_\nu - f^*_\epsilon f_\delta f^*_\alpha f_\theta f^*_\kappa f_\nu - f^*_\epsilon f_\nu f^*_\alpha f_\delta f^*_\kappa f_\theta$$

$$+ f^*_\kappa f_\nu f^*_\alpha f_\delta f^*_\epsilon f_\theta - f^*_\kappa f_\delta f^*_\alpha f_\nu f^*_\epsilon f_\theta - f^*_\kappa f_\theta f^*_\alpha f_\delta f^*_\epsilon f_\nu) \,, \tag{A.67}$$

or

$$-\frac{1}{36} \Big( + g^{-1}_{\alpha\delta} \gamma^{(4)}_{\epsilon\theta\kappa\nu} - g^{-1}_{\alpha\theta} \gamma^{(4)}_{\epsilon\delta\kappa\nu} - g^{-1}_{\alpha\nu} \gamma^{(4)}_{\epsilon\theta\kappa\delta}$$

$$+ g^{-1}_{\epsilon\theta} \gamma^{(4)}_{\alpha\delta\kappa\nu} - g^{-1}_{\epsilon\delta} \gamma^{(4)}_{\alpha\theta\kappa\nu} - g^{-1}_{\epsilon\nu} \gamma^{(4)}_{\alpha\delta\kappa\theta}$$

$$+ g^{-1}_{\kappa\nu} \gamma^{(4)}_{\alpha\delta\epsilon\theta} - g^{-1}_{\kappa\delta} \gamma^{(4)}_{\alpha\nu\epsilon\theta} - g^{-1}_{\kappa\theta} \gamma^{(4)}_{\alpha\delta\epsilon\nu} \Big) f^*_\alpha f_\delta f^*_\epsilon f_\theta f^*_\kappa f_\nu \,. \tag{A.68}$$

Factoring out inverse Green functions yields

$$+\frac{1}{36} g^{-1}_{\alpha\beta} g^{-1}_{\epsilon\zeta} g^{-1}_{\kappa\lambda} \Big( - g_{\beta\gamma} \tilde{\chi}^{(4)}_{\zeta\eta\lambda\mu} + g_{\beta\eta} \tilde{\chi}^{(4)}_{\zeta\gamma\lambda\mu} + g_{\beta\mu} \tilde{\chi}^{(4)}_{\zeta\eta\lambda\gamma}$$

$$- g_{\zeta\eta} \tilde{\chi}^{(4)}_{\beta\gamma\lambda\mu} + g_{\zeta\gamma} \tilde{\chi}^{(4)}_{\beta\eta\lambda\mu} + g_{\zeta\mu} \tilde{\chi}^{(4)}_{\beta\gamma\lambda\eta}$$

$$- g_{\lambda\mu} \tilde{\chi}^{(4)}_{\beta\gamma\zeta\eta} + g_{\lambda\gamma} \tilde{\chi}^{(4)}_{\beta\mu\zeta\eta} + g_{\lambda\eta} \tilde{\chi}^{(4)}_{\beta\gamma\zeta\mu} \Big) g^{-1}_{\gamma\delta} g^{-1}_{\eta\theta} g^{-1}_{\mu\nu} \, f^*_\alpha f_\delta f^*_\epsilon f_\theta f^*_\kappa f_\nu \,. \tag{A.69}$$

The third term on the right-hand side of Eq. A.66 is antisymmetrized using the $3! = 6$ possible permutations of the $f$'s:

$$-\frac{1}{36} g_{\alpha\delta}^{-1} g_{\epsilon\theta}^{-1} g_{\kappa\nu}^{-1} (+ f_\alpha^* f_\delta f_\epsilon^* f_\theta f_\kappa^* f_\nu - f_\alpha^* f_\delta f_\epsilon^* f_\nu f_\kappa^* f_\theta - f_\alpha^* f_\theta f_\epsilon^* f_\delta f_\kappa^* f_\nu$$
$$- f_\alpha^* f_\nu f_\epsilon^* f_\theta f_\kappa^* f_\delta + f_\alpha^* f_\nu f_\epsilon^* f_\delta f_\kappa^* f_\theta + f_\alpha^* f_\theta f_\epsilon^* f_\nu f_\kappa^* f_\delta) \,. \tag{A.70}$$

This can be rewritten as

$$-\frac{1}{36}(+ g_{\alpha\delta}^{-1} g_{\epsilon\theta}^{-1} g_{\kappa\nu}^{-1} - g_{\alpha\delta}^{-1} g_{\epsilon\nu}^{-1} g_{\kappa\theta}^{-1} - g_{\alpha\theta}^{-1} g_{\epsilon\delta}^{-1} g_{\kappa\nu}^{-1}$$
$$- g_{\alpha\nu}^{-1} g_{\epsilon\theta}^{-1} g_{\kappa\delta}^{-1} + g_{\alpha\nu}^{-1} g_{\epsilon\delta}^{-1} g_{\kappa\theta}^{-1} + g_{\alpha\theta}^{-1} g_{\epsilon\nu}^{-1} g_{\kappa\delta}^{-1}) f_\alpha^* f_\delta f_\epsilon^* f_\theta f_\kappa^* f_\nu \,, \tag{A.71}$$

or

$$\frac{1}{36} g_{\alpha\beta}^{-1} g_{\epsilon\zeta}^{-1} g_{\kappa\lambda}^{-1} (- g_{\beta\gamma} g_{\zeta\eta} g_{\lambda\mu} + g_{\beta\gamma} g_{\zeta\mu} g_{\lambda\eta} + g_{\beta\eta} g_{\zeta\gamma} g_{\lambda\mu}$$
$$+ g_{\beta\mu} g_{\zeta\eta} g_{\lambda\gamma} - g_{\beta\mu} g_{\zeta\gamma} g_{\lambda\eta} - g_{\beta\eta} g_{\zeta\mu} g_{\lambda\gamma}) g_{\gamma\delta}^{-1} g_{\eta\theta}^{-1} g_{\mu\nu}^{-1} f_\alpha^* f_\delta f_\epsilon^* f_\theta f_\kappa^* f_\nu \,. \tag{A.72}$$

The coefficient $a^{(6)}$ can thus be expressed as

$$a_{\alpha\delta\epsilon\theta\kappa\nu}^{(6)} = \frac{1}{36} \gamma_{\alpha\delta\epsilon\theta\kappa\nu}^{(6)}, \tag{A.73}$$

where $\gamma^{(6)}$ is the fully antisymmetric three-particle impurity vertex, which is defined as

$$\boxed{\gamma_{\alpha\delta\epsilon\theta\kappa\nu}^{(6)} = g_{\alpha\beta}^{-1} g_{\epsilon\zeta}^{-1} g_{\kappa\lambda}^{-1} \tilde{\chi}_{\beta\gamma\zeta\eta\lambda\mu}^{(6)} g_{\gamma\delta}^{-1} g_{\eta\theta}^{-1} g_{\mu\nu}^{-1} \,,} \tag{A.74}$$

where in turn

$$\boxed{\begin{aligned}
\tilde{\chi}_{\beta\gamma\zeta\eta\lambda\mu}^{(6)} &= \chi_{\beta\gamma\zeta\eta\lambda\mu}^{(6)} - g_{\beta\gamma} \tilde{\chi}_{\zeta\eta\lambda\mu}^{(4)} + g_{\beta\eta} \tilde{\chi}_{\zeta\gamma\lambda\mu}^{(4)} + g_{\beta\mu} \tilde{\chi}_{\zeta\eta\lambda\gamma}^{(4)} \\
&\quad - g_{\zeta\eta} \tilde{\chi}_{\beta\gamma\lambda\mu}^{(4)} + g_{\zeta\gamma} \tilde{\chi}_{\beta\eta\lambda\mu}^{(4)} + g_{\zeta\mu} \tilde{\chi}_{\beta\gamma\lambda\eta}^{(4)} \\
&\quad - g_{\lambda\mu} \tilde{\chi}_{\beta\gamma\zeta\eta}^{(4)} + g_{\lambda\gamma} \tilde{\chi}_{\beta\mu\zeta\eta}^{(4)} + g_{\lambda\eta} \tilde{\chi}_{\beta\gamma\zeta\mu}^{(4)} \\
&\quad - g_{\beta\gamma} g_{\zeta\eta} g_{\lambda\mu} + g_{\beta\gamma} g_{\zeta\mu} g_{\lambda\eta} + g_{\beta\eta} g_{\zeta\gamma} g_{\lambda\mu} \\
&\quad + g_{\beta\mu} g_{\zeta\eta} g_{\lambda\gamma} - g_{\beta\mu} g_{\zeta\gamma} g_{\lambda\eta} - g_{\beta\eta} g_{\zeta\mu} g_{\lambda\gamma} \,.
\end{aligned}} \tag{A.75}$$

It is readily verified that this expression is antisymmetric w.r.t. to exchange of any two indices coupling to either $f$ or $f^*$, e.g. the permutations $\beta \leftrightarrow \zeta$ or $\gamma \leftrightarrow \eta$. To verify this for the terms in the first three lines one needs to use the corresponding property of $\tilde{\chi}^{(4)}$. Collecting results, the dual potential is given by

$$\boxed{V[f^*, f] = -\frac{1}{4} \gamma_{\alpha\beta\gamma\delta}^{(4)} f_\alpha^* f_\beta f_\gamma^* f_\delta + \frac{1}{36} \gamma_{\alpha\beta\gamma\delta\epsilon\zeta}^{(6)} f_\alpha^* f_\beta f_\gamma^* f_\delta f_\epsilon^* f_\zeta \mp \dots} \tag{A.76}$$

## A.4　Perturbation Theory for the Dual Propagator

In this appendix, the perturbation expansion for the dual Green function is performed explicitly and expressions for dual self-energy diagrams are derived. This illustrates the particularities of the dual perturbation theory. General rules for evaluating diagrams are given in subsequent sections. The following discussion is restricted to a few lowest-order corrections to the self-energy and to diagrams containing the vertices $\gamma^{(4)}$ and $\gamma^{(6)}$ only. Greek indices on Grassmann fields comprise all quantum numbers except position, while sites are labeled by Latin indices. The index $1 \equiv \{i_1, \alpha_1\}$ is a combined index.

### A.4.1　Perturbation Expansion

The perturbation theory is constructed starting from the definition of the dual propagator

$$G^{\mathrm{d}}_{12} := -\langle f_1 f_2^* \rangle = -\frac{\tilde{\mathcal{Z}}_f}{\mathcal{Z}} \int f_1 f_2^* \, \exp(-S_{\mathrm{d}}[f^*, f]) \, \mathcal{D}[f^*, f] \,. \tag{A.77}$$

The dual action has been shown to be (Eq. A.22):

$$S^{\mathrm{d}}[f^*, f] = -\sum_{k, \alpha\beta} f_\alpha^* (G_0^{\mathrm{d}\,-1})_{\alpha\beta} f_\beta + \sum_i V_i[f_i^*, f_i] = S_0^{\mathrm{d}}[f^*, f] + \sum_i V_i[f_i^*, f_i] \,. \tag{A.78}$$

The perturbation series is generated by expanding the exponential appearing under the path integral $-(\tilde{\mathcal{Z}}_f / \mathcal{Z}) \int f_1 f_2^* \exp(-S_0^{\mathrm{d}} - \sum_i V_i) \mathcal{D}[f^*, f]$ in the interaction:

$$\exp\left( -\sum_i V_i[f_i^*, f_i] \right) = 1 - \sum_i V_i[f_i^*, f_i] + \frac{1}{2!}\left( \sum_i \sum_j V_i[f_i^*, f_i] V_j[f_j^*, f_j] \right)$$

$$- \frac{1}{3!}\left( \sum_i \sum_j \sum_k V_i[f_i^*, f_i] V_j[f_j^*, f_j] V_k[f_k^*, f_k] \right) + \dots \tag{A.79}$$

The zero-order term yields the bare dual Green function:

$$(G_0^{\mathrm{d}})_{12} = -\frac{\tilde{\mathcal{Z}}_f}{\mathcal{Z}} \int f_1 f_2^* \, \exp(-S_0^{\mathrm{d}}[f^*, f]) \, \mathcal{D}[f^*, f] \,. \tag{A.80}$$

with $S_0^{\mathrm{d}}[f^*, f] = -\sum_{\alpha\beta} f_\alpha^* (G_0^{\mathrm{d}})^{-1}_{\alpha\beta} f_\beta$. The next order gives two local contributions for the Green function $G^{\mathrm{d}}_{12}$, one from each of the vertices:

$$\left( -\frac{1}{4} \right) \sum_i \gamma^{(4)}_{i\alpha\beta\gamma\delta} \int f_1 f_2^* f_{i\alpha}^* f_{i\beta} f_{i\gamma}^* f_{i\delta} \, \exp(-S_0^{\mathrm{d}}[f^*, f]) \, \mathcal{D}[f^*, f] \,,$$

$$\left( \frac{1}{36} \right) \sum_i \gamma^{(6)}_{i\alpha\beta\gamma\delta\epsilon\zeta} \int f_1 f_2^* f_{i\alpha}^* f_{i\beta} f_{i\gamma}^* f_{i\delta} f_{i\epsilon}^* f_{i\zeta} \, \exp(-S_0^{\mathrm{d}}[f^*, f]) \, \mathcal{D}[f^*, f] \,.$$

$$\tag{A.81}$$

The second-order terms (only those involving $\gamma^{(4)}$ and $\gamma^{(6)}$) are:

$$-\frac{1}{2!}\left(-\frac{1}{4}\right)^2 \sum_i \sum_j \gamma^{(4)}_{i\alpha\beta\gamma\delta}\gamma^{(4)}_{j\kappa\lambda\mu\nu} \times$$

$$\times \int f_1 f_2^* f_{i\alpha}^* f_{i\beta} f_{i\gamma}^* f_{i\delta} f_{j\kappa}^* f_{j\lambda} f_{j\mu}^* f_{j\nu} \, \exp(-S_0^{\mathrm{d}}[f^*,f])\, \mathcal{D}[f^*,f]\,, \tag{A.82}$$

$$-\frac{1}{2!}\left(-\frac{1}{4}\right)\left(\frac{1}{36}\right) \sum_i \sum_j \gamma^{(4)}_{i\alpha\beta\gamma\delta}\gamma^{(6)}_{j\kappa\lambda\mu\nu\epsilon\zeta} \times$$

$$\times \int f_1 f_2^* f_{i\alpha}^* f_{i\beta} f_{i\gamma}^* f_{i\delta} f_{j\kappa}^* f_{j\lambda} f_{j\mu}^* f_{j\nu} f_{j\epsilon}^* f_{j\zeta} \, \exp(-S_0^{\mathrm{d}}[f^*,f])\, \mathcal{D}[f^*,f]\,, \tag{A.83}$$

$$-\frac{1}{2!}\left(-\frac{1}{4}\right)\left(\frac{1}{36}\right) \sum_i \sum_j \gamma^{(6)}_{i\kappa\lambda\mu\nu\epsilon\zeta}\gamma^{(4)}_{j\alpha\beta\gamma\delta} \times$$

$$\times \int f_1 f_2^* f_{i\kappa}^* f_{i\lambda} f_{i\mu}^* f_{i\nu} f_{i\epsilon}^* f_{i\zeta} f_{j\alpha}^* f_{j\beta} f_{j\gamma}^* f_{j\delta} \, \exp(-S_0^{\mathrm{d}}[f^*,f])\, \mathcal{D}[f^*,f]\,, \tag{A.84}$$

$$-\frac{1}{2!}\left(\frac{1}{36}\right)^2 \sum_i \sum_j \gamma^{(6)}_{i\alpha\beta\gamma\delta\epsilon\zeta}\gamma^{(6)}_{j\kappa\lambda\mu\nu\rho\eta} \times$$

$$\times \int f_1 f_2^* f_{i\alpha}^* f_{i\beta} f_{i\gamma}^* f_{i\delta} f_{i\epsilon}^* f_{i\zeta} f_{j\kappa}^* f_{j\lambda} f_{j\mu}^* f_{j\nu} f_{j\rho}^* f_{j\eta} \, \exp(-S_0^{\mathrm{d}}[f^*,f])\, \mathcal{D}[f^*,f]\,. \tag{A.85}$$

Likewise, the third-order terms are the following:

$$+\frac{1}{3!}\left(-\frac{1}{4}\right)^3 \sum_i \sum_j \sum_k \gamma^{(4)}_{i\alpha\beta\gamma\delta}\gamma^{(4)}_{j\kappa\lambda\mu\nu}\gamma^{(4)}_{k\epsilon\zeta\rho\eta} \times$$

$$\times \int f_1 f_2^* f_{i\alpha}^* f_{i\beta} f_{i\gamma}^* f_{i\delta} f_{j\kappa}^* f_{j\lambda} f_{j\mu}^* f_{j\nu} f_{k\epsilon}^* f_{k\zeta} f_{k\rho}^* f_{k\eta} \, \exp(-S_0^{\mathrm{d}}[f^*,f])\, \mathcal{D}[f^*,f]\,, \tag{A.86}$$

$$+3 \times \frac{1}{3!}\left(-\frac{1}{4}\right)^2\left(\frac{1}{36}\right) \sum_i \sum_j \sum_k \gamma^{(4)}_{i\alpha\beta\gamma\delta}\gamma^{(4)}_{j\kappa\lambda\mu\nu}\gamma^{(6)}_{k\epsilon\zeta\rho\eta\varphi\vartheta} \times$$

$$\times \int f_1 f_2^* f_{i\alpha}^* f_{i\beta} f_{i\gamma}^* f_{i\delta} f_{j\kappa}^* f_{j\lambda} f_{j\mu}^* f_{j\nu} f_{k\epsilon}^* f_{k\zeta} f_{k\rho}^* f_{k\eta} f_{k\varphi}^* f_{k\vartheta} \, \exp(-S_0^{\mathrm{d}}[f^*,f])\, \mathcal{D}[f^*,f]\,, \tag{A.87}$$

$$+3 \times \frac{1}{3!}\left(-\frac{1}{4}\right)\left(\frac{1}{36}\right)^2 \sum_i \sum_j \sum_k \gamma^{(4)}_{i\alpha\beta\gamma\delta}\gamma^{(6)}_{j\kappa\lambda\mu\nu\epsilon\zeta}\gamma^{(6)}_{k\rho\eta\varphi\vartheta\psi\omega} \times$$

$$\times \int f_1 f_2^* f_{i\alpha}^* f_{i\beta} f_{i\gamma}^* f_{i\delta} f_{j\kappa}^* f_{j\lambda} f_{j\mu}^* f_{j\nu} f_{j\epsilon}^* f_{j\zeta} f_{k\rho}^* f_{k\eta} f_{k\varphi}^* f_{k\vartheta} f_{k\psi}^* f_{k\omega} \, \exp(-S_0^{\mathrm{d}}[f^*,f])\, \mathcal{D}[f^*,f]\,,$$

$$\tag{A.88}$$

$$+\frac{1}{3!}\left(\frac{1}{36}\right)^3 \sum_i \sum_j \sum_k \gamma^{(6)}_{i\alpha\beta\gamma\delta\epsilon\zeta}\gamma^{(6)}_{j\kappa\lambda\mu\nu\rho\eta}\gamma^{(6)}_{k\varphi\vartheta\psi\omega\sigma\xi}\times$$

$$\times \int f_1 f_2^* f_{i\alpha}^* f_{i\beta} f_{i\gamma}^* f_{i\delta} f_{i\epsilon}^* f_{i\zeta} f_{j\kappa}^* f_{j\lambda} f_{j\mu}^* f_{j\nu} f_{j\rho}^* f_{j\eta} f_{k\varphi}^* f_{k\vartheta} f_{k\psi}^* f_{k\omega} f_{k\sigma}^* f_{k\xi}\,\exp(-S_0^{\mathrm{d}}[f^*,f])\,\mathcal{D}[f^*,f]\,.$$

$$(A.89)$$

## A.4.2   Self-Energy Diagrams

The first few lowest-order diagrammatic contributions to the dual self-energy are given in the following. Self-energy diagrams are obtained from the Green function diagrams by removing the external lines. As usual, only connected, irreducible diagrams are considered. Since in regular dual fermion calculations the local part of the dual Green function is taken to be zero, diagrams which contain a local propagator are omitted. An exception is diagram a), which is needed for the self-consistent determination of the hybridization function. Other diagrams can be derived in an analogous fashion using the rules given in Secs. A.4.3, A.4.4. The combinatorial prefactor of a diagram is the prefactor of the contribution times the number of pairings that lead to the same (topologically equivalent) diagram, as exemplified for diagram b). In the following, the index '0' on Green functions is omitted for brevity and because the same expressions also apply for the corresponding skeleton diagrams. As a convention, vertices are labelled counterclockwise. A contraction is defined as $f_{i\alpha}f_{j\beta}^* = -G^{\mathrm{d}}_{ij\,\alpha\beta}$.

- Diagram a)

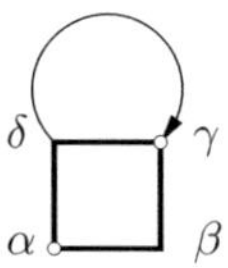


This diagram is derived from the term

$$\left(-\frac{1}{4}\right)\sum_i \gamma^{(4)}_{i\alpha\beta\gamma\delta} \int f_1 f_2^* f_{i\alpha}^* f_{i\beta} f_{i\gamma}^* f_{i\delta}\,\exp(-S_0^{\mathrm{d}}[f^*,f])\,\mathcal{D}[f^*,f]\,.$$

The pairing that corresponds to this diagram is

$$+f_1 f_2^* f_{i\alpha}^* f_{i\beta} f_{i\gamma}^* f_{i\delta} = -f_1 f_{i\alpha}^* f_{i\delta} f_{i\gamma}^* f_{i\beta} f_2^* = (-1)^4\,G^{\mathrm{d}}_{1\,i\alpha} G^{\mathrm{d}}_{i\beta\,2} G^{\mathrm{d}}_{ii\delta\gamma}\,.\qquad (A.90)$$

There are 4 pairings which give the same diagram. The corresponding correction
to the self-energy is thus

$$\Sigma^{(a)}_{ii\,\alpha\beta} = -\gamma^{(4)}_{i\alpha\beta\gamma\delta}G^{d}_{ii\delta\gamma}\,.$$

(A.91)

- Diagram b)

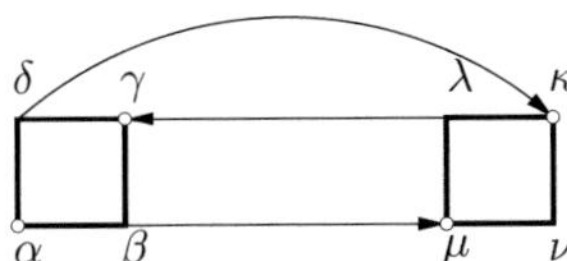


The relevant term in the perturbation expansion is:

$$-\frac{1}{2!}\left(-\frac{1}{4}\right)^{2}\sum_{i}\sum_{j}\gamma^{(4)}_{i\alpha\beta\gamma\delta}\gamma^{(4)}_{jk\lambda\mu\nu}\int f_{1}f_{2}^{*}f_{i\alpha}^{*}f_{i\beta}f_{i\gamma}^{*}f_{i\delta}f_{jk}^{*}f_{j\lambda}f_{j\mu}^{*}f_{j\nu}\times$$

$$\times\exp(-S_{0}^{d}[f^{*},f])\,\mathcal{D}[f^{*},f]\,.$$

(A.92)

The particular pairing corresponding to this diagram is:

$$+f_{1}f_{2}^{*}f_{i\alpha}^{*}f_{i\beta}f_{i\gamma}^{*}f_{i\delta}f_{jk}^{*}f_{j\lambda}f_{j\mu}^{*}f_{j\nu} = -f_{1}f_{i\alpha}^{*}f_{i\beta}f_{j\mu}^{*}f_{j\lambda}f_{i\gamma}^{*}f_{i\delta}f_{jk}^{*}f_{j\nu}f_{2}^{*}$$

$$= (-1)^{6}\,G^{d}_{1\,i\alpha}G^{d}_{ij\beta\mu}G^{d}_{ji\lambda\gamma}G^{d}_{ij\delta\kappa}G^{d}_{i\nu\,2}\,.$$

(A.93)

There are 16 different pairings that correspond to the same diagram. This can
be seen as follows: Draw two squares corresponding to the two vertices. There
are four possibilities to connect an incoming line (two on each vertex). After
attaching the incoming line to say, $\alpha$, there are two possibilities to attach the
outgoing line, since it must be connected to the other vertex to yield the desired
diagram. Connect this line to $\nu$. Now there are two more possibilities to draw a
directed line connecting the two vertices: A line going from the left to the right
vertex can only be connected to one point on the left vertex, but to two on the right
one. After this line is connected, say from $\gamma$ to $\lambda$, there is only one possibility
to connect the remaining two lines. The number of equivalent pairings is thus
$4 \cdot 2 \cdot 2 = 16$. The correction to the self-energy is hence given by

$$\Sigma^{(b)}_{ij\,\alpha\nu} = -\frac{1}{2}\gamma^{(4)}_{i\alpha\beta\gamma\delta}\gamma^{(4)}_{jk\lambda\mu\nu}G^{d}_{ij\beta\mu}G^{d}_{ji\lambda\gamma}G^{d}_{ij\delta\kappa}\,.$$

(A.94)

- Diagram c)

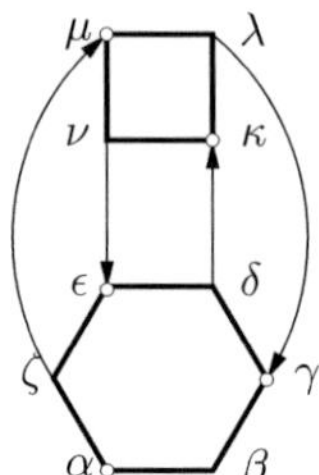

Corresponding contribution (the additional factor 2 stems from the two equivalent contributions corresponding to the two possible orders of the vertices):

$$-2 \times \frac{1}{2!}\left(-\frac{1}{4}\right)\left(\frac{1}{36}\right)\sum_i \sum_j \gamma^{(4)}_{i\alpha\beta\gamma\delta}\,\gamma^{(6)}_{j\kappa\lambda\mu\nu\epsilon\zeta} \times$$

$$\times \int f_1 f_2^* f_{i\alpha}^* f_{i\beta} f_{i\gamma}^* f_{i\delta} f_{i\epsilon}^* f_{i\zeta} f_{j\kappa}^* f_{j\lambda} f_{j\mu}^* f_{j\nu}\, \exp(-S_0^{\mathrm{d}}[f^*,f])\,\mathcal{D}[f^*,f]\,.$$

$$\text{(A.95)}$$

Corresponding pairing:

$$f_1 f_2^* f_{i\alpha}^* f_{i\beta} f_{i\gamma}^* f_{i\delta} f_{i\epsilon}^* f_{i\zeta} f_{j\kappa}^* f_{j\lambda} f_{j\mu}^* f_{j\nu} = +f_1 f_{i\alpha}^* f_{j\lambda} f_{i\gamma}^* f_{i\delta} f_{j\kappa}^* f_{j\nu} f_{i\epsilon}^* f_{i\zeta} f_{j\mu}^* f_{i\beta} f_2^*$$

$$= (-1)^6 G^{\mathrm{d}}_{1\,i\alpha} G^{\mathrm{d}}_{ji\lambda\gamma} G^{\mathrm{d}}_{ij\delta\kappa} G^{\mathrm{d}}_{ji\nu\epsilon} G^{\mathrm{d}}_{ij\zeta\mu} G^{\mathrm{d}}_{i\beta\,2}\,. \quad \text{(A.96)}$$

Number of equivalent pairings: 36. Self-energy correction:

$$\Sigma^{(c)}_{ii\,\alpha\beta} = \left(\frac{1}{4}\right)\sum_j \gamma^{(6)}_{i\alpha\beta\gamma\delta\epsilon\zeta}\,\gamma^{(4)}_{j\kappa\lambda\mu\nu}\,G^{\mathrm{d}}_{ji\lambda\gamma} G^{\mathrm{d}}_{ij\delta\kappa} G^{\mathrm{d}}_{ji\nu\epsilon} G^{\mathrm{d}}_{ij\zeta\mu}\,. \qquad \text{(A.97)}$$

- Diagram d)

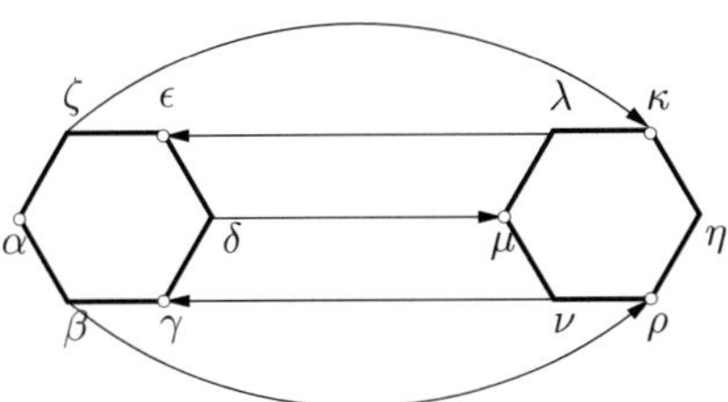


Corresponding contribution

$$-\frac{1}{2!}\left(\frac{1}{36}\right)^2 \sum_i \sum_j \gamma^{(6)}_{i\alpha\beta\gamma\delta\epsilon\zeta}\gamma^{(6)}_{jk\lambda\mu\nu\rho\eta} \int f_1 f_2^* f_{i\alpha}^* f_{i\beta} f_{i\gamma}^* f_{i\delta} f_{i\epsilon}^* f_{i\zeta} f_{jk}^* f_{j\lambda} f_{j\mu}^* f_{j\nu} f_{j\rho}^* f_{j\eta} \times$$

$$\times \exp(-S_0^{\mathrm{d}}[f^*,f])\,\mathcal{D}[f^*,f]\,. \qquad (\mathrm{A.98})$$

Corresponding pairing:

$$f_1 f_{i\alpha}^* f_{i\beta} f_{i\gamma}^* f_{i\delta} f_{i\epsilon}^* f_{i\zeta} f_{jk}^* f_{j\lambda} f_{j\mu}^* f_{j\nu} f_{j\rho}^* f_{j\eta} f_2^*$$

$$= +f_1 f_{i\alpha}^* f_{i\beta} f_{j\rho}^* f_{j\nu} f_{i\gamma}^* f_{i\delta} f_{j\mu}^* f_{j\lambda} f_{i\epsilon}^* f_{i\zeta} f_{jk}^* f_{j\eta} f_2^*$$

$$=(-1)^7 G^{\mathrm{d}}_{1\,i\alpha} G^{\mathrm{d}}_{ij\beta\rho} G^{\mathrm{d}}_{jiv\gamma} G^{\mathrm{d}}_{ij\delta\mu} G^{\mathrm{d}}_{ji\lambda\epsilon} G^{\mathrm{d}}_{ij\zeta k} G^{\mathrm{d}}_{j\eta\,2}\,. \qquad (\mathrm{A.99})$$

Number of equivalent pairings: 216. Self-energy correction:

$$\Sigma^{(d)}_{ij\,\alpha\eta} = \left(\frac{1}{12}\right)\gamma^{(6)}_{i\alpha\beta\gamma\delta\epsilon\zeta}\gamma^{(6)}_{jk\lambda\mu\nu\rho\eta} G^{\mathrm{d}}_{ij\beta\rho} G^{\mathrm{d}}_{jiv\gamma} G^{\mathrm{d}}_{ij\delta\mu} G^{\mathrm{d}}_{ji\lambda\epsilon} G^{\mathrm{d}}_{ij\zeta k}\,. \qquad (\mathrm{A.100})$$

- Diagram e)

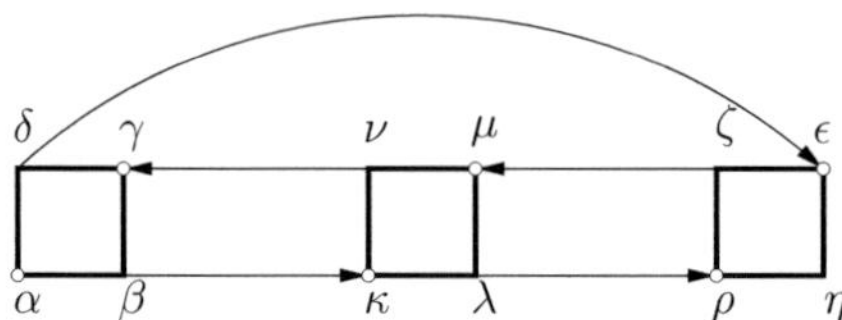

Corresponding contribution

$$+\frac{1}{3!}\left(-\frac{1}{4}\right)^3 \sum_i \sum_j \sum_k \gamma^{(4)}_{i\alpha\beta\gamma\delta}\gamma^{(4)}_{jk\lambda\mu\nu}\gamma^{(4)}_{k\epsilon\zeta\rho\eta} \int f_1 f_2^* f_{i\alpha}^* f_{i\beta} f_{i\gamma}^* f_{i\delta} f_{jk}^* f_{j\lambda} f_{j\mu}^* f_{j\nu} f_{k\epsilon}^* f_{k\zeta} f_{k\rho}^* f_{k\eta} \times$$
$$\times \exp(-S_0^{\mathrm{d}}[f^*, f])\, \mathcal{D}[f^*, f]\,.$$

$$(A.101)$$

Corresponding pairing:

$$f_1 f_{i\alpha}^* f_{i\beta} f_{i\gamma}^* f_{i\delta} f_{jk}^* f_{j\lambda} f_{j\mu}^* f_{j\nu} f_{k\epsilon}^* f_{k\zeta} f_{k\rho}^* f_{k\eta} f_2^*$$

$$= -f_1 f_{i\alpha}^* f_{i\beta} f_{jk}^* f_{j\nu} f_{i\gamma}^* f_{i\delta} f_{k\epsilon}^* f_{j\lambda} f_{k\rho}^* f_{k\zeta} f_{j\mu}^* f_{k\eta} f_2^*$$

$$= (-1)^8 G_{1\alpha}^{\mathrm{d}} G_{ij\beta\kappa}^{\mathrm{d}} G_{jiv\gamma}^{\mathrm{d}} G_{ik\delta\epsilon}^{\mathrm{d}} G_{jk\lambda\rho}^{\mathrm{d}} G_{kj\zeta\mu}^{\mathrm{d}} G_{\eta 2}^{\mathrm{d}}\,.$$

$$(A.102)$$

Number of equivalent pairings: 384. Self-energy correction:

$$\Sigma^{(e)}_{ik\,\alpha\eta} = (-1) \sum_j \gamma^{(4)}_{i\alpha\beta\gamma\delta}\gamma^{(4)}_{jk\lambda\mu\nu}\gamma^{(4)}_{k\epsilon\zeta\rho\eta} G_{ij\beta\kappa}^{\mathrm{d}} G_{jiv\gamma}^{\mathrm{d}} G_{ik\delta\epsilon}^{\mathrm{d}} G_{jk\lambda\rho}^{\mathrm{d}} G_{kj\zeta\mu}^{\mathrm{d}}\,.$$

$$(A.103)$$

- Diagram f)

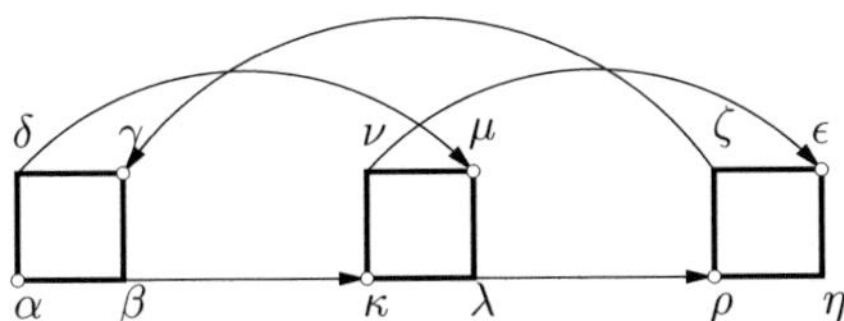

The diagram stems from the same contribution as diagram e).

Corresponding pairing:

$$f_1 f^*_{i\alpha} f_{i\beta} f^*_{i\gamma} f_{i\delta} f^*_{jk} f_{j\lambda} f^*_{j\mu} f_{j\nu} f^*_{k\epsilon} f_{k\zeta} f^*_{k\rho} f_{k\eta} f^*_2$$

$$= -f_1 f^*_{i\alpha} f_{i\beta} f^*_{jk} f_{k\zeta} f^*_{i\gamma} f_{i\delta} f^*_{j\mu} f_{j\lambda} f^*_{k\rho} f_{j\nu} f^*_{k\epsilon} f_{k\eta} f^*_2$$

$$= (-1)^8 G^{\mathrm{d}}_{1\,i\alpha} G^{\mathrm{d}}_{ij\beta k} G^{\mathrm{d}}_{ki\zeta\gamma} G^{\mathrm{d}}_{ij\delta\mu} G^{\mathrm{d}}_{jk\lambda\rho} G^{\mathrm{d}}_{jk\nu\epsilon} G^{\mathrm{d}}_{k\eta\,2} \; .$$

$$(A.104)$$

Number of equivalent pairings: 96. Self-energy correction:

$$\Sigma^{(f)}_{ik\,\alpha\eta} = \left(-\frac{1}{4}\right) \sum_j \gamma^{(4)}_{i\alpha\beta\gamma\delta} \gamma^{(4)}_{jk\lambda\mu\nu} \gamma^{(4)}_{k\epsilon\zeta\rho\eta} G^{\mathrm{d}}_{ij\beta k} G^{\mathrm{d}}_{ki\zeta\gamma} G^{\mathrm{d}}_{ij\delta\mu} G^{\mathrm{d}}_{jk\lambda\rho} G^{\mathrm{d}}_{jk\nu\epsilon} \; . \qquad (A.105)$$

Note that this diagram can be deformed as shown below by redefinition of the vertex (the prefactor remains the same). It describes renormalization of the self-energy by scattering of particle-particle pairs.

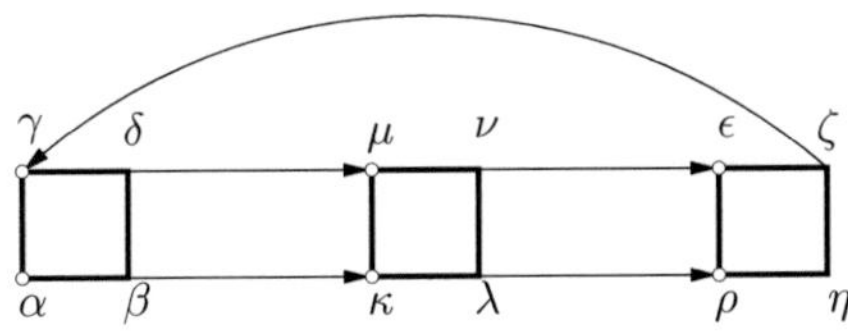

## A.4.3 Determination of Combinatorial Prefactors

The determination of the combinatorial prefactors by counting the number of equivalent pairings becomes cumbersome for diagrams at high orders. A closer analysis gives the following general rules to determine these prefactors: First consider diagrams which contain no equivalent lines (i.e. equally directed lines connecting to the same or same two vertices). This is the case for diagrams a) and e). The prefactor of such a diagram is unity at any order of the perturbation theory. In order to see this, recall that the prefactor of each vertex is $1/[(n/2)!]^2$, where $n$ is the number of edges (see Sec. A.3). This is exactly the number of possibilities to attach lines to the vertex. An additional factor $1/m!$ arises from the expansion of the exponential, where $m$ is the perturbation order. Attaching a label to each vertex of one sort (e.g. two-particle vertices) to make them distinguishable, one sees that all $m!$ permutations appear in a complete contraction. If vertices of different sorts are present in a diagram, the factor corresponding to the permutation of these vertices among themselves explicitly appears in the expansion (e.g. 2! for diagram c). Hence a diagram corresponds to the sum of $1/[(n/2)!]^2 m!$ diagrams with the same value, so that the prefactor exactly cancels.

This only holds if all ways of attaching the lines or permuting the vertices yield a different, *distinguishable* diagram. If for example two vertices are connected by $k$ equivalent lines, this reduces the number of distinguishable diagrams (pairings) by the number of permutations of these lines, since a permutation yields the identical, distinguishable diagram. Hence the prefactor is cancelled only up to a factor $1/k!$ for each set of $k$ equivalent lines connecting the same two vertices. For example, there are two equivalent lines going from left to right and three parallel lines from right to left in diagram d). Hence the prefactor is $1/2! \cdot 1/3! = 1/12$. For the vacuum to vacuum diagrams contributing to the Luttinger-Ward functional $\Phi$ used in chapter 11 one needs to account for additional symmetry factors. Noting that for the generic $n$-th order diagram depicted in Fig. A.1, each of the $2n$ cyclic permutations of the sequences $(1, 2, \ldots n)$ and $(n, \ldots, 2, 1)$ corresponds to the same distinguishable diagram, the symmetry factor of this diagram is $1/(2n)$. The symmetry factor is obviously unity for self-energy diagrams. The diagrammatic rules for the dual propagator are similar to those for Hugenholtz diagrams [177].

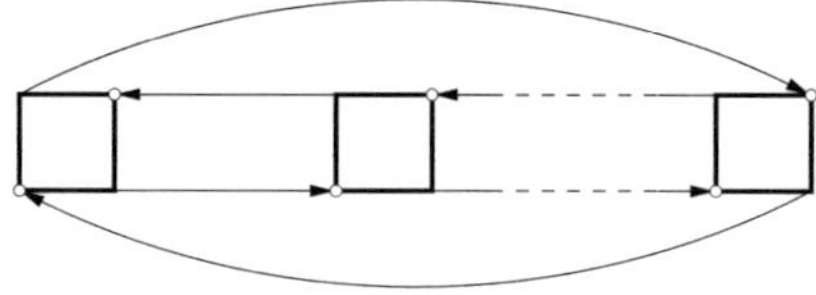

*Figure A.1:* Generic $n$-th order ring diagram contributing to the Luttinger-Ward functional $\Phi$.

### A.4.4 Determination of the Sign

Here the rules to determine the sign of a given diagram are given for the case where the dual potential is truncated after the two-particle interaction term.

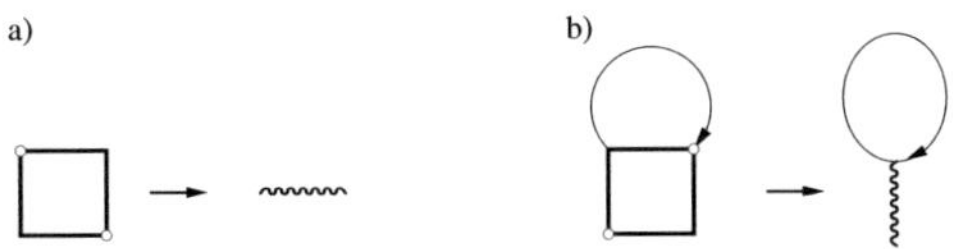

A priori (i.e. regardless of the particular pairing), the contribution to a diagram for Green's function (or the self-energy) has positive sign. A sign of $(-1)^n$, where $n$ is the order or the number of vertices arises from the expansion of the exponential. This sign cancels at any order due to the negative prefactor $-1/4$ for each vertex. In each contribution, the Grassmann numbers appear as $-ff^*f^*f\ldots f^*f$, where the sign is due to the definition of Green's function, $G = -\langle ff^* \rangle$. Reversing the order of pairs to form a complete contraction and recalling that a contraction of two Grassmann numbers is defined as $ff^* = -G$, one sees that a diagram has positive sign. An overall sign of $(-1)^{n_L}$ of a diagram arises due to $n_L$ closed fermion loops, as in standard perturbation theory. It is however not obvious how to count the loops for a given diagram in the antisymmetrized technique. This can be resolved by comparing with the *un*symmetrized technique, where the interaction has the form $U_{\alpha\beta\gamma\delta} f_\alpha^* f_\beta f_\gamma^* f_\delta = U\delta_{\alpha\beta}\delta_{\gamma\delta} f_\alpha^* f_\beta f_\gamma^* f_\delta$, which can be represented by a wiggly line as in a) in the figure above. Since the order of Grassmann variables is the same for the interaction $\gamma_{\alpha\beta\gamma\delta} f_\alpha^* f_\beta f_\gamma^* f_\delta$, the sign of a diagram is obtained by replacing the square by a wiggly line as shown in b) and counting the number of closed loops (a single loop in this example). The relative orientation of the line and the square must be kept fixed, e.g. both the line and the square in b) have been rotated counterclockwise by $\pi/2$ with respect to a).

## A.5 Vertex $\gamma^{(6)}$

For a memory-efficient implementation of $\gamma^{(6)}$, the vertex is stored for five independent frequencies only. Using the fully antisymmetric structure, only two different spin-configurations need to be stored in memory for the paramagnetic case, namely $\uparrow\uparrow\uparrow\uparrow\uparrow\uparrow$ and $\uparrow\uparrow\uparrow\uparrow\downarrow\downarrow$. All other non-zero spin configurations can be obtained from these two by a suitable permutation of the frequency (and orbital) arguments. In general, the vertex has $2^6 = 64$ distinct spin-configurations. However, taking spin-conservation for scattering processes into account, these are limited to a total of 20 non-vanishing, non-equivalent contributions (or 10, respectively, for the paramagnetic case).

## A.6 Relation Between $G$ and $G^{\mathrm{d}}$

In this section a relation between the Green functions of lattice and dual fermions is established. Start from the two equivalent representations of the partition function, Eqs. A.13, A.16:

$$\mathcal{Z} = \int \exp(-S[c^*, c])\mathcal{D}[c^*, c]$$
$$= \mathcal{Z}_f \int \int \exp(-S[c^*, c; f^*, f])\mathcal{D}[c^*, c; f^*, f] , \tag{A.106}$$

where $\mathcal{Z}_f = \det[g(\Delta - h)g] = \tilde{\mathcal{Z}}_f/C$ with $C = \prod_i \mathcal{Z}_{\mathrm{imp}}^{(i)}$. One may obtain the Green function by viewing the bare Hamiltonian as a general matrix and regarding it as the source (instead of introducing additional source terms). The Green function is then obtained by functional derivative of the partition function with respect to the transpose Hamiltonian:

$$-\frac{\delta \mathcal{Z}[h]}{\delta h_{\delta\gamma}} = -\frac{\delta}{\delta h_{\delta\gamma}} \int \exp\Big(\sum_{\alpha\beta} c_\alpha^* [(i\omega + \mu)\mathbb{1} - h]_{\alpha\beta} c_\beta - H_{\mathrm{int}}[c^*, c]\Big)\mathcal{D}[c^*, c]$$

$$= \int \sum_{\alpha\beta} c_\alpha^* \underbrace{\frac{\delta h_{\alpha\beta}}{\delta h_{\delta\gamma}}}_{\delta_{\alpha\delta}\delta_{\beta\gamma}} c_\beta \exp(-S[c^*, c])\mathcal{D}[c^*, c]$$

$$= -\int c_\gamma c_\delta^* \exp(-S[c^*, c])\mathcal{D}[c^*, c] = \mathcal{Z} G_{\gamma\delta} . \tag{A.107}$$

Hence $G_{\gamma\delta} = (-1/\mathcal{Z})\delta \mathcal{Z}/\delta h_{\delta\gamma}$. Now consider the right-hand side of Eq. A.106:

$$-\frac{1}{\mathcal{Z}}\frac{\delta \mathcal{Z}[h]}{\delta h_{\lambda\kappa}} = -\frac{1}{\mathcal{Z}}\left[\left(\frac{\delta \mathcal{Z}_f[h]}{\delta h_{\lambda\kappa}}\right)\frac{\mathcal{Z}}{\mathcal{Z}_f} - \mathcal{Z}_f \frac{\delta}{\delta h_{\lambda\kappa}}\left(\frac{\mathcal{Z}}{\mathcal{Z}_f}[h]\right)\right] , \tag{A.108}$$

where the short-hand notation $\mathcal{Z}/\mathcal{Z}_f = \int \int \exp(-S[c^*, c; f^*, f])\mathcal{D}[c^*, c; f^*, f]$ was used. The functional derivative $\delta \mathcal{Z}_f/\delta h_{\lambda\kappa} = \delta/\delta h_{\lambda\kappa} \det[g(\Delta - h)g]$ is evaluated using an identity from linear algebra for the differential of a determinant of a general matrix $A$: $\mathrm{d}\det A = \det A \, \mathrm{Tr}[A^{-1}\mathrm{d}A]$. This gives

$$-\frac{\delta \mathcal{Z}_f}{\delta h_{\lambda\kappa}} = \det[g(\Delta - h)g]\,\mathrm{Tr}\left[g^{-1}(\Delta - h)^{-1}g^{-1}\frac{g\,\delta h\,g}{\delta h_{\lambda\kappa}}\right]$$

$$= \mathcal{Z}_f \left[g^{-1}(\Delta - h)^{-1}g^{-1}\right]_{\delta\alpha} g_{\alpha\beta}\frac{\delta h_{\beta\gamma}}{\delta h_{\lambda\kappa}}g_{\gamma\delta}$$

$$= \mathcal{Z}_f\, g_{\delta\epsilon}^{-1}\left[(\Delta - h)^{-1}\right]_{\epsilon\zeta} g_{\zeta\alpha}^{-1}\, g_{\alpha\lambda}\, g_{\kappa\delta} = \mathcal{Z}_f \left[(\Delta - h)^{-1}\right]_{\kappa\lambda} . \tag{A.109}$$

The second term on the right-hand side of (A.108) is rewritten as follows:

$$-\mathcal{Z}_f \frac{\delta}{\delta h_{\lambda\kappa}}\left(\frac{\mathcal{Z}}{\mathcal{Z}_f}[h]\right) = -\mathcal{Z}_f \frac{\delta}{\delta h_{\lambda\kappa}} \int \int \exp(-S_{\mathrm{site}}[c^*,c;f^*,f])\times$$
$$\times \exp\left(-\sum_{\alpha\beta} f_\alpha^* \left[g^{-1}(\Delta-h)^{-1}g^{-1}\right]_{\alpha\beta} f_\beta\right)\mathcal{D}[c^*,c;f^*,f]$$
$$= \mathcal{Z}_f \int \int \left\{ f_\alpha^* \frac{\delta}{\delta h_{\lambda\kappa}}\left[g^{-1}(\Delta-h)^{-1}g^{-1}\right]_{\alpha\beta} f_\beta \right\} \times$$
$$\times \exp(-S[c^*,c;f^*,f])\mathcal{D}[c^*,c;f^*,f] \ .$$
$$(A.110)$$

The term in curly brackets is evaluated using an identity for the variation of the inverse of a matrix with respect to a parameter:

$$\frac{\partial A^{-1}}{\partial a_{\lambda\kappa}} = -A^{-1}\frac{\partial A}{\partial a_{\lambda\kappa}}A^{-1} \ . \tag{A.111}$$

Applying this to the inverse matrix $(\Delta-h)^{-1}$ yields

$$f_\alpha^* \frac{\delta}{\delta h_{\lambda\kappa}}\left[g^{-1}(\Delta-h)^{-1}g^{-1}\right]_{\alpha\beta} f_\beta$$
$$= f_\alpha^*\left[g^{-1}(\Delta-h)^{-1}\right]_{\alpha\gamma}\frac{\delta h_{\gamma\delta}}{\delta h_{\lambda\kappa}}\left[(\Delta-h)^{-1}g^{-1}\right]_{\delta\beta} f_\beta$$
$$= -\left[g^{-1}(\Delta-h)^{-1}\right]_{\kappa\beta} f_\beta f_\alpha^*\left[(\Delta-h)^{-1}g^{-1}\right]_{\alpha\lambda} \ . \tag{A.112}$$

Inserting this back into (A.110) and using the definition of the dual Green function[2]

$$G_{\kappa\lambda}^d = -\frac{\mathcal{Z}_f}{\mathcal{Z}} \int \int f_\kappa f_\lambda^* \exp(-S[c^*,c;f^*,f])\,\mathcal{D}[c^*,c;f^*,f] \ , \tag{A.113}$$

yields

$$-\mathcal{Z}_f\frac{\delta}{\delta h_{\lambda\kappa}}\left(\frac{\mathcal{Z}}{\mathcal{Z}_f}[h]\right) = \mathcal{Z}\left\{\left[(\Delta-h)^{-1}g^{-1}\right] G^{\mathrm{d}}\left[g^{-1}(\Delta-h)^{-1}\right]\right\} \ . \tag{A.114}$$

Inserting the results (A.109, A.114) back into Eq. A.108, leads to the following matrix relation between the dual and lattice Green functions:

$$\boxed{G = (\Delta-h)^{-1} + \left[(\Delta-h)^{-1}g^{-1}\right] G^{\mathrm{d}}\left[g^{-1}(\Delta-h)^{-1}\right] \ .} \tag{A.115}$$

---

[2]Note the different prefactor $\mathcal{Z}_f/\mathcal{Z}$ compared to (A.77), since the lattice fermions have not been integrated out from the action $S[c^*,c;f^*,f]$.

## A.7   Relation Between $\chi$ and $\chi^{\mathrm{d}}$

The two-particle Green function of the lattice fermions is obtained as the second functional derivative with respect to the transpose Hamiltonian. Using the results of the previous section, one has

$$
\begin{aligned}
\frac{1}{\mathcal{Z}} \frac{\delta^2 \mathcal{Z}[h]}{\delta h_{\nu\mu} \delta h_{\lambda\kappa}} &= -\frac{1}{\mathcal{Z}} \int c_\lambda^* c_\kappa \frac{\delta}{\delta h_{\nu\mu}} \exp(-S[c^*,c]) \, \mathcal{D}[c^*,c] \\
&= -\frac{1}{\mathcal{Z}} \int c_\lambda^* c_\kappa \frac{\delta}{\delta h_{\nu\mu}} \exp\Big( \sum_{\alpha\beta} c_\alpha^* [(i\omega + \mu)\mathbb{1} - h]_{\alpha\beta} c_\beta + H_{\mathrm{int}}[c^*,c] \Big) \mathcal{D}[c^*,c] \\
&= \frac{1}{\mathcal{Z}} \int c_\lambda^* c_\kappa \sum_{\alpha\beta} c_\alpha^* \frac{\delta h_{\alpha\beta}}{\delta h_{\nu\mu}} c_\beta \exp(-S[c^*,c]) \, \mathcal{D}[c^*,c] \\
&= \frac{1}{\mathcal{Z}} \int c_\kappa c_\lambda^* c_\mu c_\nu^* \exp(-S[c^*,c]) \, \mathcal{D}[c^*,c] \\
&= \langle c_\kappa c_\lambda^* c_\mu c_\nu^* \rangle = \chi_{\kappa\lambda\mu\nu}^{(4)} \, .
\end{aligned}
\tag{A.116}
$$

Now consider the expression obtained by functional derivative of the partition function after introducing the dual variables:

$$
\frac{1}{\mathcal{Z}} \frac{\delta}{\delta h_{\nu\mu}} \frac{\delta \mathcal{Z}}{\delta h_{\lambda\kappa}} = \frac{1}{\mathcal{Z}} \frac{\delta}{\delta h_{\mu\nu}} (-\mathcal{Z} G_{\kappa\lambda}) = -\frac{1}{\mathcal{Z}} \left( \frac{\delta \mathcal{Z}}{\delta h_{\nu\mu}} \right) G_{\kappa\lambda} - \frac{\delta G_{\kappa\lambda}}{\delta h_{\nu\mu}} = G_{\mu\nu} G_{\kappa\lambda} - \frac{\delta G_{\kappa\lambda}}{\delta h_{\nu\mu}} \, . \tag{A.117}
$$

The last term to the right is given by

$$
\frac{\delta G_{\kappa\lambda}}{\delta h_{\nu\mu}} = \frac{\delta}{\delta h_{\nu\mu}} \left( \left[ (\Delta - h)^{-1} \right]_{\kappa\lambda} + \left[ [(\Delta - h)^{-1} g^{-1}] G^{\mathrm{d}} [g^{-1}(\Delta - h)^{-1}] \right]_{\kappa\lambda} \right) \, . \tag{A.118}
$$

The first term of this expression is evaluated as

$$
\begin{aligned}
\frac{\delta}{\delta h_{\nu\mu}} \left( \left[ (\Delta - h)^{-1} \right]_{\kappa\lambda} \right) &= + \left[ (\Delta - h)^{-1} \right]_{\kappa\alpha} \frac{\delta h_{\alpha\beta}}{\delta h_{\nu\mu}} \left[ (\Delta - h)^{-1} \right]_{\beta\lambda} \\
&= + \left[ (\Delta - h)^{-1} \right]_{\kappa\nu} \left[ (\Delta - h)^{-1} \right]_{\mu\lambda} \, .
\end{aligned}
\tag{A.119}
$$

And the second term in (A.118) is

$$
\begin{aligned}
\frac{\delta}{\delta h_{\nu\mu}} \left[ [(\Delta - h)^{-1} g^{-1}] G^{\mathrm{d}} [g^{-1}(\Delta - h)^{-1}] \right]_{\kappa\lambda} &= \left( \frac{\delta}{\delta h_{\nu\mu}} [(\Delta - h)^{-1} g^{-1}] \right) G^{\mathrm{d}} [g^{-1}(\Delta - h)^{-1}] \\
&+ [(\Delta - h)^{-1} g^{-1}] G^{\mathrm{d}} \left( \frac{\delta}{\delta h_{\nu\mu}} [g^{-1}(\Delta - h)^{-1}] \right) \\
&+ [(\Delta - h)^{-1} g^{-1}] \left( \frac{\delta G^{\mathrm{d}}}{\delta h_{\nu\mu}} \right) [g^{-1}(\Delta - h)^{-1}] \, .
\end{aligned}
\tag{A.120}
$$

Now evaluate the right-hand side of (A.120) term by term. Using Eq. A.119, the first one gives

$$\left(\frac{\delta}{\delta h_{\nu\mu}}[(\Delta-h)^{-1}]_{\kappa\delta}\,g_{\delta\alpha}^{-1}\right)G_{\alpha\beta}^{d}\,[g^{-1}(\Delta-h)^{-1}]_{\beta\lambda}$$

$$= +[(\Delta-h)^{-1}]_{\kappa\nu}[(\Delta-h)^{-1}]_{\mu\delta}g_{\delta\alpha}^{-1}G_{\alpha\beta}^{d}\,[g^{-1}(\Delta-h)^{-1}]_{\beta\lambda}$$

$$= +[(\Delta-h)^{-1}]_{\kappa\nu}\left[(\Delta-h)^{-1}g^{-1}G^{d}\,g^{-1}(\Delta-h)^{-1}\right]_{\mu\lambda}\,. \tag{A.121}$$

Likewise, the second evaluates to

$$[(\Delta-h)^{-1}g^{-1}]_{\kappa\alpha}G_{\alpha\beta}^{d}\left(g_{\beta\gamma}^{-1}\frac{\delta}{\delta h_{\nu\mu}}[(\Delta-h)^{-1}]_{\gamma\lambda}\right)$$

$$= +\left[(\Delta-h)^{-1}g^{-1}G^{d}\,g^{-1}(\Delta-h)^{-1}\right]_{\kappa\nu}[(\Delta-h)^{-1}]_{\mu\lambda}\,. \tag{A.122}$$

The third term is somewhat more involved. First evaluate the functional derivative of the dual Green function given in Eq. A.113:

$$\left(\frac{\delta G_{\alpha\beta}^{d}}{\delta h_{\nu\mu}}\right) = -\frac{\mathcal{Z}_{f}}{\mathcal{Z}}\iint f_{\alpha}f_{\beta}^{*}\frac{\delta}{\delta h_{\nu\mu}}\exp(-S[c^{*},c;f^{*},f])\mathcal{D}[c^{*},c;f^{*},f] + \left(\frac{\delta}{\delta h_{\nu\mu}}\frac{\mathcal{Z}_{f}}{\mathcal{Z}}\right)\left(G_{\alpha\beta}^{d}\frac{\mathcal{Z}}{\mathcal{Z}_{f}}\right). \tag{A.123}$$

Performing the same steps as in Eqs. A.110 and A.112, the first term on the right-hand side becomes

$$\frac{\mathcal{Z}_{f}}{\mathcal{Z}}\iint f_{\alpha}f_{\beta}^{*}\,f_{\gamma}^{*}\frac{\delta}{\delta h_{\nu\mu}}[g^{-1}(\Delta-h)^{-1}g^{-1}]_{\gamma\delta}f_{\delta}\exp(-S[c^{*},c;f^{*},f])\mathcal{D}[c^{*},c;f^{*},f]$$

$$= \frac{-\mathcal{Z}_{f}}{\mathcal{Z}}\iint f_{\alpha}f_{\beta}^{*}f_{\delta}f_{\gamma}^{*}[g^{-1}(\Delta-h)^{-1}]_{\gamma\nu}[(\Delta-h)^{-1}g^{-1}]_{\mu\delta}\exp(-S[c^{*},c;f^{*},f])\mathcal{D}[c^{*},c;f^{*},f]$$

$$= -[(\Delta-h)^{-1}g^{-1}]_{\mu\delta}\chi_{\alpha\beta\delta\gamma}^{d}\,[g^{-1}(\Delta-h)^{-1}]_{\gamma\nu}\,. \tag{A.124}$$

Here the dual two-particle Green function has been defined as

$$\chi_{\alpha\beta\delta\gamma}^{d} = \frac{\mathcal{Z}_{f}}{\mathcal{Z}}\int\int f_{\alpha}f_{\beta}^{*}f_{\delta}f_{\gamma}^{*}\exp(-S[c^{*},c;f^{*},f])\mathcal{D}[c^{*},c;f^{*},f]\,. \tag{A.125}$$

The second term in Eq. A.123 contains

$$\frac{\delta}{\delta h_{\nu\mu}}\frac{\mathcal{Z}_{f}}{\mathcal{Z}} = \frac{1}{\mathcal{Z}^{2}}\left(\left(\frac{\delta\mathcal{Z}_{f}}{\delta h_{\nu\mu}}\right)\mathcal{Z} - \left(\frac{\delta\mathcal{Z}}{\delta h_{\nu\mu}}\right)\mathcal{Z}_{f}\right)$$

$$= \frac{1}{\mathcal{Z}^{2}}\left(\left(-\mathcal{Z}_{f}[(\Delta-h)^{-1}]_{\mu\nu}\right)\mathcal{Z} + [\mathcal{Z}\,G_{\mu\nu}]\mathcal{Z}_{f}\right)$$

$$= -\frac{\mathcal{Z}_{f}}{\mathcal{Z}}\left([(\Delta-h)^{-1}]_{\mu\nu} - G_{\mu\nu}\right)\,. \tag{A.126}$$

Expressing the lattice Green function in terms of the dual Green function according to Eq. A.115 finally yields

$$
\left(G^{\mathrm{d}}_{\alpha\beta}\frac{\mathcal{Z}}{\mathcal{Z}_f}\right)\frac{\delta}{\delta h_{\nu\mu}}\left(\frac{\mathcal{Z}_f}{\mathcal{Z}}\right) = -G^{\mathrm{d}}_{\alpha\beta}\left\{[(\Delta-h)^{-1}]_{\mu\nu} - G_{\mu\nu}\right\}
$$
$$
= G^{\mathrm{d}}_{\alpha\beta}\left\{\left[(\Delta-h)^{-1}g^{-1}\right]G^{\mathrm{d}}\left[g^{-1}(\Delta-h)^{-1}\right]\right\}_{\mu\nu} . \tag{A.127}
$$

Gathering the partial results from Eqs. A.115, A.119, A.121, A.122 and A.124, one finds the following relation for the lattice and dual two-particle Green function:

$$
\begin{aligned}
\chi_{\kappa\lambda\mu\nu} = {} & \frac{1}{\mathcal{Z}}\frac{\delta^2\mathcal{Z}}{\delta h_{\lambda\kappa}\delta h_{\nu\mu}} = G_{\kappa\lambda}G_{\mu\nu} + \frac{\delta G_{\kappa\lambda}}{\delta h_{\nu\mu}}\\
= {} & + \left[(\Delta-h)^{-1}\right]_{\kappa\lambda}\left[(\Delta-h)^{-1}\right]_{\mu\nu}\\
& + \left[(\Delta-h)^{-1}\right]_{\kappa\lambda}\left[(\Delta-h)^{-1}g^{-1}G^{\mathrm{d}}g^{-1}(\Delta-h)^{-1}\right]_{\mu\nu}\\
& + \left[(\Delta-h)^{-1}g^{-1}G^{\mathrm{d}}g^{-1}(\Delta-h)^{-1}\right]_{\kappa\lambda}\left[(\Delta-h)^{-1}\right]_{\mu\nu}\\
& + \left[(\Delta-h)^{-1}g^{-1}G^{\mathrm{d}}g^{-1}(\Delta-h)^{-1}\right]_{\kappa\lambda}\left[(\Delta-h)^{-1}g^{-1}G^{\mathrm{d}}g^{-1}(\Delta-h)^{-1}\right]_{\mu\nu}\\
& - \left[(\Delta-h)^{-1}\right]_{\kappa\nu}\left[(\Delta-h)^{-1}\right]_{\mu\lambda}\\
& - [(\Delta-h)^{-1}]_{\kappa\nu}\left[(\Delta-h)^{-1}g^{-1}G^{\mathrm{d}}g^{-1}(\Delta-h)^{-1}\right]_{\mu\lambda}\\
& - \left[(\Delta-h)^{-1}g^{-1}G^{\mathrm{d}}g^{-1}(\Delta-h)^{-1}\right]_{\kappa\nu}[(\Delta-h)^{-1}]_{\mu\lambda}\\
& + [(\Delta-h)^{-1}g^{-1}]_{\kappa\alpha}[g^{-1}(\Delta-h)^{-1}]_{\beta\lambda}\chi^{\mathrm{d}}_{\alpha\beta\delta\gamma}[(\Delta-h)^{-1}g^{-1}]_{\mu\delta}[g^{-1}(\Delta-h)^{-1}]_{\gamma\nu}\\
& - \left\{\left[(\Delta-h)^{-1}g^{-1}\right]G^{\mathrm{d}}\left[g^{-1}(\Delta-h)^{-1}\right]\right\}_{\kappa\lambda}\left\{\left[(\Delta-h)^{-1}g^{-1}\right]G^{\mathrm{d}}\left[g^{-1}(\Delta-h)^{-1}\right]\right\}_{\mu\nu} .
\end{aligned}
$$
$$\tag{A.128}$$

Here the first four lines stem from the direct product of lattice Green functions. The following line is the result from Eq. A.119, the following two lines are the first two lines of the right-hand side of (A.120) and the last two lines originate from the last line of (A.120). The fourth and last line of this equation cancel. The rest can be simplified considerably by introducing the antisymmetrized direct matrix product

$$
[A \otimes B]_{\kappa\lambda\mu\nu} := A_{\kappa\lambda}B_{\mu\nu} - A_{\kappa\nu}B_{\mu\lambda} . \tag{A.129}
$$

With this definition, the result becomes

$$
\begin{aligned}
\chi_{\kappa\lambda\mu\nu} = {} & \left\{\left[(\Delta - h)^{-1}\right] \otimes \left[(\Delta - h)^{-1}\right]\right\}_{\kappa\lambda\mu\nu} \\
& + \left\{\left[(\Delta - h)^{-1}\right] \otimes \left[(\Delta - h)^{-1} g^{-1} G^{\mathrm{d}} g^{-1}(\Delta - h)^{-1}\right]\right\}_{\kappa\lambda\mu\nu} \\
& + \left\{\left[(\Delta - h)^{-1} g^{-1} G^{\mathrm{d}} g^{-1}(\Delta - h)^{-1}\right] \otimes \left[(\Delta - h)^{-1}\right]\right\}_{\kappa\lambda\mu\nu} \\
& + [(\Delta - h)^{-1} g^{-1}]_{\kappa\alpha}[g^{-1}(\Delta - h)^{-1}]_{\beta\lambda}\chi^{\mathrm{d}}_{\alpha\beta\delta\gamma}[(\Delta - h)^{-1} g^{-1}]_{\mu\delta}[g^{-1}(\Delta - h)^{-1}]_{\gamma\nu} .
\end{aligned}
$$

$$\tag{A.130}$$

A particularly simple relation is obtained by considering the nontrivial part of the two-particle Green function:

$$
\begin{aligned}
\chi_{\kappa\lambda\mu\nu} - [G \otimes G]_{\kappa\lambda\mu\nu} = {} & [(\Delta - h)^{-1} g^{-1}]_{\kappa\alpha}[g^{-1}(\Delta - h)^{-1}]_{\beta\lambda}\chi^{\mathrm{d}}_{\alpha\beta\delta\gamma}\times \\
& \times [(\Delta - h)^{-1} g^{-1}]_{\mu\delta}[g^{-1}(\Delta - h)^{-1}]_{\gamma\nu} \\
& - \left\{[(\Delta - h)^{-1} g^{-1} G^{\mathrm{d}} g^{-1}(\Delta - h)^{-1}] \otimes \right. \\
& \left. \otimes [(\Delta - h)^{-1} g^{-1} G^{\mathrm{d}} g^{-1}(\Delta - h)^{-1}]\right\}_{\kappa\lambda\mu\nu} .
\end{aligned}
$$

$$\tag{A.131}$$

Combining the two terms by separating the dual Green functions from the rest in the last two lines yields the final relation

$$
\boxed{
\begin{aligned}
\chi_{\kappa\lambda\mu\nu} - [G \otimes G]_{\kappa\lambda\mu\nu} = {} & [(\Delta - h)^{-1} g^{-1}]_{\kappa\alpha}[g^{-1}(\Delta - h)^{-1}]_{\beta\lambda}\left[\chi^{\mathrm{d}} - G^{\mathrm{d}} \otimes G^{\mathrm{d}}\right]_{\alpha\beta\delta\gamma}\times \\
& \times [(\Delta - h)^{-1} g^{-1}]_{\mu\delta}[g^{-1}(\Delta - h)^{-1}]_{\gamma\nu} .
\end{aligned}
}
$$

$$\tag{A.132}$$

It is useful to define

$$
L_{\alpha\beta} := [(\Delta - h)^{-1} g^{-1}]_{\alpha\beta} \qquad R_{\alpha\beta} := [g^{-1}(\Delta - h)^{-1}]_{\alpha\beta} . \tag{A.133}
$$

Then (A.132) can be reexpressed as

$$
\chi_{\kappa\lambda\mu\nu} - [G \otimes G]_{\kappa\lambda\mu\nu} = L_{\kappa\alpha}L_{\mu\delta}\left[\chi^{\mathrm{d}} - G^{\mathrm{d}} \otimes G^{\mathrm{d}}\right]_{\alpha\beta\delta\gamma} R_{\beta\lambda}R_{\gamma\nu} . \tag{A.134}
$$

From this relation, one can obtain an equivalent one relating the vertices of lattice and dual fermions. The lattice (dual) vertex is related to the non-trivial part of the (dual) two-particle Green function according to

$$
\chi^{(\mathrm{d})}_{\kappa\lambda\mu\nu} - \left[G^{(\mathrm{d})} \otimes G^{(\mathrm{d})}\right]_{\kappa\lambda\mu\nu} = G^{(\mathrm{d})}_{\kappa\alpha} G^{(\mathrm{d})}_{\beta\lambda} \Gamma^{(\mathrm{d})}_{\alpha\beta\delta\gamma} G^{(\mathrm{d})}_{\mu\delta} G^{(\mathrm{d})}_{\gamma\nu} . \tag{A.135}
$$

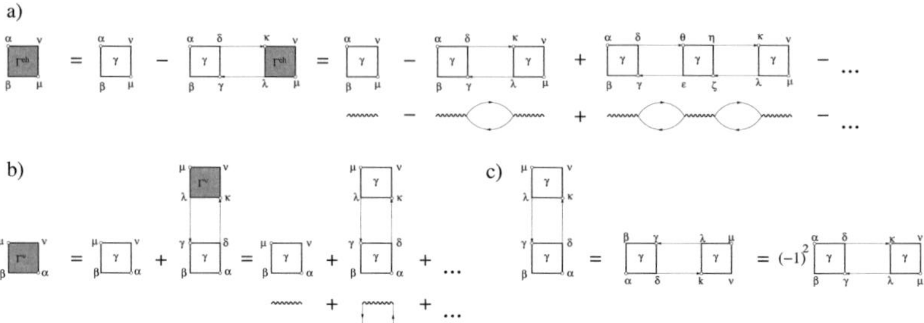

*Figure A.2:* Diagrammatic representation of the Bethe-Salpeter equations in the (horizontal) electron-hole (a) and vertical channel (b). The diagrams which originate from replacing the vertex by a density-density interaction in order to determine the sign of the prefactors, are also shown. c) illustrates that an additional sign arises when switching between channels.

If one additionally defines

$$L'_{\alpha\beta} := [G^{-1}(\Delta - h)^{-1}g^{-1}G^{d}]_{\alpha\beta} \overset{A.29}{=} -(\mathbb{1} + \Sigma^{d}g)^{-1} \tag{A.136}$$

and

$$R'_{\alpha\beta} := [G^{d}g^{-1}(\Delta - h)^{-1}G^{-1}]_{\alpha\beta} = -(\mathbb{1} + g\Sigma^{d})^{-1}\,, \tag{A.137}$$

the relation between the vertex of dual and lattice fermions becomes

$$\boxed{\Gamma_{\kappa\lambda\mu\nu} = L'_{\kappa\alpha}L'_{\mu\delta}\Gamma^{d}_{\alpha\beta\delta\gamma}R'_{\beta\lambda}R'_{\gamma\nu}\,.} \tag{A.138}$$

## A.8   LDFA Equations

In this section, the LDFA equations are derived in general form. They are based on the Schwinger-Dyson equation and an approximation to the vertex, which is constructed from the Bethe-Salpeter equations in the horizontal and vertical channels.

### A.8.1   Bethe-Salpeter Equations

It is useful to consider the structure of the Bethe-Salpeter equations in the antisymmetrized theory in some detail. The Bethe-Salpeter equation (BSE) for the (horizontal) electron-hole channel shown in Fig. A.2 a) reads (the index (4) on the two-particle vertex is omitted)

$$\Gamma^{eh}_{\alpha\beta\mu\nu} = \gamma_{\alpha\beta\mu\nu} - \gamma_{\alpha\beta\gamma\delta}G_{\delta\kappa}G_{\lambda\gamma}\Gamma^{eh}_{\kappa\lambda\mu\nu}\,. \tag{A.139}$$

The BSE for the vertical channel according to Fig. A.2 b) is given by

$$\Gamma^{\mathrm{v}}_{\mu\beta\alpha\nu} = \gamma_{\mu\beta\alpha\nu} + \gamma_{\gamma\beta\alpha\delta}G_{\delta\kappa}G_{\lambda\gamma}\Gamma^{\mathrm{v}}_{\mu\lambda\kappa\nu} \ . \tag{A.140}$$

In the antisymmetric theory the horizontal and vertical channels are equivalent. This is obvious from the topology of the diagrams. It is hence possible to express the vertex in the vertical channel through the one in the horizontal channel. In order to see this and how the different sign in Eqs. A.139, A.140 emerges, consider, for example, the first-order contribution to the vertical channel:

$$\delta\Gamma^{\mathrm{v}\,(1)}_{\mu\beta\alpha\nu} = \gamma_{\mu\beta\alpha\nu} = -\gamma_{\alpha\beta\mu\nu} = -\delta\Gamma^{\mathrm{eh}\,(1)}_{\alpha\beta\mu\nu} \ . \tag{A.141}$$

This follows from the antisymmetry property of $\gamma$. The second order vertex correction is (cf. Fig. A.2 b) )

$$\delta\Gamma^{\mathrm{v}\,(2)}_{\mu\beta\alpha\nu} = \gamma_{\gamma\beta\alpha\delta}G_{\delta\kappa}G_{\lambda\gamma}\gamma_{\mu\lambda\kappa\nu} \ . \tag{A.142}$$

This can be transformed into a diagram for the horizontal channel by rotating it clockwise by $\pi/2$ and then mirror it about the horizontal axis as shown in Fig. A.2 c). The mirror-operation changes the sense of rotation of the labeling of each vertex. Restoring the original sense of rotation by exchanging the two incoming endpoints (marked by the open circle) yields a minus sign for each vertex in the diagram. Hence the overall sign change in general is $(-1)^n$, where $n$ is the number of vertices in the diagram. For the second order correction, the resulting expression in terms of the vertex correction to the horizontal channel reads

$$\delta\Gamma^{\mathrm{v}\,(2)}_{\mu\beta\alpha\nu} = \gamma_{\gamma\beta\alpha\delta}G_{\delta\kappa}G_{\lambda\gamma}\gamma_{\mu\lambda\kappa\nu} = (-1)^2\gamma_{\alpha\beta\gamma\delta}G_{\delta\kappa}G_{\lambda\gamma}\gamma_{\kappa\lambda\mu\nu} = -\delta\Gamma^{\mathrm{eh}\,(2)}_{\alpha\beta\mu\nu} \ . \tag{A.143}$$

The additional alternating sign that arises by changing between the channels is canceled by the alternating sign in the electron-hole channel. The correspondence between the diagrams in different channels is hence valid for all orders, whence it follows that

$$\Gamma^{\mathrm{v}}_{\mu\beta\alpha\nu} = -\Gamma^{\mathrm{eh}}_{\alpha\beta\mu\nu} \ , \qquad \Gamma^{\mathrm{v}}_{\alpha\nu\mu\beta} = -\Gamma^{\mathrm{eh}}_{\alpha\beta\mu\nu} \ . \tag{A.144}$$

Hence the vertex in the vertical channel is transformed into the vertex in the electron-hole channel by exchanging two incoming endpoints and reversing the sign. The second equality corresponds to the exchange of outgoing endpoints and holds by symmetry.

## A.8.2 Approximation to the Vertex

An approximation to the vertex can be constructed as the sum of vertices obtained from the Bethe-Salpeter equations in all three channels including a double counting correction. In the following, only the electron-hole and vertical channels are considered for

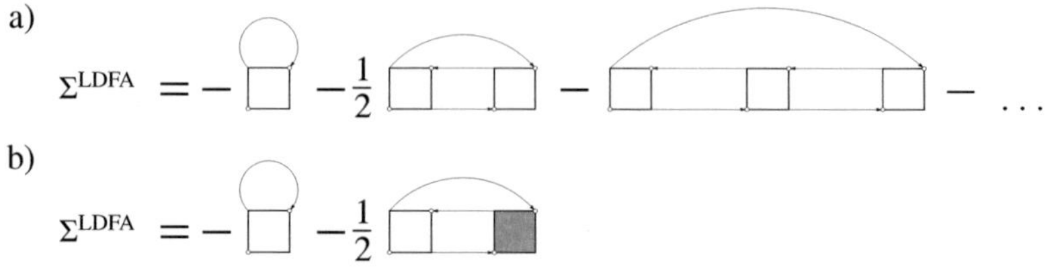

a)

b)

*Figure A.3:* Diagrammatic representation of the ladder approximation to the dual self-energy (a) and the Schwinger-Dyson equation (b).

simplicity. Adding the contributions from the two channels is motivated by the need to regain crossing symmetry. The approximation to the full vertex is given by

$$\Gamma_{\kappa\lambda\mu\nu} = \Gamma^{\mathrm{eh}}_{\kappa\lambda\mu\nu} + \Gamma^{\mathrm{v}}_{\kappa\lambda\mu\nu} - \gamma_{\kappa\lambda\mu\nu} \,. \tag{A.145}$$

As was shown in the previous section, the vertex from the vertical channel can be expressed in terms of the electron-hole vertex. It can thus be written as

$$\Gamma_{\kappa\lambda\mu\nu} = \Gamma^{\mathrm{eh}}_{\kappa\lambda\mu\nu} - \Gamma^{\mathrm{eh}}_{\mu\lambda\kappa\nu} - \gamma_{\kappa\lambda\mu\nu} \,. \tag{A.146}$$

This vertex has the property that it obeys partial crossing symmetry:

$$\Gamma_{\mu\lambda\kappa\nu} = \Gamma^{\mathrm{eh}}_{\mu\lambda\kappa\nu} - \Gamma^{\mathrm{eh}}_{\kappa\lambda\mu\nu} - \gamma_{\mu\lambda\kappa\nu} = -\Gamma_{\kappa\lambda\mu\nu} \,, \tag{A.147}$$

i.e. it changes sign when the endpoints corresponding to incoming lines (indices $\mu$, $\kappa$) are interchanged.

## A.8.3    Schwinger-Dyson Equation

The diagrammatic representation of the Schwinger-Dyson equation (SDE) is shown in Fig. A.3. It can formally be obtained using the equation of motion. According to the diagrammatic rules derived in Secs. A.4.3, A.4.4, it reads

$$\Sigma_{\alpha\nu} = -\gamma_{\alpha\nu\gamma\delta}G_{\delta\gamma} - \frac{1}{2}\gamma_{\alpha\beta\gamma\delta}G_{\beta\mu}G_{\lambda\gamma}G_{\delta\kappa}\Gamma_{\kappa\lambda\mu\nu} \,. \tag{A.148}$$

Inserting the approximation to $\Gamma$, Eq. A.146, into the SDE yields

$$\Sigma_{\alpha\nu} = -\gamma_{\alpha\nu\gamma\delta}G_{\delta\gamma} - \frac{1}{2}\gamma_{\alpha\beta\gamma\delta}G_{\beta\mu}G_{\lambda\gamma}G_{\delta\kappa}\Gamma^{\mathrm{eh}}_{\kappa\lambda\mu\nu} + \frac{1}{2}\gamma_{\alpha\beta\gamma\delta}G_{\beta\mu}G_{\lambda\gamma}G_{\delta\kappa}\Gamma^{\mathrm{eh}}_{\mu\lambda\kappa\nu}$$

$$+ \frac{1}{2}\gamma_{\alpha\beta\gamma\delta}G_{\beta\mu}G_{\lambda\gamma}G_{\delta\kappa}\gamma_{\kappa\lambda\mu\nu} \,. \tag{A.149}$$

Note that diagrammatically this corresponds to rotating the vertex in the electron-hole channel with respect to Fig. A.2 a) (cf. also Fig. A.4 a). Renaming $\mu \leftrightarrow \kappa$ and $\beta \leftrightarrow \delta$ in the second term gives

$$\frac{1}{2}\gamma_{\alpha\beta\gamma\delta}G_{\beta\mu}G_{\lambda\gamma}G_{\delta\kappa}\Gamma^{\mathrm{eh}}_{\mu\lambda\kappa\nu} = \frac{1}{2}\gamma_{\alpha\delta\gamma\beta}G_{\delta\kappa}G_{\lambda\gamma}G_{\beta\mu}\Gamma^{\mathrm{eh}}_{\kappa\lambda\mu\nu} = -\frac{1}{2}\gamma_{\alpha\beta\gamma\delta}G_{\delta\kappa}G_{\lambda\gamma}G_{\beta\mu}\Gamma^{\mathrm{eh}}_{\kappa\lambda\mu\nu} \ . \quad (\mathrm{A}.150)$$

where in the last step the antisymmetry property of $\gamma$ was used. Hence the second and third term are equal. The SDE becomes

$$\Sigma_{\alpha\nu} = -\gamma_{\alpha\nu\gamma\delta}G_{\delta\gamma} - \gamma_{\alpha\beta\gamma\delta}G_{\beta\mu}G_{\lambda\gamma}G_{\delta\kappa}\Gamma^{\mathrm{eh}}_{\kappa\lambda\mu\nu} + \frac{1}{2}\gamma_{\alpha\beta\gamma\delta}G_{\beta\mu}G_{\lambda\gamma}G_{\delta\kappa}\gamma_{\kappa\lambda\mu\nu}$$

$$= -\gamma_{\alpha\nu\gamma\delta}G_{\delta\gamma} - \frac{1}{2}\gamma_{\alpha\beta\gamma\delta}G_{\beta\mu}G_{\lambda\gamma}G_{\delta\kappa}\left(2\Gamma^{\mathrm{eh}}_{\kappa\lambda\mu\nu} - \gamma_{\kappa\lambda\mu\nu}\right) \ . \quad (\mathrm{A}.151)$$

## A.8.4  Ladder Approximation

In order to show that the SDE together with the approximation for the vertex generates the ladder approximation to the self-energy, first consider the dual self-energy up to third order in $\gamma$:

$$\Sigma^{(3)}_{\alpha\nu} = -\gamma_{\alpha\lambda\gamma\delta}G_{\delta\gamma} - \frac{1}{2}\gamma_{\alpha\beta\gamma\delta}G_{\beta\mu}G_{\lambda\gamma}G_{\delta\kappa}\gamma_{\kappa\lambda\mu\nu} - \gamma_{\alpha\beta\gamma\delta}G_{\beta\epsilon}G_{\zeta\mu}G_{\lambda\eta}G_{\theta\gamma}\gamma_{\epsilon\zeta\eta\theta}\gamma_{\kappa\lambda\mu\nu}G_{\delta\kappa} \ . \quad (\mathrm{A}.152)$$

The vertex up to second order in $\gamma$ is

$$\Gamma^{\mathrm{eh}}_{\kappa\lambda\mu\nu} = \gamma_{\kappa\lambda\mu\nu} - \gamma_{\kappa\lambda\epsilon\theta}G_{\theta\eta}G_{\zeta\epsilon}\gamma_{\eta\zeta\mu\nu} \ . \quad (\mathrm{A}.153)$$

Inserting this into the SDE, Eq. A.151, gives the following expression:

$$\Sigma_{\alpha\nu} = -\gamma_{\alpha\nu\gamma\delta}G_{\delta\gamma} - \frac{1}{2}\gamma_{\alpha\beta\gamma\delta}G_{\beta\mu}G_{\lambda\gamma}G_{\delta\kappa}\gamma_{\kappa\lambda\mu\nu} + \gamma_{\alpha\beta\gamma\delta}G_{\beta\mu}G_{\lambda\gamma}G_{\delta\kappa}\gamma_{\kappa\lambda\epsilon\theta}G_{\theta\eta}G_{\zeta\epsilon}\gamma_{\eta\zeta\mu\nu} \ . \quad (\mathrm{A}.154)$$

The first two terms obviously resemble the first two given in Eq. A.152. It remains to be shown that the identity holds for the last term. To see this, it is very instructive to perform operations on the diagrams. Deform the self-energy diagram as shown in Fig. A.4 b). The ladder is rotated and mirrored about the horizontal axis. By the same arguments as in Sec. A.8.1, this changes the sign of each vertex. The resulting diagram hence corresponds to the expression

$$(-1)^2\gamma_{\alpha\beta\gamma\delta}G_{\beta\mu}G_{\lambda\gamma}G_{\delta\kappa}\gamma_{\kappa\theta\epsilon\lambda}G_{\theta\eta}G_{\zeta\epsilon}\gamma_{\mu\zeta\eta\nu} \ . \quad (\mathrm{A}.155)$$

In order to arrive at the final result, one needs to make use of the antisymmetry property of the vertex $\gamma$ in order to deform the diagram into one with the same topology as the

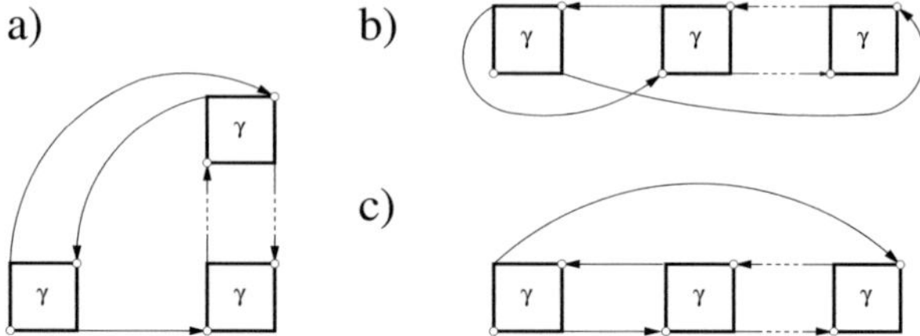

*Figure A.4:* a) Deformations of the self-energy diagram originating from the $n$-th-order vertex correction. The sign of the diagram in a) is $(-1)^{n+2}$ (a factor $-1$ stems from the Schwinger-Dyson equation and a factor $(-1)^{n+1}$ from the Bethe-Salpeter equation in the horizontal channel which is rotated counterclockwise by $\pi/2$ with respect to Fig. A.2 a) ). Deforming this to the one shown in b) yields a factor of $(-1)^n$. Exchanging endpoints of the left vertex to give diagram c) contributes another factor $(-1)$, so that the self-energy diagrams have negative sign at every order.

corresponding self-energy diagram, Fig. A.4 c). This introduces an additional minus sign, so that this term becomes

$$- \gamma_{\alpha\delta\gamma\beta} G_{\beta\mu} G_{\lambda\gamma} G_{\delta\kappa} \gamma_{\kappa\theta\epsilon\lambda} G_{\theta\eta} G_{\zeta\epsilon} \gamma_{\mu\zeta\eta\nu} \, . \tag{A.156}$$

This expression must be the same as the third order term in the self-energy, Eq. A.152. This is formally seen by first renaming variables: $\beta \leftrightarrow \delta$, $\kappa \leftrightarrow \eta$ and $\lambda \leftrightarrow \zeta$. This yields

$$- \gamma_{\alpha\beta\gamma\delta} G_{\delta\mu} G_{\zeta\gamma} G_{\beta\eta} \gamma_{\eta\theta\epsilon\zeta} G_{\theta\kappa} G_{\lambda\epsilon} \gamma_{\mu\lambda\kappa\nu} \, . \tag{A.157}$$

Then renaming $\kappa \leftrightarrow \mu$, $\zeta \leftrightarrow \theta$ and $\epsilon \leftrightarrow \eta$ finally gives

$$- \gamma_{\alpha\beta\gamma\delta} G_{\delta\kappa} G_{\theta\gamma} G_{\beta\epsilon} \gamma_{\epsilon\zeta\eta\theta} G_{\zeta\mu} G_{\lambda\eta} \gamma_{\kappa\lambda\mu\nu} \, , \tag{A.158}$$

which is the same expression as the last term in Eq. A.152. By the rules of the perturbation theory, any ladder diagram has a prefactor of $(-1)$, except for the second order, which has a prefactor of $-(1/2)$. According to the foregoing, these are correctly reproduced up to third order by the SDE and the above approximation to the vertex. In order to see that the correct sign is reproduced at any order, note that a diagram in the BSE comes with a sign of $(-1)^{n+1}$, where $n$ is the number of vertices in the diagram. Inserting this into the SDE yields a prefactor of $(-1)^{n+2}$. Deforming this diagram into the diagram b) of Fig. A.4 yields a factor $(-1)^n$, so that the diagram has positive sign. Changing endpoints of the first vertex yields the correct prefactor $(-1)$.

Using the SDE and the above approximation for the vertex is needed to obtain the correct prefactors of the diagrams in the ladder. From the structure of the ladder diagrams depicted in Fig. A.3 one can see that the self-energy in the ladder approximation can be constructed using for example the vertex in the vertical channel depicted in Fig. A.2 b) by connecting the endpoints $\beta$, $\mu$ by a dual Green function: $\Sigma_{\alpha\nu} = -\Gamma^{\mathrm{v}}_{\mu\beta\alpha\nu} G^{\mathrm{d}}_{\beta\mu}$. Alternatively, this can be expressed in terms of the horizontal channel using the equivalence of the two channels, Eq. A.144. The resulting diagram series has the same structure as the ladder series, but leads to overcounting of the second order contribution, which assumes a prefactor $(-1)$. The correct prefactor $-1/2$ is obtained by explicitly subtracting the second order contribution $-(1/2)\gamma_{\gamma\beta\alpha\delta} G_{\delta\kappa} G_{\lambda\gamma} \gamma_{\mu\lambda\kappa\nu} G_{\beta\mu}$, thus recovering A.151.

# Appendix B

# Asymptotics for Bare Susceptibility

The bubble diagram

$$\chi^0(i\Omega_m) = -\frac{1}{\beta} \sum_{n=-\infty}^{\infty} G(i\omega_n)\, G[i(\omega_n + \Omega_m)] \tag{B.1}$$

takes on negative values for large $m$ if the Green function is artificially set to zero starting from a cutoff frequency $\omega_N$. To avoid this, define the Green function as

$$G^{\mathrm{def}}(i\omega_n) = \begin{cases} G^{\mathrm{num}}(i\omega_n) & =: \; G_n^{\mathrm{num}} & n \in [-N, N-1] \\[2mm] G^{\mathrm{as}}(i\omega_n) & =: \; G_n^{\mathrm{as}} = \frac{1}{i\omega_n} & \text{else} \end{cases}, \tag{B.2}$$

where $G^{\mathrm{num}}$ is the part that is numerically known up to the cutoff, from where the Green function is taken to be equal to $G = 1/i\omega_n$. Within the interval $[-N-m, N-1]$, the bubble can be evaluated using $G$ as defined in Eq. B.2. The susceptibility is written in the form

$$\chi_m = -\frac{1}{\beta} \sum_{n=-N-m}^{N-1} G_n^{\mathrm{def}} G_{n+m}^{\mathrm{def}} - \frac{1}{\beta} \sum_{n=-\infty}^{-N-m-1} G_n^{\mathrm{as}} G_{n+m}^{\mathrm{as}} - \frac{1}{\beta} \sum_{n=N}^{\infty} G_n^{\mathrm{as}} G_{n+m}^{\mathrm{as}}. \tag{B.3}$$

The purely asymptotic terms are evaluated to be

$$-\frac{1}{\beta} \sum_{n=-\infty}^{\infty} G_n^{\mathrm{as}} G_{n+m}^{\mathrm{as}} + \frac{1}{\beta} \sum_{n=-N-m}^{N-1} G_n^{\mathrm{as}} G_{n+m}^{\mathrm{as}}. \tag{B.4}$$

Using the definition of the Matsubara frequencies, $\omega_n = (2n+1)\pi/\beta$, the infinite sum is

$$\frac{\beta}{\pi^2} \sum_{n=-\infty}^{\infty} \frac{1}{(2n+1)^2} = \frac{\beta}{4}. \tag{B.5}$$

for $m = 0$ and

$$\frac{\beta}{\pi^2} \sum_{n=-\infty}^{\infty} \frac{1}{2n+1} \frac{1}{2(n+m)+1} = \frac{\beta}{\pi^2} \sum_{n=-\infty}^{\infty} \frac{1}{2m} \left( \frac{1}{2n+1} - \frac{1}{2(n+m)+1} \right) = 0 \, . \qquad \text{(B.6)}$$

for $m \neq 0$. Taken together, the susceptibility is evaluated as

$$\chi_m = \frac{\beta}{4} \delta_{m,0} - \frac{1}{\beta} \sum_{n=-N-m}^{N-1} \left( G_n^{\text{def}} G_{n+m}^{\text{def}} - G_n^{\text{as}} G_{n+m}^{\text{as}} \right) \, , \qquad \text{(B.7)}$$

where $G$ is defined as in Eq. B.2.

# Appendix C

# Fourier Transforms

The following Fourier transforms are often encountered when dealing with two-particle
quantities. For a function of two arguments, the Fourier transform reads

$$
G(\omega, \omega') = \int_0^\beta d\tau \int_0^\beta d\tau' G(\tau, \tau') e^{i(\omega\tau - \omega'\tau')} . \tag{C.1}
$$

Using time translation invariance,

$$
\begin{aligned}
G(\omega, \omega') &= \int_0^\beta d\tau \int_0^\beta d\tau' G(\tau - \tau', 0) e^{i(\omega(\tau-\tau') + (\omega-\omega')\tau')} \\
&= \int_0^\beta d\tau' \int_{-\tau'}^{\beta-\tau'} d\tilde{\tau}\, G(\tilde{\tau}, 0) e^{i\omega\tilde{\tau}} e^{i(\omega-\omega')\tau'} = \int_0^\beta d\tilde{\tau}\, G(\tilde{\tau}, 0) e^{i\omega\tilde{\tau}} \beta\delta_{\omega\omega'} ,
\end{aligned} \tag{C.2}
$$

where it was used that due to the $2\beta$-periodicity of $G(\tilde{\tau}, 0)$ the integral over $\tilde{\tau}$ is independent of $\tau'$ and $\int d\tau' e^{i(\omega-\omega')\tau'} = \beta\delta_{\omega,\omega'}$. The Fourier transform for the two-particle Green
function is

$$
G(\omega_1, \omega_2, \omega_3, \omega_4) = \int_0^\beta d\tau_1 \int_0^\beta d\tau_2 \int_0^\beta d\tau_3 \int_0^\beta d\tau_4 G(\tau_1, \tau_2, \tau_3, \tau_4) e^{i(\omega_1\tau_1 - \omega_2\tau_2 + \omega_3\tau_3 - \omega_4\tau_4)} \tag{C.3}
$$

Using time translation invariance, energy conservation requires that $\omega_1 + \omega_3 = \omega_2 + \omega_4$.
Introducing the bosonic frequency $\Omega := \omega_1 - \omega_2 = \omega_4 - \omega_3$ and using the short-hand
notation

$$
G(\omega, \omega', \Omega) := G(\omega + \Omega, \omega, \omega', \omega' + \Omega) , \tag{C.4}
$$

one has the useful identity

$$
G(\omega, \omega', -\Omega) = G^*(-\omega, -\omega', \Omega) . \tag{C.5}
$$

A quantity that is frequently encountered is $F(\tau) := G(\tau, \tau, 0, 0)$. One has

$$
\begin{aligned}
F(\tau) &= \frac{1}{\beta^4} \sum_{\substack{\omega_1,\omega_2 \\ \omega_3,\omega_4}} G(\omega_1, \omega_2, \omega_3, \omega_4) e^{-i(\omega_1-\omega_2)\tau} \\
&= \frac{1}{\beta^3} \sum_{\Omega} \sum_{\omega_2,\omega_3} G(\omega_2 + \Omega, \omega_2, \omega_3, \omega_3 + \Omega) e^{-i\Omega\tau} \, .
\end{aligned}
\tag{C.6}
$$

The inverse transform is

$$
F(\Omega) = \int d\tau \, F(\tau) e^{i\Omega\tau} = \frac{1}{\beta^2} \sum_{\omega,\omega'} G(\omega + \Omega, \omega, \omega', \omega' + \Omega) \, .
\tag{C.7}
$$

# Appendix D

# CTQMC Formulae

## D.1  Fast Updates

In the CTQMC algorithm, ratios of determinants need to be computed between matrices which differ only by a few rows or columns. This can be done efficiently using the fast update formulae derived below. Computing the determinant ratio is much more efficient and numerically more stable than the calculation of the determinants themselves. Let $\hat{g}_{(k)}$ be the matrix that contains bare Green functions $g_{ij} = g_{\alpha\beta}(\tau_i - \tau_j)$ according to the particular configuration at perturbation order $k$. The matrix $\hat{g}_{(k+1)}$ results from adding a row and a column to the matrix $\hat{g}_{(k)}$. Without loss of generality, the row and colum are assumed to be the last ones; otherwise they can be commuted through without changing the determinant. The key idea is to introduce the matrix

$$\Delta = \hat{g}_{(k+1)} - \hat{g}_{(k)} = \left( \begin{array}{ccc|c} 0 & \cdots & 0 & g_{1,k+1} \\ \vdots & \ddots & \vdots & \vdots \\ 0 & \cdots & 0 & g_{k,k+1} \\ \hline g_{k+1,1} & \cdots & g_{k+1,k} & g_{k+1,k+1} - 1 \end{array} \right), \tag{D.1}$$

where $\hat{g}_{(k)}$ is extended to be a $k + 1 \times k + 1$-matrix, with the last row and column filled with zeros and the $k + 1, k + 1$- element equal to one. With this definition, one can write

$$\hat{g}_{(k+1)} = \hat{g}_{(k)} + \Delta = \hat{g}_{(k)}(\mathbb{1} + \hat{g}_{(k)}^{-1}\Delta) . \tag{D.2}$$

Here $\mathbb{1}$ denotes the $k + 1 \times k + 1$ unit matrix. The desired determinant ratio is hence given by

$$\frac{\det \hat{g}_{(k+1)}}{\det \hat{g}_{(k)}} = \det(\mathbb{1} + \hat{g}_{(k)}^{-1}\Delta) . \tag{D.3}$$

This can be computed efficiently, given that the inverse $M = \hat{g}^{-1}$ of the matrix $\hat{g}$ is known. The inversion of $\hat{g}$ is computationally costly (it requires $O(k^3)$ operations).

Since the updated elements of the matrix $M$ can be computed efficiently as shown below, and the Green function can be directly measured from this matrix, $M$ rather than $\hat{g}$ is manipulated and stored in memory. Assuming summation over repeated indices, the relevant matrix for the determinant ratio can be written as

$$(\mathbb{1} + M^{(k)}\Delta) = \left(\begin{array}{ccc|c} 1 & \cdots & 0 & M_{1,j}g_{j,k+1} \\ \vdots & \ddots & \vdots & \vdots \\ 0 & \cdots & 1 & M_{k,j}g_{j,k+1} \\ \hline g_{k+1,1} & \cdots & g_{k+1,k} & g_{k+1,k+1} \end{array}\right). \tag{D.4}$$

The index $j$ runs from 1 to $k$. The determinant is readily evaluated to be (expand, e.g. along the last row)

$$r = \frac{\det \hat{g}_{(k+1)}}{\det \hat{g}_{(k)}} = \det(\mathbb{1}_{(k)} + M^{(k)}\Delta) = g_{k+1,k+1} - g_{k+1,i}M^{(k)}_{i,j}g_{j,k+1}. \tag{D.5}$$

This equation defines the measurement of Green's function. More generally, the measurement of Green's function obtained by removing row $a$ and column $b$ is given by

$$\boxed{g_{a,b} = \frac{\det \hat{g}_{(k+1)}}{\det \hat{g}_{(k)}} = g_{a,b} - g_{a,i}M^{(k)}_{i,j}g_{j,b}.} \tag{D.6}$$

The sign that arises by exchanging two rows or columns is compensated by permuting the corresponding Grassmann numbers in the numerator of the determinant ratio, Eq. 3.62. By inversion of Eq. D.2, the elements of the matrix $M^{(k+1)}$ are given by $M^{(k+1)} = (\mathbb{1} + M^{(k)}\Delta)^{-1}M^{(k)}$. The inverse of Eq. D.4 is straightforwardly found using block matrix inversion:

$$\left(\begin{array}{cc} \mathbb{1} & B \\ C & D \end{array}\right)^{-1} = \left(\begin{array}{cc} \mathbb{1} + B(D - CB)^{-1}C & -B(D - CB)^{-1} \\ -(D - CB)^{-1}C & (D - CB)^{-1} \end{array}\right). \tag{D.7}$$

Comparing to (D.4) and using (D.5), one finds that $D - CB = r$. Multiplying the result by $M^{(k)}$ using block matrix multiplication leads to

$$M^{(k+1)} = \left(\begin{array}{cc} \mathbb{1} & B \\ C & D \end{array}\right)^{-1}\left(\begin{array}{cc} M^{(k)} & 0 \\ 0 & 1 \end{array}\right) = \left(\begin{array}{cc} M^{(k)} + B(D - CB)^{-1}CM^{(k)} & -B(D - CB)^{-1} \\ -(D - CB)^{-1}CM^{(k)} & (D - CB)^{-1} \end{array}\right). \tag{D.8}$$

Applying this to D.4 yields the fast update formulas. Defining $L_{i,j} = M_{i,l}G_{l,j}$ and $R_{i,j} = G_{i,l}M_{l,j}$, one has $B_i = L_{i,k+1}$, $[CM^{(k)}]_j = R_{j,k+1}$ and the elements of the matrix $M^{(k+1)}$ are updated as

$$M^{(k+1)} = \left(\begin{array}{c|c} M_{ij} + L_{i,k+1}R_{k+1,j}/r & -L_{i,k+1}/r \\ \hline -R_{k+1,j}/r & 1/r \end{array}\right). \tag{D.9}$$

Obviously, the update for the matrix elements $M_{ij}^{(k+1)}$ of the updated matrix for $i, j = 1, \ldots, k$ can be expressed in terms of the elements in row and column $k + 1$:

$$M_{ij}^{(k+1)} = M_{ij}^{(k)} + \frac{M_{i\,k+1}^{(k+1)} M_{k+1\,j}^{(k+1)}}{M_{k+1\,k+1}^{(k+1)}} \, , \quad i, j = 1, \ldots, k \, . \tag{D.10}$$

This also implies that the matrix elements of the matrix $M_{(k-1)}$, which results from removing row and column $k$, are restored via

$$M_{ij}^{(k-1)} = M_{ij}^{(k)} - \frac{M_{ik}^{(k)} M_{kj}^{(k)}}{M_{kk}^{(k)}} \, , \quad i, j = 1, \ldots, k - 1 \, . \tag{D.11}$$

## D.2   Determinant Ratio for $2n$-Point Correlators

The formulae for the determinant ratio of general $n$-particle correlation functions are derived here in analogy to (D.3). The determinant ratio relevant for the measurement of a $2n$-point correlation function is given by the ratio of determinants of matrices

$$\frac{\det \hat{g}_{(k+n)}}{\det \hat{g}_{(k)}} = \det(\mathbb{1}_{(k)} + \hat{g}_{(k)}^{-1} \Delta) \, . \tag{D.12}$$

Here the matrix $\Delta$ is given by

$$\Delta = \hat{g}_{(k+n)} - \hat{g}_{(k)} = \left( \begin{array}{ccc|ccc} 0 & \cdots & 0 & g_{1,k+1} & \cdots & g_{1,k+n} \\ \vdots & \ddots & \vdots & & \vdots & \\ 0 & \cdots & 0 & g_{k,k+1} & \cdots & g_{k,k+n} \\ \hline g_{k+1,1} & \cdots & g_{k+1,k} & g_{k+1,k+1} - 1 & \cdots & g_{k+1,k+n} \\ \vdots & & \vdots & \vdots & & \vdots \\ g_{k+n,1} & \cdots & g_{k+n,k} & g_{k+n,k+1} & \cdots & g_{k+n,k+n} - 1 \end{array} \right) . \tag{D.13}$$

The matrix $\hat{g}_{(k)}$ is extended to be the $k + n \times k + n$-matrix

$$\hat{g}_{(k)} \rightarrow \left( \begin{array}{c|c} \hat{g}_{(k)} & 0 \\ \hline 0 & \mathbb{1}_{(n)} \end{array} \right) . \tag{D.14}$$

Then

$$(\mathbb{1} + M^{(k)} \Delta) = \left( \begin{array}{ccc|ccc} 1 & \cdots & 0 & M_{1,j} g_{j,k+1} & \cdots & M_{1,j} g_{j,k+n} \\ \vdots & \ddots & \vdots & \vdots & & \vdots \\ 0 & \cdots & 1 & M_{k,j} g_{j,k+1} & \cdots & M_{k,j} g_{j,k+n} \\ \hline g_{k+1,1} & \cdots & g_{k+1,k} & g_{k+1,k+1} & \cdots & g_{k+1,k+n} \\ \vdots & & \vdots & \vdots & & \vdots \\ g_{k+n,1} & \cdots & g_{k+n,k} & g_{k+n,k+1} & \cdots & g_{k+n,k+n} \end{array} \right) . \tag{D.15}$$

Now use the following identity for the determinant of a general $2 \times 2$ block matrix:

$$\det \begin{pmatrix} A & B \\ C & D \end{pmatrix} = \det(A) \det(D - CA^{-1}B) . \tag{D.16}$$

For the special case $A = 1$ one has to evaluate $\det(D - CB)$. Application to Eq. D.15 yields the determinant ratio in the form

$$\frac{\det \hat{g}_{(k+n)}}{\det \hat{g}_{(k)}} = \det \begin{pmatrix} g_{k+1,k+1} - g_{k+1,i}M_{i,j}g_{j,k+1} & \cdots & g_{k+1,k+n} - g_{k+1,i}M_{i,j}g_{j,k+n} \\ \vdots & & \vdots \\ g_{k+n,k+1} - g_{k+n,i}M_{i,j}g_{j,k+1} & \cdots & g_{k+n,k+n} - g_{k+n,i}M_{i,j}g_{j,k+n} \end{pmatrix} . \tag{D.17}$$

Measurements for $2n$-point correlators are thus expressed in terms of measurements for the single-particle Green function.

# Bibliography


[1] R. E. Prange and S. M. Girvin, *The Quantum Hall Effect* (Springer, New York, 1997).

[2] G. R. Stewart, *Heavy-fermion systems*, Rev. Mod. Phys. **56**, 755 (1984).

[3] H. v. Löhneysen, A. Rosch, M. Vojta, and P. Wölfle, *Fermi-liquid instabilities at magnetic quantum phase transitions*, Rev. Mod. Phys. **79**, 1015 (2007).

[4] A. C. Hewson, *The Kondo Problem to Heavy Fermions* (Cambridge Univ. Press, Cambridge, 1993).

[5] P. W. Anderson, *The Theory of Superconductivity in High-$T_c$ Cuprates* (Princeton Univ. Press, Princeton, 1997).

[6] D. J. Scalapino, *The case for $d_{x^2-y^2}$ pairing in the cuprate superconductors*, Physics Reports **250**, 329 (1995).

[7] S. Biermann, A. Poteryaev, A. I. Lichtenstein, and A. Georges, *Dynamical Singlets and Correlation-Assisted Peierls Transition in $VO_2$*, Phys. Rev. Lett. **94**, 026404 (2005).

[8] Y. Kamihara, T. Watanabe, M. Hirano, and H. Hosono, *Iron-Based Layered Superconductor $La[O_{1-x}F_x]FeAs$ ($x = 0.05$-$0.12$) with $T_c = 26\,K$*, J. Am. Chem. Soc. **130**, 3296 (2008).

[9] M. Imada, A. Fujimori, and Y. Tokura, *Metal-insulator transitions*, Rev. Mod. Phys. **70**, 1039 (1998).

[10] V. I. Anisimov, J. Zaanen, and O. K. Andersen, *Band theory and Mott insulators: Hubbard U instead of Stoner I*, Phys. Rev. B **44**, 943 (1991).

[11] F. Aryasetiawan and O. Gunnarsson, *The GW method*, Reports on Progress in Physics **61**, 237 (1998).

[12] G. Kotliar, S. Y. Savrasov, K. Haule, V. S. Oudovenko, O. Parcollet, and C. A. Marianetti, *Electronic structure calculations with dynamical mean-field theory*, Rev. Mod. Phys. **78**, 865 (2006).

[13] G. Kotliar and D. Vollhardt, *Strongly Correlated Materials: Insights From Dynamical Mean-Field Theory*, Physics Today **57**, 53 (2004).

[14] N. F. Mott, *Metal-Insulator Transitions* (Taylor and Francis, London, 1974).

[15] A. Georges, G. Kotliar, W. Krauth, and M. J. Rozenberg, *Dynamical mean-field theory of strongly correlated fermion systems and the limit of infinite dimensions*, Rev. Mod. Phys. **68**, 13 (1996).

[16] W. Metzner and D. Vollhardt, *Correlated Lattice Fermions in $d = \infty$ Dimensions*, Phys. Rev. Lett. **62**, 324 (1989).

[17] E. Müller-Hartmann, *Self-consistent perturbation theory of the Anderson model: ground state properties*, Z. Phys. B **57**, 281 (1984).

[18] V. I. Anisimov, A. I. Poteryaev, M. A. Korotin, A. O. Anokhin, and G. Kotliar, *First-principles calculations of the electronic structure and spectra of strongly correlated systems: dynamical mean-field theory*, J. Phys.: Condensed Matter **9**, 7359 (1997a).

[19] A. I. Lichtenstein and M. I. Katsnelson, *Ab initio calculations of quasiparticle band structure in correlated systems: LDA++ approach*, Phys. Rev. B **57**, 6884 (1998).

[20] A. I. Lichtenstein, M. I. Katsnelson, and G. Kotliar, *Finite-Temperature Magnetism of Transition Metals: An ab initio Dynamical Mean-Field Theory*, Phys. Rev. Lett. **87**, 067205 (2001).

[21] A. I. Poteryaev, A. I. Lichtenstein, and G. Kotliar, *Nonlocal Coulomb Interactions and Metal-Insulator Transition in $Ti2O3$: A Cluster LDA + DMFT Approach*, Phys. Rev. Lett. **93**, 086401 (2004).

[22] T. Saha-Dasgupta, A. Lichtenstein, M. Hoinkis, S. Glawion, M. Sing, R. Claessen, and R. Valenti, *Cluster dynamical mean-field calculations for Ti-OCl*, New Journal of Physics **9**, 380 (2007).

[23] J. Schäfer, D. Schrupp, E. Rotenberg, K. Rossnagel, H. Koh, P. Blaha, and R. Claessen, *Electronic Quasiparticle Renormalization on the Spin Wave Energy Scale*, Phys. Rev. Lett. **92**, 097205 (2004).

[24] M. Eschrig and M. R. Norman, *Neutron Resonance: Modeling Photoemission and Tunneling Data in the Superconducting State of Bi2S r2CaCu2O8 + δ*, Phys. Rev. Lett. **85**, 3261 (2000).

[25] E. Schachinger, J. J. Tu, and J. P. Carbotte, *Angle-resolved photoemission spectroscopy and optical renormalizations: Phonons or spin fluctuations*, Phys. Rev. B **67**, 214508 (2003).

[26] R. Claessen, M. Sing, U. Schwingenschlögl, P. Blaha, M. Dressel, and C. S. Jacobsen, *Spectroscopic Signatures of Spin-Charge Separation in the Quasi-One-Dimensional Organic Conductor TTF-TCNQ*, Phys. Rev. Lett. **88**, 096402 (2002).

[27] A. Schiller and K. Ingersent, *Systematic $1/d$ Corrections to the Infinite-Dimensional Limit of Correlated Lattice Electron Models*, Phys. Rev. Lett. **75**, 113 (1995).

[28] M. V. Sadovskii, I. A. Nekrasov, E. Z. Kuchinskii, T. Pruschke, and V. I. Anisimov, *Pseudogaps in strongly correlated metals: A generalized dynamical mean-field theory approach*, Phys. Rev. B **72**, 155105 (2005).

[29] S. Pairault, D. Sénéchal, and A.-M. S. Tremblay, *Strong-Coupling Expansion for the Hubbard Model*, Phys. Rev. Lett. **80**, 5389 (1998).

[30] S. K. Sarker, *A new functional integral formalism for strongly correlated Fermi systems*, Journal of Physics C: Solid State Physics **21**, L667 (1988).

[31] T. D. Stanescu and G. Kotliar, *Strong coupling theory for interacting lattice models*, Phys. Rev. B **70**, 205112 (2004).

[32] A. I. Lichtenstein and M. I. Katsnelson, *Antiferromagnetism and d-wave superconductivity in cuprates: A cluster dynamical mean-field theory*, Phys. Rev. B **62**, R9283 (2000).

[33] G. Kotliar, S. Y. Savrasov, G. Pálsson, and G. Biroli, *Cellular Dynamical Mean Field Approach to Strongly Correlated Systems*, Phys. Rev. Lett. **87**, 186401 (2001).

[34] M. Potthoff, M. Aichhorn, and C. Dahnken, *Variational Cluster Approach to Correlated Electron Systems in Low Dimensions*, Phys. Rev. Lett. **91**, 206402 (2003).

[35] T. Maier, M. Jarrell, T. Pruschke, and M. H. Hettler, *Quantum cluster theories*, Rev. Mod. Phys. **77**, 1027 (2005).

[36] V. Y. Irkhin, A. A. Katanin, and M. I. Katsnelson, *Robustness of the Van Hove Scenario for High-Tc Superconductors*, Phys. Rev. Lett. **89**, 076401 (2002).

[37] E. Y. Loh, J. E. Gubernatis, R. T. Scalettar, S. R. White, D. J. Scalapino, and R. L. Sugar, *Sign problem in the numerical simulation of many-electron systems*, Phys. Rev. B **41**, 9301 (1990).

[38] C. Slezak, M. Jarrell, T. Maier, and J. Deisz, *Multi-scale Extensions to Quantum Cluster Methods for Strongly Correlated Electron Systems*, arXiv:0603421, unpublished (2006).

[39] A. Toschi, A. A. Katanin, and K. Held, *Dynamical vertex approximation: A step beyond dynamical mean-field theory*, Phys. Rev. B **75**, 045118 (2007).

[40] H. Kusunose, *Influence of Spatial Correlations in Strongly Correlated Electron Systems: Extension to Dynamical Mean Field Approximation*, Journal of the Physical Society of Japan **75**, 054713 (2006).

[41] A. N. Rubtsov, M. I. Katsnelson, and A. I. Lichtenstein, *Dual fermion approach to nonlocal correlations in the Hubbard model*, Phys. Rev. B **77**, 033101 (2008).

[42] T. O. Wehling, I. Grigorenko, A. I. Lichtenstein, and A. V. Balatsky, *Phonon-Mediated Tunneling into Graphene*, Phys. Rev. Lett. **101**, 216803 (2008).

[43] R. M. Martin, *Electronic Structure, Basic Theory and Practical Methods* (Cambridge University Press, Cambridge, 2004).

[44] O. K. Andersen, *Linear methods in band theory*, Phys. Rev. B **12**, 3060 (1975).

[45] J. Hubbard, *Electron Correlations in Narrow Energy Bands*, Proceedings of the Royal Society of London. Series A. Mathematical and Physical Sciences **276**, 238 (1963).

[46] J. Hubbard, *Electron Correlations in Narrow Energy Bands. II. The Degenerate Band Case*, Proceedings of the Royal Society of London. Series A. Mathematical and Physical Sciences **277**, 237 (1964a).

[47] E. H. Lieb and F. Y. Wu, *Absence of Mott Transition in an Exact Solution of the Short-Range, One-Band Model in One Dimension*, Phys. Rev. Lett. **20**, 1445 (1968).

[48] E. H. Lieb and F. Y. Wu, *The one-dimensional Hubbard model: a reminiscence*, Physica A: Statistical Mechanics and its Applications **321**, 1 (2003), statphys-Taiwan-2002: Lattice Models and Complex Systems.

[49] V. Mastropietro, *Rigorous Proof of Luttinger Liquid Behavior in the 1d Hubbard Model*, Journal of Statistical Physics **121**, 373 (2005).

[50] A. Macridin, M. Jarrell, T. Maier, P. R. C. Kent, and E. D'Azevedo, *Pseudogap and Antiferromagnetic Correlations in the Hubbard Model*, Phys. Rev. Lett. **97**, 036401 (2006).

[51] A. Koga, N. Kawakami, T. M. Rice, and M. Sigrist, *Orbital-Selective Mott Transitions in the Degenerate Hubbard Model*, Phys. Rev. Lett. **92**, 216402 (2004).

[52] T. Ohashi, N. Kawakami, and H. Tsunetsugu, *Mott Transition in Kagomé Lattice Hubbard Model*, Phys. Rev. Lett. **97**, 066401 (2006).

[53] J. W. Negele and H. Orland, *Quantum Many-Particle Systems* (Westview Press, Boulder, 1998).

[54] P. W. Anderson, *Localized Magnetic States in Metals*, Phys. Rev. **124**, 41 (1961).

[55] P. W. Anderson, *Localized Magnetic States and Fermi-Surface Anomalies in Tunneling*, Phys. Rev. Lett. **17**, 95 (1966).

[56] A. N. Rubtsov, V. V. Savkin, and A. I. Lichtenstein, *Continuous-time quantum Monte Carlo method for fermions*, Phys. Rev. B **72**, 035122 (2005).

[57] S. Weiss, J. Eckel, M. Thorwart, and R. Egger, *Iterative real-time path integral approach to nonequilibrium quantum transport*, Phys. Rev. B **77**, 195316 (2008).

[58] M. Caffarel and W. Krauth, *Exact diagonalization approach to correlated fermions in infinite dimensions: Mott transition and superconductivity*, Phys. Rev. Lett. **72**, 1545 (1994).

[59] J. Hubbard, *Electron Correlations in Narrow Energy Bands. III. An Improved Solution*, Proceedings of the Royal Society of London. Series A. Mathematical and Physical Sciences **281**, 401 (1964b).

[60] J. M. Luttinger and J. C. Ward, *Ground-State Energy of a Many-Fermion System. II*, Phys. Rev. **118**, 1417 (1960).

[61] G. Baym and L. P. Kadanoff, *Conservation Laws and Correlation Functions*, Phys. Rev. **124**, 287 (1961).

[62] E. Müller-Hartmann, *Correlated fermions on a lattice in high dimensions*, Z. Phys. B **74**, 507 (1989).

[63] A. Fuhrmann, S. Okamoto, H. Monien, and A. J. Millis, *Fictive-impurity approach to dynamical mean-field theory: A strong-coupling investigation*, Phys. Rev. B **75**, 205118 (2007).

[64] S. Okamoto, A. J. Millis, H. Monien, and A. Fuhrmann, *Fictive impurity models: An alternative formulation of the cluster dynamical mean-field method*, Phys. Rev. B **68**, 195121 (2003).

[65] G. Biroli and G. Kotliar, *Cluster methods for strongly correlated electron systems*, Phys. Rev. B **65**, 155112 (2002).

[66] M. Potthoff, *Self-energy-functional approach to systems of correlated electrons*, Eur. Phys. J. B **32**, 429 (2003).

[67] M. Fleck, A. I. Liechtenstein, A. M. Oleś, L. Hedin, and V. I. Anisimov, *Dynamical Mean-Field Theory for Doped Antiferromagnets*, Phys. Rev. Lett. **80**, 2393 (1998).

[68] B. Amadon, F. Lechermann, A. Georges, F. Jollet, T. O. Wehling, and A. I. Lichtenstein, *Plane-wave based electronic structure calculations for correlated materials using dynamical mean-field theory and projected local orbitals*, Phys. Rev. B **77**, 205112 (2008).

[69] T. Miyake and F. Aryasetiawan, *Screened Coulomb interaction in the maximally localized Wannier basis*, Phys. Rev. B **77**, 085122 (2008).

[70] V. I. Anisimov, F. Aryasetiawan, and A. I. Lichtenstein, *First-principles calculations of the electronic structure and spectra of strongly correlated systems: the LDA + U method*, Journal of Physics: Condensed Matter **9**, 767 (1997b).

[71] C. Z. Mooney, *Monte Carlo simulation (Quantitative Applications in the Social Sciences)* (Sage, Thousand Oaks, California, 1997), ISBN 0-8039-5943-5.

[72] D. J. Scalapino and R. L. Sugar, *Method for Performing Monte Carlo Calculations for Systems with Fermions*, Phys. Rev. Lett. **46**, 519 (1981).

[73] R. Blankenbecler, D. J. Scalapino, and R. L. Sugar, *Monte Carlo calculations of coupled boson-fermion systems. I*, Phys. Rev. D **24**, 2278 (1981).

[74] J. E. Hirsch, *Two-dimensional Hubbard model: Numerical simulation study*, Phys. Rev. B **31**, 4403 (1985).

[75] J. E. Hirsch and R. M. Fye, *Monte Carlo Method for Magnetic Impurities in Metals*, Phys. Rev. Lett. **56**, 2521 (1986).

[76] A. N. Rubtsov and A. I. Lichtenstein, *Continuous-time quantum Monte Carlo method for fermions: Beyond auxiliary field framework*, JETP Lett. **80**, 61 (2004).

[77] A. N. Rubtsov, *Quantum Monte Carlo determinantal algorithm without Hubbard-Stratonovich transformation: a general consideration*, arXiv:0302228, unpublished (2003).

[78] P. Werner, A. Comanac, L. de' Medici, M. Troyer, and A. J. Millis, *Continuous-Time Solver for Quantum Impurity Models*, Phys. Rev. Lett. **97**, 076405 (2006).

[79] P. Werner and A. J. Millis, *Hybridization expansion impurity solver: General formulation and application to Kondo lattice and two-orbital models*, Phys. Rev. B **74**, 155107 (2006).

[80] K. Haule, *Quantum Monte Carlo impurity solver for cluster dynamical mean-field theory and electronic structure calculations with adjustable cluster base*, Phys. Rev. B **75**, 155113 (2007).

[81] M. Troyer and U.-J. Wiese, *Computational Complexity and Fundamental Limitations to Fermionic Quantum Monte Carlo Simulations*, Phys. Rev. Lett. **94**, 170201 (2005).

[82] N. Metropolis, A. W. Rosenbluth, M. N. Rosenbluth, A. H. Teller, and E. Teller, *Equation of State Calculations by Fast Computing Machines*, The Journal of Chemical Physics **21**, 1087 (1953).

[83] W. K. Hastings, *Monte Carlo sampling methods using Markov chains and their applications*, Biometrika **57**, 97 (1970).

[84] H. Fehske, R. Schneider, and A. Weiße, *Computational Many-Particle Physics*, vol. 739 of *Lecture Notes in Physics* (Springer, Berlin, 2008), ISBN 3-540-74685-4.

[85] B. A. Berg, *Markov Chain Monte Carlo Simulations and Their Statistical Analysis* (World Scientific Publishing Co., Singapore, 2004), ISBN 981-238935-0.

[86] J. E. Hirsch, *Discrete Hubbard-Stratonovich transformation for fermion lattice models*, Phys. Rev. B **28**, 4059 (1983).

[87] J. Sherman and W. J. Morrison, *Adjustment of an Inverse Matrix Corresponding to a Change in One Element of a Given Matrix*, The Annals of Mathematical Statistics **21**, 124 (1950).

[88] N. Blümer, *Multigrid Hirsch-Fye quantum Monte Carlo method for dynamical mean-field theory*, arXiv:0801.1222, unpublished (2008).

[89] N. Blümer, *Numerically exact Green functions from Hirsch-Fye quantum Monte Carlo simulations*, arXiv:0712,1290, unpublished (2007).

[90] J. Yoo, S. Chandrasekharan, R. K. Kaul, D. Ullmo, and H. U. Baranger, *On the sign problem in the Hirsch & Fye algorithm for impurity problems*, Journal of Physics A: Mathematical and General **38**, 10307 (2005).

[91] S. Sakai, R. Arita, K. Held, and H. Aoki, *Quantum Monte Carlo study for multi-orbital systems with preserved spin and orbital rotational symmetries*, Phys. Rev. B **74**, 155102 (2006).

[92] N. V. Prokof'ev, B. V. Svistunov, and I. S. Tupitsyn, *Exact quantum Monte Carlo process for the statistics of discrete systems*, JETP Lett. **64**, 911 (1996).

[93] A. A. Abrikosov, L. P. Gor'kov, and I. E. Dzyaloshinskii, *Methods of Quantum Field Theory in Statistical Physics* (Pergamon Press, New York, 1965).

[94] E. Gorelov, Ph.D. thesis, University of Hamburg (2007).

[95] E. Gull, Ph.D. thesis, ETH Zurich (2008).

[96] F. Wang and D. P. Landau, *Efficient, Multiple-Range Random Walk Algorithm to Calculate the Density of States*, Phys. Rev. Lett. **86**, 2050 (2001a).

[97] F. Wang and D. P. Landau, *Determining the density of states for classical statistical models: A random walk algorithm to produce a flat histogram*, Phys. Rev. E **64**, 056101 (2001b).

[98] G. D. Mahan, *Many-particle Physics* (Kluwer Academic Publishers, New York, 2000).

[99] J. E. Gubernatis, J. E. Hirsch, and D. J. Scalapino, *Spin and charge correlations around an Anderson magnetic impurity*, Phys. Rev. B **35**, 8478 (1987).

[100] *Plot Digitizer*, version 2.4.1, available at `http://plotdigitizer.sourceforge.net/`.

[101] C. Herkt-Januschek, Ph.D. thesis, University of Hamburg (2008).

[102] V. J. Emery, *Theory of the quasi-one-dimensional electron gas with strong on-site interactions*, Phys. Rev. B **14**, 2989 (1976).

[103] A. M. Läuchli and P. Werner, *Krylov implementation of the hybridization expansion impurity solver and application to 5-orbital models*, Phys. Rev. B **80**, 235117 (2009).

[104] E. Gull, P. Werner, A. Millis, and M. Troyer, *Performance analysis of continuous-time solvers for quantum impurity models*, Phys. Rev. B **76**, 235123 (2007).

[105] P. Werner, T. Oka, and A. J. Millis, *Diagrammatic Monte Carlo simulation of nonequilibrium systems*, Phys. Rev. B **79**, 035320 (2009).

[106] M. Jarrell and J. E. Gubernatis, *Bayesian inference and the analytic continuation of imaginary-time quantum Monte Carlo data*, Physics Reports **269**, 133 (1996).

[107] A. S. Mishchenko, N. V. Prokof'ev, A. Sakamoto, and B. V. Svistunov, *Diagrammatic quantum Monte Carlo study of the Fröhlich polaron*, Phys. Rev. B **62**, 6317 (2000).

[108] H. J. Vidberg and J. W. Serene, *Solving the Eliashberg Equations by Means of N-Point Padé Approximants*, J. Low Temp. Phys. **29**, 179 (1977).

[109] A. Yamamoto, P. A. Sharma, Y. Okamoto, A. Nakao, H. A. Katori, S. Niitaka, D. Hashizume, and H. Takagi, *Metal–Insulator Transition in a Pyrochlore-type Ruthenium Oxide, $Hg_2Ru_2O_7$*, Journal of the Physical Society of Japan **76**, 043703 (2007).

[110] M. Schmidt, W. Ratcliff, P. G. Radaelli, K. Refson, N. M. Harrison, and S. W. Cheong, *Spin Singlet Formation in $MgTi2O4$: Evidence of a Helical Dimerization Pattern*, Phys. Rev. Lett. **92**, 056402 (2004).

[111] M. Shaz, S. van Smaalen, L. Palatinus, M. Hoinkis, M. Klemm, S. Horn, and R. Claessen, *Spin-Peierls transition in TiOCl*, Phys. Rev. B **71**, 100405 (2005).

[112] M. Hase, I. Terasaki, and K. Uchinokura, *Observation of the spin-Peierls transition in linear Cu2+ (spin-1/2) chains in an inorganic compound $CuGeO3$*, Phys. Rev. Lett. **70**, 3651 (1993).

[113] A. Georges and G. Kotliar, *Hubbard model in infinite dimensions*, Phys. Rev. B **45**, 6479 (1992).

[114] J. Kondo, *Resistance Minimum in Dilute Magnetic Alloys*, Progress of Theoretical Physics **32**, 37 (1964).

[115] A. O. Anokhin, V. Y. Irkhin, and M. I. Katsnelson, *On the theory of the Mott transition in the paramagnetic phase*, J. Phys.: Condensed Matter **3**, 1475 (1991).

[116] D. R. Hamann, *Path Integral Theory of Magnetic Alloys*, Phys. Rev. B **2**, 1373 (1970).

[117] G. Moeller, V. Dobrosavljević, and A. E. Ruckenstein, *RKKY interactions and the Mott transition*, Phys. Rev. B **59**, 6846 (1999).

[118] A. Fuhrmann, D. Heilmann, and H. Monien, *From Mott insulator to band insulator: A dynamical mean-field theory study*, Phys. Rev. B **73**, 245118 (2006).

[119] S. S. Kancharla and S. Okamoto, *Band insulator to Mott insulator transition in a bilayer Hubbard model*, Phys. Rev. B **75**, 193103 (2007).

[120] T. Yoshikawa and M. Ogata, *Role of frustration and dimensionality in the Hubbard model on the stacked square lattice: Variational cluster approach*, Phys. Rev. B **79**, 144429 (2009).

[121] K. Bouadim, G. G. Batrouni, F. Hébert, and R. T. Scalettar, *Magnetic and transport properties of a coupled Hubbard bilayer with electron and hole doping*, Phys. Rev. B **77**, 144527 (2008).

[122] T. Miyazaki, I. Nakamura, and D. Yoshioka, *Bilayer Heisenberg model studied by the Schwinger-boson Gutzwiller-projection method*, Phys. Rev. B **53**, 12206 (1996).

[123] R. Bulla, T. A. Costi, and T. Pruschke, *Numerical renormalization group method for quantum impurity systems*, Rev. Mod. Phys. **80**, 395 (2008).

[124] L. Kouwenhoven and L. Glazman, *Revival of the Kondo effect*, Physics World **14**, 33 (2001).

[125] O. Y. Kolesnychenko, R. de Kort, M. I. Katsnelson, A. I. Lichtenstein, and H. van Kempen, *Real-space imaging of an orbital Kondo resonance on the Cr(001) surface*, Nature **415**, 507 (2002).

[126] L. Borda, *Kondo screening cloud in a one-dimensional wire: Numerical renormalization group study*, Phys. Rev. B **75**, 041307 (2007).

[127] I. Affleck, L. Borda, and H. Saleur, *Friedel oscillations and the Kondo screening cloud*, Phys. Rev. B **77**, 180404 (2008).

[128] L. Borda, M. Garst, and J. Kroha, *Kondo cloud and spin-spin correlations around a partially screened magnetic impurity*, Phys. Rev. B **79**, 100408 (2009).

[129] K. R. Patton, H. Hafermann, S. Brener, A. I. Lichtenstein, and M. I. Katsnelson, *Measuring spin and charge correlations via tunneling-current conductance fluctuations*, arXiv:0902,0225, unpublished (2008).

[130] G. E. Santoro and G. F. Giuliani, *Impurity spin susceptibility of the Anderson model: A perturbative approach*, Phys. Rev. B **44**, 2209 (1991).

[131] C. F. Hirjibehedin, C. P. Lutz, and A. J. Heinrich, *Spin Coupling in Engineered Atomic Structures*, Science **312**, 1021 (2006).

[132] J. Fernandez-Rossier, *Theory of single spin inelastic tunneling spectroscopy*, arXiv:0901.4839, unpublished (2009).

[133] A. N. Rubtsov, *Quality of the mean-field approximation: A low-order generalization yielding realistic critical indices for three-dimensional Ising-class systems*, Phys. Rev. B **66**, 052107 (2002).

[134] A. N. Rubtsov, *Small parameter for lattice models with strong interaction*, arXiv:0601333, unpublished (2006).

[135] H. Hafermann, S. Brener, A. N. Rubtsov, M. I. Katsnelson, and A. I. Lichtenstein, *Cluster dual fermion approach to nonlocal correlations*, JETP Lett. p. 677 (2007).

[136] S. Brener, H. Hafermann, A. N. Rubtsov, M. I. Katsnelson, and A. I. Lichtenstein, *Dual fermion approach to susceptibility of correlated lattice fermions*, Phys. Rev. B **77**, 195105 (2008).

[137] G. Li, H. Lee, and H. Monien, *Determination of the lattice susceptibility within the dual fermion method*, Phys. Rev. B **78**, 195105 (2008).

[138] H. Lee, G. Li, and H. Monien, *Hubbard model on the triangular lattice using dynamical cluster approximation and dual fermion methods*, Phys. Rev. B **78**, 205117 (2008).

[139] A. N. Rubtsov, M. I. Katsnelson, A. I. Lichtenstein, and A. Georges, *Dual fermion approach to the two-dimensional Hubbard model: Antiferromagnetic fluctuations and Fermi arcs*, Phys. Rev. B **79**, 045133 (2009).

[140] H. Hafermann, G. Li, A. N. Rubtsov, M. I. Katsnelson, A. I. Lichtenstein, and H. Monien, *Efficient Perturbation Theory for Quantum Lattice Models*, Phys. Rev. Lett. **102**, 206401 (2009a).

[141] D. Boies, C. Bourbonnais, and A. M. S. Tremblay, *One-Particle and Two-Particle Instability of Coupled Luttinger Liquids*, Phys. Rev. Lett. **74**, 968 (1995).

[142] S. Pairault, D. Sénéchal, and A.-M. Tremblay, *Strong-coupling perturbation theory of the Hubbard model*, The European Physical Journal B **16**, 85 (2000).

[143] G. Baym, *Self-Consistent Approximations in Many-Body Systems*, Phys. Rev. **127**, 1391 (1962).

[144] *Intel ® Math Kernel Library*, http://software.intel.com/en-us/intel-mkl/.

[145] P. Monthoux and D. J. Scalapino, *Self-consistent $d_{x^2-y^2}$ pairing in a two-dimensional Hubbard model*, Phys. Rev. Lett. **72**, 1874 (1994).

[146] U. Schollwöck, *The density-matrix renormalization group*, Rev. Mod. Phys. **77**, 259 (2005).

[147] M. Ferrero, P. S. Cornaglia, L. D. Leo, O. Parcollet, G. Kotliar, and A. Georges, *Valence bond dynamical mean-field theory of doped Mott insulators with nodal/antinodal differentiation*, EPL **85**, 57009 (2009).

[148] Y. Z. Zhang and M. Imada, *Pseudogap and Mott transition studied by cellular dynamical mean-field theory*, Phys. Rev. B **76**, 045108 (2007).

[149] H. Park, K. Haule, and G. Kotliar, *Cluster Dynamical Mean Field Theory of the Mott Transition*, Phys. Rev. Lett. **101**, 186403 (2008).

[150] M. Balzer, B. Kyung, D. Sénéchal, A.-M. S. Tremblay, and M. Potthoff, *First-order Mott transition at zero temperature in two dimensions: Variational plaquette study*, EPL **85**, 17002 (2009).

[151] V. Y. Irkhin and M. I. Katsnelson, *Current carriers in a quantum two-dimensional antiferromagnet*, J. Phys.: Condensed Matter **3**, 6439 (1991).

[152] B. Stroustrup, *The C++ programming language* (Addison-Wesley, Boston, 2006).

[153] M. Balzer, W. Hanke, and M. Potthoff, *Mott transition in one dimension: Benchmarking dynamical cluster approaches*, Phys. Rev. B **77**, 045133 (2008).

[154] M. Capone, M. Civelli, S. S. Kancharla, C. Castellani, and G. Kotliar, *Cluster-dynamical mean-field theory of the density-driven Mott transition in the one-dimensional Hubbard model*, Phys. Rev. B **69**, 195105 (2004).

[155] H. Hafermann, S. Brener, A. N. Rubtsov, M. I. Katsnelson, and A. I. Lichtenstein, *Understanding the electronic structure and magnetism of correlated nanosystems*, J. Phys.: Condensed Matter **21**, 064248 (2009b).

[156] X. Dai, K. Haule, and G. Kotliar, *Strong-coupling solver for the quantum impurity model*, Phys. Rev. B **72**, 045111 (2005).

[157] E. Koch, G. Sangiovanni, and O. Gunnarsson, *Sum rules and bath parametrization for quantum cluster theories*, Phys. Rev. B **78**, 115102 (2008).

[158] *GSL – GNU Scientific Library*, http://www.gnu.org/software/gsl/.

[159] S. Hochkeppel, F. F. Assaad, and W. Hanke, *Dynamical-quantum-cluster approach to two-particle correlation functions in the Hubbard model*, Phys. Rev. B **77**, 205103 (2008).

[160] A. B. Migdal, *Theory of finite Fermi Systems and applications to atomic nuclei* (Interscience Publishers, New York, 1967).

[161] P. Nozières, *Theory of interacting Fermi systems* (Benjamin, New York, 1964).

[162] N. E. Bickers, *Analysis of electron parquet equations: an eigenvector-based solution algorithm*, unpublished.

[163] N. E. Bickers and S. R. White, *Conserving approximations for strongly fluctuating electron systems. II. Numerical results and parquet extension*, Phys. Rev. B **43**, 8044 (1991).

[164] N. Bulut, D. J. Scalapino, and S. R. White, *Bethe-Salpeter eigenvalues and amplitudes for the half-filled two-dimensional Hubbard model*, Phys. Rev. B **47**, 14599 (1993).

[165] T. A. Maier, M. S. Jarrell, and D. J. Scalapino, *Structure of the Pairing Interaction in the Two-Dimensional Hubbard Model*, Phys. Rev. Lett. **96**, 047005 (2006).

[166] C.-H. Pao and N. E. Bickers, *Superconductivity in the two-dimensional Hubbard model: One-particle correlation functions*, Phys. Rev. B **51**, 16310 (1995).

[167] P. A. Lee, N. Nagaosa, and X.-G. Wen, *Doping a Mott insulator: Physics of high-temperature superconductivity*, Rev. Mod. Phys. **78**, 17 (2006).

[168] P. W. Anderson, *Is There Glue in Cuprate Superconductors?*, Science **317**, 1705 (2007).

[169] T. A. Maier, D. Poilblanc, and D. J. Scalapino, *Dynamics of the Pairing Interaction in the Hubbard and t-J Models of High-Temperature Superconductors*, Phys. Rev. Lett. **100**, 237001 (2008).

[170] *ARPACK – Arnoldi Package*, http://www.caam.rice.edu/software/ARPACK/.

[171] E. Dagotto, *Correlated electrons in high-temperature superconductors*, Rev. Mod. Phys. **66**, 763 (1994).

[172] A. A. Katanin, A. Toschi, and K. Held, *Comparing pertinent effects of anti-ferromagnetic fluctuations in the two and three dimensional Hubbard model*, arXiv:0808.0689, unpublished (2006).

[173] A. Auerbach, ed., *Interacting Electrons and Quantum Magnetism* (Springer, New York, 1998).

[174] N. E. Bickers, D. J. Scalapino, and S. R. White, *Conserving Approximations for Strongly Correlated Electron Systems: Bethe-Salpeter Equation and Dynamics for the Two-Dimensional Hubbard Model*, Phys. Rev. Lett. **62**, 961 (1989).

[175] N. E. Bickers and D. J. Scalapino, *Conserving approximations for strongly fluctuating electron systems. I. Formalism and calculational approach*, Annals of Physics **193**, 206 (1989).

[176] V. Janiš, *Stability of self-consistent solutions for the Hubbard model at intermediate and strong coupling*, Phys. Rev. B **60**, 11345 (1999).

[177] N. Hugenholtz, *Perturbation theory of large quantum systems*, Physica **23**, 481 (1957).

[178] A. Moreo, *Magnetic susceptibility of the two-dimensional Hubbard model*, Phys. Rev. B **48**, 3380 (1993).

[179] C. Huscroft, M. Jarrell, T. Maier, S. Moukouri, and A. N. Tahvildarzadeh, *Pseudogaps in the 2D Hubbard Model*, Phys. Rev. Lett. **86**, 139 (2001).

[180] P. R. C. Kent, M. Jarrell, T. A. Maier, and T. Pruschke, *Efficient calculation of the antiferromagnetic phase diagram of the three-dimensional Hubbard model*, Phys. Rev. B **72**, 060411 (2005).

[181] A. Toschi, private communication (2009).

[182] M. I. Katsnelson, V. Y. Irkhin, L. Chioncel, A. I. Lichtenstein, and R. A. de Groot, *Half-metallic ferromagnets: From band structure to many-body effects*, Rev. Mod. Phys. **80**, 315 (2008).

[183] Y. Nagaoka, *Ferromagnetism in a Narrow, Almost Half-Filled s Band*, Phys. Rev. **147**, 392 (1966).

[184] V. Y. Irkhin and M. I. Katsnelson, *Spin waves in narrow band ferromagnet*, Journal of Physics C: Solid State Physics **18**, 4173 (1985).

# List of Abbreviations and Acronyms

| | | | | |
|---|---|---|---|---|
| 2PGF | two-particle Green function | | GGA | generalized gradient approximation |
| AF | antiferromagnetic | | GMR | giant magnetoresistance |
| AFI | antiferromagnetic instability | | GSL | GNU scientific library |
| AIM | Anderson impurity model | | HST | Hubbard-Stratonovich transformation |
| APW | augmented plane waves | | IPT | iterated perturbation theory |
| ARPES | angle resolved photoemission spectroscopy | | LDA | local density approximation |
| BSE | Bethe-Salpeter equation | | LDFA | ladder dual fermion approach |
| CDF | cluster dual fermion | | LMTO | linear muffin-tin orbitals |
| CDFA | cluster dual fermion approach | | MaxEnt | maximum entropy method |
| CDMFT | cellular DMFT | | MB | megabyte |
| CTQMC | continuous-time quantum Monte Carlo | | MC | Monte Carlo |
| DCA | dynamical cluster approximation | | MIT | metal-insulator transition |
| DF | dual fermion | | MPI | message passing interface |
| DFA | dual fermion approach | | NCA | non-crossing approximation |
| DFT | density functional theory | | NN | nearest neighbor |
| DMFT | dynamical mean-field theory | | NNN | next-nearest neighbor |
| DMRG | density matrix renormalization group | | NRG | numerical renormalization group |
| DOS | density of states | | PG | pseudogap |
| DΓA | dynamical vertex approximation | | QMC | quantum Monte Carlo |
| ED | exact diagonalization | | RKKY | Ruderman Kittel Kasuya Yosida |
| FFT | fast Fourier transform | | RPA | random phase approximation |
| FLAPW | full potential linearized augmented plane waves | | SDE | Schwinger-Dyson equation |
| FLEX | fluctuation-exchange approximation | | SFT | self-energy functional theory |
| FLOP | floating point operation | | SPERT | superperturbation theory |
| FM | ferromagnetic | | STM | scanning-tunneling microscope |
| GB | gigabyte | | VCA | variational cluster approach |
| | | | VLA | variational lattice approach |
| | | | openMP | open multi-processing |

# List of Publications

## Refereed Journal Articles


6. Kelly R. Patton, Hartmut Hafermann, Sergej Brener, Alexander I. Lichtenstein and Mikhail I. Katsnelson, *Probing the Kondo screening cloud via tunneling-current conductance fluctuations*, Phys. Rev. B **80**, 212403 (2009).

5. H. Hafermann, G. Li, A. N. Rubtsov, M. I. Katsnelson, A. I. Lichtenstein and H. Monien, *Efficient Perturbation Theory for Quantum Lattice Models*, Phys. Rev. Lett. **102**, 206401 (2009).

4. H. Hafermann, M. I. Katsnelson and A. I. Lichtenstein, *Metal-insulator transition by suppression of spin fluctuations*, EPL **85**, 37006 (2009).

3. H. Hafermann, C. Jung, S. Brener, M. I. Katsnelson, A. N. Rubtsov and A. I. Lichtenstein, *Superperturbation solver for quantum impurity models*, EPL **85**, 27007 (2009).

2. S. Brener, H. Hafermann, A. N. Rubtsov, M. I. Katsnelson and A. I. Lichtenstein, *Dual fermion approach to susceptibility of correlated lattice fermions*, (PRB Editors' Suggestion), Phys. Rev. B **77**, 195105 (2008).

1. H. Hafermann, S. Brener, A. N. Rubtsov, M. I. Katsnelson and A. I. Lichtenstein, *Cluster dual fermion approach to nonlocal correlations*, JETP Lett. **86**, 677 (2007) [Письма в ЖЭТФ **86**, 769 (2007)].

# Conference Proceedings

(peer reviewed articles)

2. H. Hafermann, S. Brener, A. N. Rubtsov, M. I. Katsnelson and A. I. Lichtenstein, Proceedings of the 2nd International Conference on Quantum Simulators and Design - Tokyo, Japan, May 31 – June 3, 2008.
   *Understanding the electronic structure and magnetism of correlated nanosystems*, J. Phys.: Condens. Matter **21**, 064248 (2009).

1. H. Hafermann, M. Kecker, S. Brener, A. N. Rubtsov, M. I. Katsnelson and A. I. Lichtenstein, Proceedings of the International Conference on Quantum Phenomena in Complex Matter (Stripes08) - Erice, Italy, July 26 – August 1, 2008.
   *Dual Fermion Approach to High-Temperature Superconductivity*, J. Supercond. Nov. Magn. **22**, 45 (2008).


# Unrelated Work

The following peer reviewed publications have been prepared alongside the work reported in this thesis, but the topic of these papers is otherwise unrelated.

2. H. Hafermann, R. Bručas, I. L. Soroka, M. I. Katsnelson, D. Iuşan, B. Sanyal, O. Eriksson and B. Hjörvarsson, *Competing anisotropies in bcc $Fe_{81}Ni_{19}/Co(001)$ superlattices*, Appl. Phys. Lett. **94**, 073102 (2009).

1. R. Bručas, H. Hafermann, I. L. Soroka, D. Iuşan, B. Sanyal, M. I. Katsnelson, O. Eriksson and B. Hjörvarsson, *Magnetic anisotropy and evolution of ground-state domain structures in bcc $Fe_{81}Ni_{19}/Co(001)$ superlattices*, Phys. Rev. B **78**, 024421 (2008).

# Danksagung

An dieser Stelle möchte ich all jenen, die zum Gelingen dieser Arbeit beigetragen haben, meinen Dank aussprechen. Dies gilt vor allem meinem Betreuer Alexander Lichtenstein für seine Geduld in der Anfangsphase, die stete Unterstützung, viele erhellende Diskussionen und Denkanstöße, das mir entgegengebrachte Vertrauen und die vielen Freiräume meinen eigenen Ansätzen nachzugehen. Nicht zuletzt danke ich ihm auch dafür, dass er die Verlängerung meines Vertrages ermöglicht und mir in der Phase des Zusammenschreibens den Rücken freigehalten hat. Hervorheben möchte ich auch die Ermöglichung der Teilnahme an zahlreichen Konferenzen, Workshops und Summerschools und vor allem erfolgreicher Kollaborationen. All diese Faktoren haben maßgeblich zum Erfolg der Arbeit beigetragen.

Ich möchte mich sehr herzlich bei Michael Potthoff für die zügige Erstellung des Zweitgutachtens und für anregende Diskussionen bedanken. Ferner danke ich ihm sowie Matthias Balzer für den offenen Austausch von Daten.

Alexey Rubtsov danke ich für die Kollaboration, wertvolle Gespräche, die Bereitstellung seiner Codes und die Einladung an die Staatliche Universität Moskau. Bei Mikhail Katsnelson möchte ich mich bedanken für die langjährige Zusammenarbeit, für neue Ideen, die Ermöglichung meines mehrwöchigen Aufenthaltes in Nimwegen und für die nicht selbstverständliche Möglichkeit, mein Projekt nach der Doktorarbeit weiterführen zu können. Ich danke Philipp Werner für die Kollaboration, die Übernahme des Gutachtens der Disputation, die Einladung nach Zürich mit der Gelegenheit zu einem Vortrag an der ETH, sowie ihm und Emanuel Gull für das bereitwillige zur Verfügung stellen der Quanten Monte Carlo Codes und Emanuel für das LaTeX-Template. Ich möchte ferner Herrn Monien für sehr nützliche Diskussionen, auch beim Erstellen des Manuskriptes für die Publikation, und besonders für das Einweihen in zahlreiche nützliche Konzepte von C++ danken, die vielseitige Verwendung gefunden haben. Seine Idee, die Konvergenzeigenschaften mittels der Eigenwerte zu untersuchen und seine immerwährende Bereitschaft sich Zeit zu nehmen, wenn ich auf meinem Weg von Heidelberg nach Hamburg vorbeikam, waren sehr hilfreich. Außerdem möchte ich Václav Janiš und Alexander Shick für die Gastfreundschaft in Prag danken, sowie für zahlreiche wertvolle Diskussionen mit Václav Janiš sowie Matouš Ringel und dessen Kommentar „Wenn sich das Vorzeichen nicht ändert, dann verstehst du deine Störungstheorie nicht!..." Frank Lechermann danke ich für zahlreiche Diskussion und die Zusammenarbeit bei der Anwendung der Methode dualer Fermionen auf reale Systeme, die leider noch keinen Eingang in die vorliegende Arbeit gefunden hat, aber sicherlich in Kürze Früchte tragen wird. Herrn Scharnberg danke ich für kritische Fragen und die Zusammenarbeit an einem Nebenprojekt. Für die Einladung ins kalifornische Livermore und die Gelegenheit, den Großrechner vor Ort nutzen zu können, möchte ich Alexander Landa danken. Darüberhinaus danke ich Frithjof Anders, Andrey Antipov, Leonid Glazmann, Igor Krivenko, Gang Li, Andrei Ruban und Alessandro Toschi für nützliche Gespräche, sowie Andrei Mishchenko für das Programm zur analytischen Fortsetzung. Ich möchte mich darüberhinaus für die finanzielle Unterstützung durch den Sonderforschungsbereich 668 (Teilprojekt A3) der DFG und bei Herrn Wiesendanger für die unkomplizierte Verlängerung meines Vertrages bedanken.

Mein besonderer Dank gilt auch der gesamten Arbeitsgruppe th1-li für die angenehme und freundschaftliche Atmosphäre, Heiterkeit im Büroalltag und auf Konferenzen und Unterstützung jeglicher Art. Insbesondere danke ich meinem Büronachbarn Tim Wehling für den Austausch über unsere recht unterschiedlichen Arbeitsthemen und kritische Bemerkungen zu Teilen des Manuskriptes, Evgeny Gorelov für die Starthilfe mit dem CTQMC Code und Sergey Brener für unzählige Diskussionen, viele neue Einsichten und den Einfluss der russischen Denkweise, den er auf mich ausgeübt und der sich sicherlich an der ein oder anderen Stelle in dieser Arbeit niedergeschlagen hat. Kelly Patton danke ich für die gute Zusammenarbeit, zahlreiche Diskussionen, für die Illustration und darüberhinaus für die vielen Feierabendgespräche in Altona, für Anregungen und vor allem für das Korrekturlesen des gesamten Manuskriptes. Christoph Jung danke ich für die sehr effektive Teamarbeit an unserem gemeinsamen Projekt und auch für das Lesen von Teilen dieser Arbeit. Nicht zuletzt sei den Diplomanden Claudius Herkt-Januschek und Martin Kecker gedankt für das Testen und Anwenden der Codes. Diese Arbeit auf dem Gebiet der rechnergestützen Physik wäre ohne die Unterstützung des PHYSnet Rechenzentrums kaum denkbar gewesen. Vor allem Bodo Krause-Kyora danke ich für schnelle, unkonventionelle und absolut unentbehrliche Hilfe bei Computerproblemen, vor allem für seine Bereitschaft sogar am Wochenende Probleme zu beheben, die mich mehrfach gerettet und ihn wahrscheinlich gelehrt hat, niemals seine private Handynummer herauszugeben. Herrn Mercado danke ich für die kompetente und unkomplizierte Hilfe bei Verwaltungsfragen, sowie Frau Sen und Frau Schmidtke für Hilfe bei Formalitäten.

All meinen Freunden in und außerhalb Hamburgs danke ich für die schönen gemeinsamen Erlebnisse und Aufheiterungen, insbesondere der Altona-WG für ein Dach über dem Kopf nach meinem Umzug, für die kulinarische Unterstützung in Heidelberg und allen für das Verständnis, wenn ich mich zeitweise rar gemacht habe. Henning Labuhn danke ich für die Gestaltung des Buchumschlags. Meiner Familie, insbesondere meinen Eltern bin ich überaus dankbar für den Rückhalt und die Unterstützung während meines Studiums, welche diese Arbeit erst möglich gemacht hat. Inga, ich danke Dir dafür dass Du immer für mich da warst, nicht zu vergessen die Illustrationen, aber vor allem für unendlich viel Geduld.

Nimwegen  
im Juni 2009

*Hartmut Hafermann*